Fractal Geometry in Pure and Applied Mathematics

CONTEMPORARY MATHEMATICS

828

Fractal Geometry in Pure and Applied Mathematics

AMS-SMF-EMS Special Session
Fractal Geometry in Pure and Applied Mathematics
July 5–9, 2022
Grenoble, France

Hafedh Herichi
Maria Rosaria Lancia
Therese Basa Landry
Anna Rozanova-Pierrat
Steffen Winter
Editors

AMS AMERICAN MATHEMATICAL SOCIETY

2020 *Mathematics Subject Classification.* Primary 28A80, 26A33, 28A75, 35K05, 35R30, 46E36, 05C50, 60K37.

Library of Congress Cataloging-in-Publication Data

Names: Herichi, Hafedh editor

Title: Fractal geometry in pure and applied mathematics / Hafedh Herichi, Maria Rosaria Lancia, Therese Basa Landry, Anna Rozanova-Pierrat, Steffen Winter, editors.

Description: Providence, Rhode Island : American Mathematical Society, [2025] | Series: Contemporary mathematics, 0271-4132 ; volume 828 | "Special Session at the Joint International Meeting of the AMS-SMF-EMS, July 18–22, 2022 Grenoble, France"–Title page.

Identifiers: LCCN 2025010631 | ISBN 9781470474621 paperback | ISBN 9781470481063 ebook

Subjects: LCSH: Fractals | Differential equations, Partial | AMS: Measure and integration – Classical measure theory – Fractals | Real functions – Functions of one variable – Fractional derivatives and integrals | Measure and integration – Classical measure theory – Length, area, volume, other geometric measure theory | Partial differential equations – Parabolic equations and systems – Heat equation | Partial differential equations – Miscellaneous topics – Inverse problems | Combinatorics – Graph theory – Graphs and linear algebra (matrices, eigenvalues, etc.) | Probability theory and stochastic processes – Special processes – Processes in random environments

Classification: LCC QC20.7.F73 F71745 2025

LC record available at https://lccn.loc.gov/2025010631

∞ The paper used in this book is acid-free and falls within the guidelines
established to ensure permanence and durability.
Visit the AMS home page at `https://www.ams.org/`

10 9 8 7 6 5 4 3 2 1   30 29 28 27 26 25

To Benoît B. Mandelbrot, the fractalist; a brilliant mathematician and a visionary, in honor and remembrance of his 100th birthday.

Contents

Preface

At the Joint International Meeting of the AMS-SMF-EMS held in Grenoble, France, from July 5-9, 2022, we organized a special session titled "*Fractal Geometry in Pure and Applied Mathematics*". The presentations at this session sparked a lively exchange of ideas, leading to the decision to make the discussed subjects accessible to a broader mathematical community. The topics of the lectures and talks presented at the special session capture some of the recent developments in the field of fractal geometry and its applications in pure and applied mathematics.

To capture the essence of the inspiring and productive atmosphere of the meeting, we invited the speakers to contribute to this volume. Many of them enthusiastically agreed, resulting in a collection of papers that reflect the diversity and depth of the field. Some papers present original research, while others offer expository summaries of recent findings, placing them in context.

We highlight some key results and mention some connections: The Sierpinski gasket is certainly one of the benchmark examples in fractal geometry. New concepts and theories are often developed and tested first on this set and its variants. Several contributions in this volume are related in one way or another to this classical self-similar set. In "*Time fractional equations on Sierpinski gasket like fractals*", Raffaela Capitanelli studies non-Brownian diffusion processes on such fractals. Sierpinski gasket-like fractals can also be obtained by randomizing the construction of the fractal, and this approach is taken by D. Fontaine, D.J. Kelleher, and A. Teplyaev in "*Green's function and eigenfunctions on random Sierpinski gaskets*", where randomized Dirichlet forms are used to construct nonstandard Laplacians. They study how the randomness is reflected in the Green's function and eigenfunctions of the new operators.

Fractal structures such as the Sierpinski gasket also arise as limiting objects of sequences of graphs. In "*Spectra of three-peg Hanoi towers graphs,*" Brett Hungar, Gamal Mograby, Madison Phelps, Luke G. Rogers, and Jonathan Wheeler compare Laplacians on two sequences of planar graphs approximating the Sierpinski gasket, using spectral decimation techniques to infer properties of one from the other.

Domains with fractal boundaries are models of rough objects, relevant for instance in the study of energy exchange, and widely used in the applied sciences. In this volume, diffusion phenomena across fractal interfaces are studied. In "*Existence of optimal shapes for heat diffusions across irregular interfaces*", Gabriel Claret and Anna Rozanova-Pierrat study diffusion between two domains connected via a general fractal boundary, by developing Biegert's theory of trace operators on a class of non-Lipschitz boundaries. Looking for shapes which maximize the speed of heat energy transfer, they prove that maximizers typically have a fractal boundary, confirming ideas of the physicist De Gennes from 1982. In the article

"*Inverse problems in irregular domains: approximation via Mosco convergence*" Simone Creo, Maria Rosaria Lancia, Gianluca Mola, and Silvia Romanelli obtained one of the first results for some inverse problems on fractal-boundary domains and suitable approximations by smoother domains in an abstract framework. After proving the well-posedness, they show, via Mosco convergence, that the solution to the original problem arises as a suitable limit of solutions on the approximating domains, being highly relevant for numerical approximation.

Many fractal spaces are locally non-Euclidean. The theory of metric measure spaces provides another framework in which differential calculus can be extended beyond the Riemannian setting. For metric measure spaces which have a doubling measure and support a Poincaré-type inequality, there exists a notion of Sobolev space. In "*Homogeneous Newton-Sobolev spaces in metric measure spaces and their Banach space properties*", Nageswari Shanmugalingam demonstrates that such spaces also exhibit Banach space properties which are useful for analyzing PDEs and variational problems.

In "*Iterated fractal drums ∼ some new perspectives: polyhedral measures, atomic decompositions and Morse theory*", Claire David and Michel L. Lapidus present new ideas for defining differential operators on very irregular fractals by extending classical trace methods due to Jonsson and Wallin. Working in the setting of the Weierstrass curve, they develop their new theory, which relates atomic decompositions, trace theorems and complex dimensions via polyhedral measures related to particular polygonal approximations of the Weierstrass curve. This also leads to ideas for some Morse theory on fractals.

We hope that this volume will enable mathematicians and scientists alike to expand their knowledge in the field of fractal geometry. To ensure high quality, all articles have been peer-reviewed by experts of the respective topic and double-checked by the editors. We aimed to make these proceedings accessible and interesting to a wide audience; including young researchers, graduate students, post doctorates, as well as experts in the field, and scientists from other fields. Our hope is that through this volume, knowledge will be shared, communication be supported, collaboration be initiated, and community be established among researchers with connections to fractal geometry.

Hafedh Herichi
Maria Rosaria Lancia
Therese Basa Landry
Anna Rozanova-Pierrat
Steffen Winter
Editors

Contemporary Mathematics
Volume **828**, 2025
https://doi.org/10.1090/conm/828/16591

Time fractional equations on Sierpinski gasket like fractals

Raffaela Capitanelli

ABSTRACT. The present paper deals with solutions of time fractional equations on Sierpinski gasket like fractals. Aim is to show that a heat kernel estimate can be related either to a time fractional diffusion on the Sierpinski gasket or to a diffusion on a variant of the Sierpinski gasket.

1. Introduction

In recent years, anomalous diffusion phenomena have been extensively studied as they occur in diverse fields including biology, physics, chemistry, engineering like, for example, oxygen in human lungs, ions near rough electrodes, reactive species in porous catalysts (see, for instance, [**6**], [**21**], [**23**], [**30**] and the references therein).

All these phenomena are characterized by the fact that the corresponding mean squared displacement is proportional to a power of time.

This anomalous behaviour can be caused by the presence of *non-homogeneous* media like fractals and by the action of *non-Brownian* type processes like time changed processes.

Merging the theory of time changed processes with the theory of fractal sets is a challenging task and aim of the present paper is just to investigate time changed processes on a class of self-similar fractals. More precisely, the present paper deals with heat kernel estimates for time fractional equations on Sierpinski gasket like fractals.

Literature on time fractional Cauchy problems is very extensive both from the probability and the analysis point of view. We mention only few works dealing with basic and fundamental aspects [**1**], [**5**], [**15**], [**16**], [**20**], [**25**] but we are not exhaustive certainly. In [**8**], we have established existence, uniqueness and convergence results of solutions to general time fractional equations by giving the probabilistic representations. In particular, we study fractional Cauchy problem with Robin condition on the pre-fractal boundary obtaining asymptotic results for the corresponding fractional diffusions with Robin, Neumann and Dirichlet boundary conditions

2020 *Mathematics Subject Classification.* Primary 26A33; Secondary 35R11, 60J50, 35A70.
Key words and phrases. Time changes, fractional operators, time-fractional equations, fractals.

The author is member of GNAMPA (INdAM) and is partially supported by Grants Ateneo "Sapienza" 2022-RM1221816C364975 "Fractal and Fractional" and PRIN 2022 (APRIDACAS) Anomalous Phenomena on Regular and Irregular Domains: Approximating Complexity for the Applied Sciences Funded by EU - Next Generation EU CUP B53D23009540006 - CP 2022XZSAFN - PNRR M4.C2.1.1.

on the fractal domain in [**10**]. We point out that the non-local time-operator we deal with is very general and covers a huge class of non-local (convolution type) operators. From the probabilistic point of view, non-local time-operator are related to time changed Brownian motion where the time change is given by an inverse to a subordinator defined by a symbol which is a Bernstein function. Such operators have been considered also in the papers [**11**] and [**28**]. Furthermore in [**13**] and [**14**] the authors proved sharp two-sided heat kernel estimates for fundamental solutions of general time fractional equations in metric measure spaces. At this point, we quote from page 149 of the fundamental book [**29**] *Differential equations on fractals. A tutorial* of R. Strichartz "*The study of the heat equation, the heat kernel and heat kernel estimates has long been an important part of analysis for a wide variety of Laplacian. In recent decades it has taken on added significance because of many works showing that, starting with heat kernel estimates, it is possible to derive many desirable consequences. We might paraphrase this philosophy as: "Whatever you want, first get heat kernel estimates.""*

Then our aim is to obtain heat kernel estimates related to time changed diffusions on Sierpinski gasket type fractals (see Theorem 4.1). As byproduct we show that the same heat kernel estimate can be related either to a time fractional diffusion on Sierpinski gasket or a diffusion on a variant of Sierpinski gasket (see Theorem 4.2) In particular we discuss as a process on irregular domains can be considered as a suitable time fractional Cauchy problem on a *simpler* (irregular) domain.

The plan of the paper is the following.

First we introduce the base process X and the corresponding Dirichlet form and we define the time changes, the fractional operators and the fractional equations governing these time changed processes in Section 2.

In Section 3 we recall the definitions and the main properties of Sierpinski gasket $SG(2)$ and of the variant of Sierpinski gasket $SG(3)$.

Finally, in Section 4 we obtain heat kernel estimates related to time changed diffusions on such fractals and we conclude with some remarks on such estimates.

2. Time changed processes

In this section we introduce the setting for the base process X and the corresponding time changes of the base process (see, for instance, [**12**] and [**18**]).

Let E be a locally compact, separable metric space and $E_\partial = E \cup \{\partial\}$ be the one-point compactification of E. Let $\mathcal{B}(E)$ be the σ-field of the Borel sets in E where $\mathcal{B}_\partial$ is the σ-field in E_∂.

We consider the symmetric Markov process $X = \{X_t, t \geq 0\}$ on $(E, \mathcal{B}(E))$ with infinitesimal generator $(A, D(A))$ and with transition function $p(t, x, B)$ on $[0, \infty) \times E \times \mathcal{B}(E)$. The point ∂ is the cemetery point for X and a function f on E can be extended to E_∂ by setting $f(\partial) = 0$. The associated semigroup is uniquely defined by

$$P_t f(x) := \int_E f(y) p(t, x, dy) = \mathbf{E}_x[f(X_t)], \quad f \in C_\infty(E)$$

with $X_0 = x \in E$ where $\mathbf{E}_x$ denote the mean value with respect to the probability measure

$$\mathbf{P}_x(X_t \in dy) = p(t, x, dy)$$

and C_∞ is the set of continuous function $C(E)$ on E such that $f(x) \to 0$ as $x \to \partial$.

The Dirichlet form associated with (the non-positive definite, self-adjoint operator) A is defined by $\mathcal{E}(u,v) = (\sqrt{-A}u, \sqrt{-A}v)$ with domain $D(\mathcal{E}) = D(\sqrt{-A})$.

We point out that X is equivalent to an m-symmetric Hunt process whose Dirichlet form $(\mathcal{E}, D(\mathcal{E}))$ is on $L^2(E)$ (see the books [**12, 18**]). Without restrictions we assume that the form is regular ([**18**, page 143]).

Now we introduce the time changes of the base process X.

We consider the subordinator $H = \{H_t, t \geq 0\}$ that is a non-decreasing Lèvy process for which

$$\mathbf{E}_0[\exp(-\lambda H_t)] = \exp(-t\Phi(\lambda))$$

where Φ is the symbol of H. We recall that there is a 1-to-1 correspondence between subordinators and Bernstein functions (see the books [**7, 27**]). The symbol Φ we consider is a Bernstein function with the following representation

$$\Phi(\lambda) = \int_0^\infty \left(1 - e^{-\lambda z}\right) \Pi(dz), \quad \lambda \geq 0 \tag{2.1}$$

where Π on $(0,\infty)$ with $\int_0^\infty (1 \wedge z)\Pi(dz) < \infty$ is the associated *Lévy measure.* We focus only on strictly increasing subordinators with infinite measures that is $\Pi((0,\infty)) = +\infty$.

We point out that by introducing the *tail of the Lévy measure*

$$\overline{\Pi} = \Pi((z,\infty))$$

we have that

$$\frac{\Phi(\lambda)}{\lambda} = \int_0^\infty e^{-\lambda z}\overline{\Pi}(z)dz. \tag{2.2}$$

The same symbol Φ can be associated to the inverse L of H , that is

$$L = \{L_t, t \geq 0\}$$

defined as

$$L_t = \inf\{s \geq 0 \, : \, H_s > t\}, \quad t \geq 0.$$

We assume that $H_0 = 0$, $L_0 = 0$. By definition, we also have that

$$\mathbf{P}_0(H_t < s) = \mathbf{P}_0(L_s > t), \quad s,t > 0. \tag{2.3}$$

Both random times H, L are not decreasing and the inverse process L turns out to be a continuous process as we focus only on strictly increasing subordinators with infinite measures.

We now introduce the fractional operators and the fractional equations governing the time changed process $X^L = \{X \circ L_t, t \geq 0\}$, that is the base process $X = \{X_t, t \geq 0\}$ with the time change L characterized by the symbol Φ. Let $M > 0$ and $w \geq 0$. Let $\mathcal{M}_w$ be the set of (piecewise) continuous function on $[0,\infty)$ of exponential order w such that $|u(t)| \leq Me^{wt}$. Denote by $\widetilde{u}$ the Laplace transform of u. Then, we define the operator

$$\mathfrak{D}_t^\Phi : \mathcal{M}_w \mapsto \mathcal{M}_w$$

such that

$$\int_0^\infty e^{-\lambda t}\mathfrak{D}_t^\Phi u(t)\, dt = \Phi(\lambda)\widetilde{u}(\lambda) - \frac{\Phi(\lambda)}{\lambda}u(0), \quad \lambda > w$$

where Φ is given in (2.1). Since u is exponentially bounded, the integral $\widetilde{u}$ is absolutely convergent for $\lambda > w$. We point out that

$$\Phi(\lambda)\widetilde{u}(\lambda) - \frac{\Phi(\lambda)}{\lambda}u(0) = (\lambda\widetilde{u}(\lambda) - u(0))\,\frac{\Phi(\lambda)}{\lambda}. \tag{2.4}$$

Then the fractional derivative $\mathfrak{D}_t^\Phi$ can be written as a convolution involving the ordinary derivative and the inverse transform of (2.2) iff $u \in \mathcal{M}_w \cap C([0,\infty), \mathbb{R}_+)$ and $u' \in \mathcal{M}_w$, that is,

$$\mathfrak{D}_t^\Phi u(t) = \int_0^t u'(s)\overline{\Pi}(t-s)ds.$$

We notice that when $\Phi(\lambda) = \lambda$ we have that a.s. $H_t = t$ and $L_t = t$ and the operator $\mathfrak{D}_t^\Phi u(t)$ is the ordinary time derivative.

In the case $\Phi(\lambda) = \lambda^\beta$, that is $\Phi(\lambda)$ is the symbol of a stable subordinator, the operator $\mathfrak{D}_t^\Phi$ becomes the Caputo fractional derivative

$$\mathfrak{D}_t^\Phi u(t) = \frac{1}{\Gamma(1-\beta)}\int_0^t \frac{u'(s)}{(t-s)^\beta}ds \tag{2.5}$$

with $u'(s) = du/ds$ and Γ is the Gamma function, that is, $\Gamma(\beta) = \int_0^\infty t^{\beta-1}e^{-s}ds$.

We point out that in the case $\Phi(\lambda) = (\lambda+\eta)^\beta - \eta^\beta$, with $\eta \geq 0$ and $\beta \in (0,1)$, the operator $\mathfrak{D}_t^\Phi$ becomes the so called tempered fractional derivative

$$\mathfrak{D}_t^\Phi u(t) = \int_0^t u'(s)\overline{\Pi}(t-s)\,ds.$$

with

$$\overline{\Pi}(s) = \frac{\beta}{\Gamma(1-\beta)}\int_s^\infty e^{-\eta z}z^{-\beta-1}\,dz.$$

For other explicit representation of the operator $\mathfrak{D}_t^\Phi$ see also the works [**11**], [**28**] and the references therein.

Now we introduce time fractional equation related to the base process X with generator $(A, D(A))$ introduced above.

More precisely, we consider the following time fractional equation

$$\mathfrak{D}_t^\Phi u = Au, \qquad u_0 = f \in D(A). \tag{2.6}$$

The fractional Cauchy problem has been investigated by many authors by considering Caputo derivative and only recently by taking into account more general operators.

The probabilistic representation of the solution to (2.6) can be written in terms of the time changed process X^L, that is,

$$u(t,x) = \mathbf{E}_x[f(X_t^L)] = \int_0^\infty P_s f(x)\,\mathbf{P}_0(L_t \in ds), \quad x \in E,\, t > 0. \tag{2.7}$$

We point out that (2.7) is not a semigroup, indeed the random time L is not Markovian and therefore, the composition X^L is not a Markov process.

We recall the following theorem (proved in [**8**]) where existence and uniqueness of strong solutions to general time fractional equations have been obtained for Feller processes (not necessarily Feller diffusions)

THEOREM 2.1. *The function* (2.7) *is the unique strong solution in* $L^2(E)$ *to* (2.6) *in the sense that:*

(1) $\varphi : t \mapsto u(t,\cdot)$ *is such that* $\varphi \in C([0,\infty), \mathbb{R}_+)$ *and* $\varphi' \in \mathcal{M}_0$,

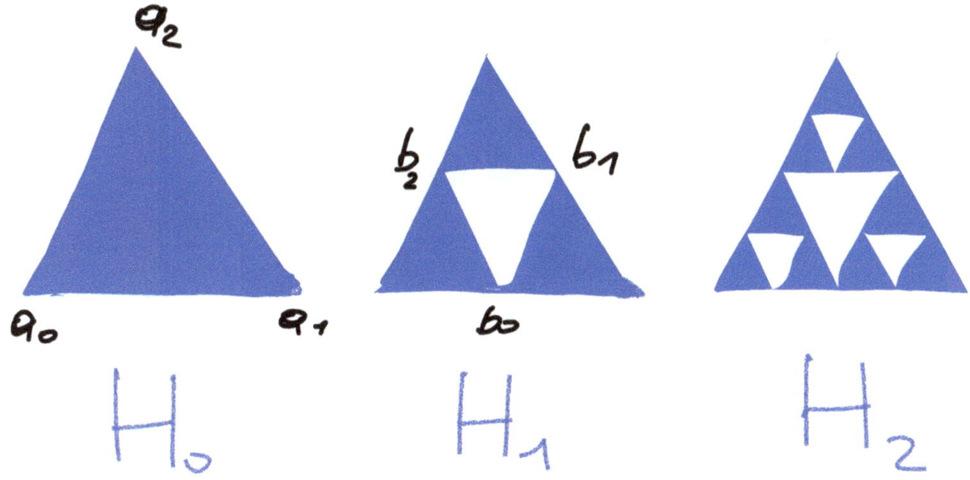

FIGURE 1. H_0, H_1, H_2 for SG(2)

(2) $\vartheta : x \mapsto u(\cdot, x)$ *is such that* $\vartheta, A\vartheta \in D(A)$,
(3) $\forall\, t > 0$, $\mathfrak{D}_t^{\Phi} u(t,x) = Au(t,x)$ *holds a.e in* E
(4) $\forall\, x \in E$, $u(t,x) \to f(x)$ *as* $t \downarrow 0$.

We point out that the result in Theorem 2.1 has been proved in the general setting of Feller process (right-continuous with no discontinuity other than jumps) following a very different approach with respect to [**11**], [**13**], [**14**].

3. Sierpinski gasket type fractals

In this section we recall the construction of Sierpinski gasket type fractals. We point out that Sierpinski gasket type fractals have two main geometrical features: such structures exist in all Euclidean dimensions and they exhibit self-similarity at all scales. In the following we consider Sierpinski gasket type fractals in the plane (see [**2**], [**3**], [**19**], [**22**], [**24**], [**26**], [**29**] and the reference therein).

We start by considering the standard Sierpinski gasket $SG(2)$.

Let $G_0 = \{a_0, a_1, a_2\}$ with $a_0 = (0,0)$, $a_1 = (1,0)$ and $a_2 = (\frac{1}{2}, \frac{\sqrt{3}}{2})$ be the vertices of the unit triangle in $\mathbb{R}^2$, and let H_0 be the closed convex hull of G_0. The construction of the Sierpinski gasket $SG(2)$ can be obtained by the following Cantor-type subtraction procedure. Let $\{b_0, b_1, b_2\}$ be the midpoints of the 3 sides of G_0, that is $b_0 = (\frac{1}{2}, 0)$, $b_1 = (\frac{3}{4}, \frac{\sqrt{3}}{2})$ and $b_2 = (\frac{1}{4}, \frac{\sqrt{3}}{4})$ and let A be the interior of the triangle with vertices $\{b_0, b_1, b_2\}$. Let $H_1 = H_0 \setminus A$, so H_1 consists of 3 closed upward facing triangles, each of side $\frac{1}{2}$. Now by repeating the operation on each of these triangles we obtain the set H_2 consisting of 9 upward closed upward facing triangles each of side 2^{-2} (see figure 1). Continuing in this way, we obtain a decreasing sequence of closed non-empty sets H_n, and set

$$SG(2) = \cap_{n=0}^{\infty} H_n.$$

We point out that the Sierpinski gasket can be obtained from the general theory of self-similar set as the unique compact set invariant with respect to the set of the 3

FIGURE 2. The third step of SG(2)

following contractive similarities with contraction factor $\frac{1}{2}$

$$\psi_i(x) = a_i + \frac{1}{2}(x - a_i), \qquad i = 0, 1, 2$$

that is

$$SG(2) = \bigcup_{i_1,\ldots,i_n=1}^{4} \psi_i(SG(2))$$

(see figure 2).

We introduce the following notation for set $A \subset \mathbb{R}^2$ and for a family of N contractive similarities: we define, for arbitrary n-tuples of indices $i_1, \ldots, i_n \in \{1, \ldots, N\}$, $\psi_{i_1 \ldots i_n} := \psi_{i_1} \circ \cdots \circ \psi_{i_n}$, $A_{i_1 \ldots i_n} := \psi_{i_1 \ldots i_m}(A)$, $A^{(0)} = A$ and

$$A^{(n)} = \bigcup_{i_1,\ldots,i_n=1}^{N} A_{i_1 \ldots i_n}.$$

If we set $V_\star = \cup_{n \geq 0}(G_0)^{(n)}$; the set $\bar{V_\star}$, that is, the closure in $\mathbb{R}^2$ of $V_\star$ is again the Sierpinski gasket $SG(2)$.

Moreover, as the open set condition holds, there exists an unique invariant measure μ with the respect the same self-similar family (see [**19**]) which is given, after normalization, by the the restriction to K of the d_f-dimensional Hausdorff measure of $\mathbb{R}^2$, that is,

$$\mu = \frac{H^{d_f}(\cdot)\lfloor K}{(H^{d_f}(K))}$$

where $d_f = \frac{\log 4}{\log 3}$.

On this set it is possible to construct self-similar energy forms by using Kigami's idea to approximate the fractal energy form by a sequence of renormalized energy forms on finite graphs (see [**22**]). More precisely, we define, for $u : V_\star \to \mathbb{R}$,

$$\mathcal{E}_n^{(2)}[u] = \frac{1}{2}\left(\frac{5}{3}\right)^n \sum_{i_1,\ldots,i_n=1}^{3} \sum_{\xi,\eta \in G_0} |u(\psi_{i_1 \ldots i_n}(\xi)) - u(\psi_{i_1 \ldots i_n}(\eta))|^2.$$

By using the decimation principle, the sequence $\mathcal{E}_n^{(p)}$ is non-decreasing, and by defining for $u : V_\star \to \mathbb{R}$

$$\mathcal{E}^{(2)}[u] = \lim_{n\to\infty} \mathcal{E}_n^{(2)}[u],$$

the set

$$\mathcal{F}_\star^{(2)} = \{u : V_\star \to \mathbb{R} : \mathcal{E}^{(2)}[u] < \infty\}$$

does not degenerate to a space containing only constant functions. Moreover each $f \subset \mathcal{F}_\star^{(2)}$ can be uniquely extended in $C(SG(2))$ and so we define the space

$$\mathcal{F}^{(2)} = \{u \in C(SG(2)) : \mathcal{E}^{(2)}[f] < \infty\},$$

where $\mathcal{E}^{(2)}[u] := \mathcal{E}^{(2)}[u|_{V^\star}]$. Hence $\mathcal{F}^{(2)} \subset C(SG(2)) \subset L^2(\mu)$. By polarization, we obtain that $(\mathcal{E}^{(2)}, \mathcal{F}^{(2)})$ is a local regular Dirichlet form on $L^2(\mu)$; moreover it is self-similar

$$\mathcal{E}^{(2)}[u] = \sum_{i=1}^{4} \frac{5}{3} \mathcal{E}^{(2)}[u \circ \psi_i].$$

As $(\mathcal{E}^{(2)}, \mathcal{F}^{(2)})$ is a closed form on $L^2(\mu)$, there exists a unique self-adjoint, non-positive operator Δ on $L^2(\mu)$ - with domain $D(\Delta)$ - such that

$$\mathcal{E}^{(2)}(u,v) = -\int_{SG(2)} (\Delta u) v \, d\mu.$$

The operator Δ is called the Laplacian on the Sierpinski gasket $SG(2)$.

In a similar way we can construct a variant of Sierpinski gasket $SG(3)$ by the same procedure seen before by subdividing now the triangle into 9 smaller triangles, and deleting the 3 "downward facing" ones. More precisely, the construction of the variant Sierpinski gasket $SG(3)$ can be obtained by starting again from the triangle G_0. Let $\{c_0, c_1\}$ be the points that divide the side with vertices $\{a_0, a_1\}$ of G_0 in third equal parts, that is $c_0 = (\frac{1}{3}, 0)$ $c_1 = (\frac{2}{3}, 0)$; let $\{c_2, c_3\}$ be the points that divide the side with vertices $\{a_1, a_2\}$ of G_0 in third equal parts, that is $c_3 = (\frac{2}{3}, \sqrt{3}/3)$ $c_2 = (5/6, \sqrt{3}/6)$, let $\{c_4, c_5\}$ be the points that divide the side with vertices $\{a_0, a_2\}$ of G_0 in third equal parts, that is $c_5 = (\frac{1}{6}, \sqrt{3}/6)$ $c_4 = (1/3, \sqrt{3}/3)$. Moreover we consider the point $c_6 = (1/2, \sqrt{3}/6)$. Let A_1 be the interior of the triangle with vertices $\{c_0, c_6, c_5\}$; let A_2 be the interior of the triangle with vertices $\{c_1, c_2, c_6\}$ and let A_3 be the interior of the triangle with vertices $\{c_6, c_3, c_4\}$. Let $H_1 = H_0 \setminus (A_1 \cup A_2 \cup A_3)$, so H_1 consists of 6 closed upward facing triangles, each of side $\frac{1}{3}$. Now repeat the operation on each of these triangles to obtain the set H_2 consisting of 36 upward closed upward facing triangles, each of side $\frac{1}{9}$. Continuing in this way, we obtain a decreasing sequence of closed non-empty sets H_n, and set $SG(3) = \cap_{n=0}^{\infty} H_n$.

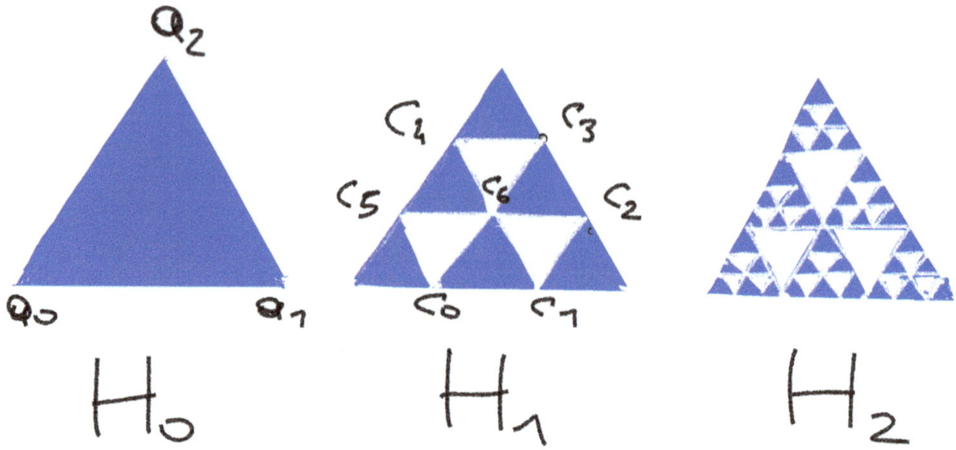

FIGURE 3. H_0, H_1, H_2 for SG(3)

Moreover, also the Sierpinski gasket $SG(3)$ can be defined as the unique compact set invariant with respect to the set of the following $N = 6$ contractive similarities with contraction factor $\frac{1}{3}$:

$$\varphi_i(x) = a_i + \frac{1}{3}(x - a_i), \qquad i = 0, 1, 2,$$

$$\varphi_3(x) = \frac{1}{3}x + c_0, \varphi_4(x) = \frac{1}{3}x + c_6, \varphi_5(x) = \frac{1}{3}x + c_5.$$

Moreover, if we define $V_0 = (G_0)^{(n)}$, $V_n = (G_0)^{(n)}$ for $n > 1$, $V_\star = \cup_{n \geq 0} V_n$ the set $K = \bar{V_\star}$ is again the Sierpinski gasket $SG(3)$.

Furthermore, as the open set condition holds with the same open triangle, we can characterize the unique Borel regular measure invariant with the respect the same self-similar family as the normalized Hausdorff measure with Hausdorff dimension $d_f = \frac{\log 6}{\log 3}$. In order to construct the energy on SG(3) we start considering for $f : V_\star \to \mathbb{R}$

$$\mathcal{E}_n^{(3)}[f] = \frac{1}{2}\left(\frac{15}{7}\right)^n \sum_{i_1,\dots,i_n=1}^{3} \sum_{\xi,\eta \in G_0} |u(\psi_{i_1\dots i_n}(\xi)) - u(\psi_{i_1\dots i_n}(\eta))|^2.$$

FIGURE 4. The third step of SG(3)

We point out that the renormalization energy factor $\rho = \frac{15}{7}$ can be obtained by the decimation procedure in the following way. Suppose u defined on G_0 by $u(a_0) = 1$ and $u(a_1) = 0$ $u(a_2) = 0$ and we want to extend u on V_1 to minimize energy that is we are seeking for the harmonic extension of the function u from V_0 to V_1. By using the symmetry, setting $u(c_0) = u(c_5) = x$, $u(c_1) = u(c_4) = y$, $u(c_2) = u(c_3) = w$, $u(c_6) = z$ we have to minimize the following energy

$$\mathcal{E}_1^{(3)}[u] = \frac{1}{2}\rho((1-x)^2+(y-w)^2+(x-y)^2+y^2+w^2+(x-z)^2+(z-y)^2+(z-w)^2).$$

and we obtain $x = \frac{8}{5}$, $y = \frac{4}{15}$ $w = \frac{4}{15}$, $z = \frac{1}{3}$.

We point out that the same values on $V_1 \setminus V_0$ can be obtained as the average of the values at the neighbouring points in V_1

The value of ρ is such that $\rho \min \mathcal{E}_1^{(3)} = \mathcal{E}_0^{(3)}$ that is $\rho 7/15 = 1$. We recall that by using symmetry, the finite ramification and self-similarity it is enough to study only the first step (see [**22**] and [**29**]). Once again, as before for the Sierpinski Gasket $SG(2)$, we can construct the associated Dirichlet form and the corresponding Laplacian for the Sierpinski Gasket $SG(3)$.

We remark that it is possible to consider other variant of Sierpinski gasket by changing in a suitable way the number of the contractive similarities and the corresponding contraction factor but, for our analysis, it is enough to focus our attention on $SG(2)$ and $SG(3)$.

4. Heat kernel estimates

In the following, we write $h(s) \simeq f(s)$ if there exist constants $c_1, c_2 > 0$ such that $c_1 f(s) \leq h(s) \leq c_2 f(s)$ for the specified range of the argument s. Similarly, we write $h(s) \approx f(s)g(s)$, if there exist constants $C_1, c_1, C_2, c_2 > 0$ such that $f(C_1 s)g(c_1 s) \leq h(s) \leq f(C_2 s)g(c_2 s)$ for the specified range of s.

Heat kernel estimates related to the *Laplacian* on the two-dimensional Sierpinski gasket $SG(2)$ were first proved in [**4**] (see also [**2**]). More precisely, it is shown by Barlow and Perkins in [4] that the following heat kernel estimate hold for $SG(2)$

$$p_0(t,x,y) \approx t^{-\frac{d}{\alpha}} \exp\left(-\left(\frac{d(x,y)}{t^{\frac{1}{\alpha}}}\right)^{\frac{\alpha}{\alpha-1}}\right) \tag{4.1}$$

with

$$d = d_f = \frac{\log 3}{\log 2} \qquad \alpha = d_w := \frac{\log 5}{\log 2} \qquad (SG(2)). \tag{4.2}$$

Similar estimates hold for other fractals with the appropriate self similar dimension $d_f = \frac{\log N}{\log \gamma}$ and walk dimension $d_w = \frac{\log \rho N}{\log \gamma}$, where N is the number of contractive similarities with contraction factor $\frac{1}{\gamma}$ and ρ is the renormalization energy factor. In particular, for the variant of Sierpinski gasket $SG(3)$ we have that estimates (4.1) holds with

$$d = d_f = \frac{\log 6}{\log 3} \qquad \alpha = d_w := \frac{\log(\frac{90}{7})}{\log 3} \qquad (SG(3)). \tag{4.3}$$

We recall that the walk dimension d_w can be expressed by the Einstein relation

$$d_w = 2\frac{d_f}{d_s}$$

where d_s is the called spectral dimension. We point out that the fractal dimension relates the volume of small balls with their radii and the walk dimension relates mean exit times from balls with their radii (see, for instance, [**2**], [**17**], [**26**] and the references therein).

We point out that for the transition density of standard Brownian motion on $\mathbb{R}^d$ satisfies with $d = d_s = d_f$ and $\alpha = d_w = 2$.

We now consider the following time fractional equation on $SG(2)$

$$\mathfrak{D}_t^{\Phi} u = \Delta u, \qquad u_0 = f \in D(\Delta). \tag{4.4}$$

in the case $\Phi(\lambda) = \lambda^\beta$, that is $\Phi(\lambda)$ is the symbol of a stable subordinator.

In the following Theorem 4.1 we establish heat kernel estimates for time fractional equation (4.4) on $SG(2)$.

THEOREM 4.1. *The fundamental solution $q(t,x,y)$ for the time fractional equation* (4.4) *on $SG(2)$ with $\Phi(\lambda) = \lambda^\beta$, $0 < \beta < 1$, satisfies the following estimates*

$$q(t,x,y) \simeq t^{\beta-1-\frac{\beta d}{\alpha}} \qquad \text{if} \qquad d(x,y) \le t^{\frac{\beta}{\alpha}}$$

and

$$q(t,x,y) \approx t^{\beta-1-\frac{\beta d}{\alpha}} \exp(-(d(x,y)/t^{\beta/\alpha})^{\frac{\alpha}{\alpha-\beta}}) \qquad \text{if} \qquad d(x,y) \ge t^{\frac{\beta}{\alpha}}$$

with $d = d_f = \frac{\log 3}{\log 2}$ *and* $\alpha = d_w := \frac{\log 5}{\log 2}$.

PROOF. The result follows from the estimates proved in [**13**] and [**14**]) in the case of β-stable subordinator with $0 < \beta < 1$. In fact, in Theorem 1.4 of [**14**] the authors derive the heat kernel estimates for time fractional equations in locally compact separable metric space E. More precisely, if the fundamental solution $p(t,x,y)$ satisfies

$$p(t,x,y) \asymp \frac{1}{t^{d/\alpha}} \exp(-\left(\frac{d(x,y)}{t^{1/\alpha}}\right)^{\alpha/(\alpha-1)}), \quad t > 0, x, y \in M, \tag{4.5}$$

with $\alpha \geq 2$ ($\mu(B(x,r)) \simeq r^d$ for all $x \in M$ and $r > 0$) then the fundamental solution $q(t,x,y)$ for the time fractional equation with $\Phi(\lambda) = \lambda^\beta$, $0 < \beta < 1$, satisfies

$$q(t,x,y) \simeq H_{\leq 1}(t, d(x,y)) \quad \text{if } d(x,y)t^{-\beta/\alpha} \leq 1,$$
$$q(t,x,y) \asymp H^{(c)}_{\geq 1}(t, d(x,y)) \quad \text{if } d(x,y)t^{-\beta/\alpha} \geq 1$$

where

$$H_{\leq 1}(t,r) = \begin{cases} t^{\beta-1-\beta d/\alpha}, & \text{if } d < 2\alpha, \\ t^{-1-\beta}\log\left(\dfrac{2t^\beta}{r^\alpha}\right), & \text{if } d = 2\alpha, \\ t^{-1-\beta}/r^{d-2\alpha}, & \text{if } d > 2\alpha, \end{cases}$$

and

$$H^{(c)}_{\geq 1}(t,r) = t^{\beta-1-\beta d/\alpha}\exp\Big(-(r^\alpha/t^\beta)^{1/(\alpha-\beta)}\Big).$$

Therefore, starting from the estimates (4.1) on the Sierpinski gasket $SG(2)$ as $d = d_f = \frac{\log 3}{\log 2} < \alpha = d_w := \frac{\log 5}{\log 2}$ we achieve the result. □

We focus our attention on-diagonal estimates. From Theorem 4.1 we obtain the following Theorem 4.2 by a straightforward calculation.

THEOREM 4.2. *The fundamental solution $q(t,x,y)$ for the time fractional equation (4.4) on $SG(2)$ with $\Phi(\lambda) = \lambda^\beta$ with*

$$\beta = \frac{1 - \frac{\log 6}{\log(\frac{90}{7})}}{1 - \frac{\log 3}{\log 5}}$$

satisfies the following estimates

$$q(t,x,x) \simeq t^{-d/\alpha} \tag{4.6}$$

with $d = \frac{\log 6}{\log 3}$ and $\alpha = \frac{\log(\frac{90}{7})}{\log 3}$.

PROOF. We consider on-diagonal heat kernel estimates for the time fractional equation on $SG(2)$

$$q(t,x,x) \simeq t^{\beta-1-\frac{\beta d}{\alpha}}$$

with $d = d_f = \frac{\log 3}{\log 2}$ and $\alpha = d_w := \frac{\log 5}{\log 2}$.

The idea of the proof is to find a value of β such that the exponent of on-diagonal heat kernel estimates for the time fractional equation on $SG(2)$ is equal to the exponent of on-diagonal heat kernel estimates for the heat equation on $SG(3)$ that is equivalent to solve the following equation

$$\beta - 1 - \beta\frac{\frac{\log 3}{\log 2}}{\frac{\log 5}{\log 2}} = -\frac{\frac{\log 6}{\log 3}}{\frac{\log(\frac{90}{7})}{\log 3}}.$$

□

Theorem 4.2 gives us a on-diagonal heat kernel estimates for the time fractional equation on $SG(2)$ that is *comparable* to on-diagonal heat kernel estimates for the heat equation on $SG(3)$. From this, we obtain that the same heat kernel estimate can be related either to a time fractional diffusion on Sierpinski gasket or a diffusion on a variant of Sierpinski gasket.

In particular, Theorem 4.2 underlines the link between the features of the medium and the anomalous diffusions which are well-described by time changed processes. Moreover, Theorem 4.2 show as a process on irregular domains like $SG(3)$ can be considered as a suitable time fractional Cauchy problem on the *simpler* Sierpinki gasket $SG(2)$.

As it is possible to define and construct Sierpinski gasket type fractals in any dimension, we can obtain examples with same heat kernel estimates in any dimension. Furthermore, it is also possible to consider homogeneous random Sierpinski gaskets or scale irregular Sierpinski gaskets as in [**3**] and we can obtain examples with same heat kernel estimates also for not self similar structures.

Then we might paraphrase the philosophy "*Whatever you want, first get heat kernel estimates*" with the following philosophy "*Whatever you get the heat kernel estimates, first decide what you want.*"

In particular, we could approach the study of processes on fractal sets by considering *time changed* processes on *simpler* domains and viceversa.

As the fractal sets and fractals operators are the result of an asymptotic process (the one induced by the infinite iterations of a family of contractive similarities), this approach could be useful for the numerical approximation of the anomalous phenomena on fractal structures (where the asymptotic feature of fractals introduces an additional approximation level to those already inherent in the numerical approximation of the classical at hand) by considering structures with less ramifications at all scales inside. By the same approach, it is also possible to study of delayed and rushed anomalous diffusions processes (see [**9**]) on trap domains and their approximation: this is a working in progress paper with M. D'Ovidio.

References

[1] Mark Allen, Luis Caffarelli, and Alexis Vasseur, *A parabolic problem with a fractional time derivative*, Arch. Ration. Mech. Anal. **221** (2016), no. 2, 603–630, DOI 10.1007/s00205-016-0969-z. MR3488533

[2] Martin T. Barlow, *Diffusions on fractals*, Lectures on probability theory and statistics (Saint-Flour, 1995), Lecture Notes in Math., vol. 1690, Springer, Berlin, 1998, pp. 1–121, DOI 10.1007/BFb0092537. MR1668115

[3] M. T. Barlow and B. M. Hambly, *Transition density estimates for Brownian motion on scale irregular Sierpinski gaskets* (English, with English and French summaries), Ann. Inst. H. Poincaré Probab. Statist. **33** (1997), no. 5, 531–557, DOI 10.1016/S0246-0203(97)80104-5. MR1473565

[4] Martin T. Barlow and Edwin A. Perkins, *Brownian motion on the Sierpiński gasket*, Probab. Theory Related Fields **79** (1988), no. 4, 543–623, DOI 10.1007/BF00318785. MR966175

[5] Emilia Bazhlekova, *The abstract Cauchy problem for the fractional evolution equation*, Fract. Calc. Appl. Anal. **1** (1998), no. 3, 255–270. MR1657572

[6] Daniel ben-Avraham and Shlomo Havlin, *Diffusion and reactions in fractals and disordered systems*, Cambridge University Press, Cambridge, 2000, DOI 10.1017/CBO9780511605826. MR1881970

[7] Jean Bertoin, *Subordinators: examples and applications*, Lectures on probability theory and statistics (Saint-Flour, 1997), Lecture Notes in Math., vol. 1717, Springer, Berlin, 1999, pp. 1–91, DOI 10.1007/978-3-540-48115-7_1. MR1746300

[8] Raffaela Capitanelli and Mirko D'Ovidio, *Fractional equations via convergence of forms*, Fract. Calc. Appl. Anal. **22** (2019), no. 4, 844–870, DOI 10.1515/fca-2019-0047. MR4023098

[9] Raffaela Capitanelli and Mirko D'Ovidio, *Delayed and rushed motions through time change*, ALEA Lat. Am. J. Probab. Math. Stat. **17** (2020), no. 1, 183–204, DOI 10.30757/alea.v17-08. MR4105292

[10] Raffaela Capitanelli and Mirko D'Ovidio, *Fractional Cauchy problem on random snowflakes*, J. Evol. Equ. **21** (2021), no. 2, 2123–2140, DOI 10.1007/s00028-021-00673-7. MR4278424

[11] Zhen-Qing Chen, *Time fractional equations and probabilistic representation*, Chaos Solitons Fractals **102** (2017), 168–174, DOI 10.1016/j.chaos.2017.04.029. MR3672008

[12] Zhen-Qing Chen and Masatoshi Fukushima, *Symmetric Markov processes, time change, and boundary theory*, London Mathematical Society Monographs Series, vol. 35, Princeton University Press, Princeton, NJ, 2012. MR2849840

[13] Zhen-Qing Chen, Panki Kim, Takashi Kumagai, and Jian Wang, *Heat kernel estimates for time fractional equations*, Forum Math. **30** (2018), no. 5, 1163–1192, DOI 10.1515/forum-2017-0192. MR3849640

[14] Zhen-Qing Chen, Panki Kim, Takashi Kumagai, and Jian Wang, *Time fractional Poisson equations: representations and estimates*, J. Funct. Anal. **278** (2020), no. 2, 108311, 48, DOI 10.1016/j.jfa.2019.108311. MR4030268

[15] Zhen-Qing Chen, Mark M. Meerschaert, and Erkan Nane, *Space-time fractional diffusion on bounded domains*, J. Math. Anal. Appl. **393** (2012), no. 2, 479–488, DOI 10.1016/j.jmaa.2012.04.032. MR2921690

[16] Samuil D. Eidelman and Anatoly N. Kochubei, *Cauchy problem for fractional diffusion equations*, J. Differential Equations **199** (2004), no. 2, 211–255, DOI 10.1016/j.jde.2003.12.002. MR2047909

[17] U. R. Freiberg, *Einstein relation on fractal objects*. Communications to SIMAI Congress, Vol. 2 (2007).

[18] Masatoshi Fukushima, Yōichi Ōshima, and Masayoshi Takeda, *Dirichlet forms and symmetric Markov processes*, De Gruyter Studies in Mathematics, vol. 19, Walter de Gruyter & Co., Berlin, 1994, DOI 10.1515/9783110889741. MR1303354

[19] John E. Hutchinson, *Fractals and self-similarity*, Indiana Univ. Math. J. **30** (1981), no. 5, 713–747, DOI 10.1512/iumj.1981.30.30055. MR625600

[20] Valentin Keyantuo, Carlos Lizama, and Mahamadi Warma, *Existence, regularity and representation of solutions of time fractional diffusion equations*, Adv. Differential Equations **21** (2016), no. 9-10, 837–886. MR3513120

[21] Anatoly A. Kilbas, Hari M. Srivastava, and Juan J. Trujillo, *Theory and applications of fractional differential equations*, North-Holland Mathematics Studies, vol. 204, Elsevier Science B.V., Amsterdam, 2006. MR2218073

[22] Jun Kigami, *Analysis on fractals*, Cambridge Tracts in Mathematics, vol. 143, Cambridge University Press, Cambridge, 2001, DOI 10.1017/CBO9780511470943. MR1840042

[23] A. N. Kochubeĭ, *Diffusion of fractional order* (Russian), Differentsial′nye Uravneniya **26** (1990), no. 4, 660–670, 733–734; English transl., Differential Equations **26** (1990), no. 4, 485–492. MR1061448

[24] Shigeo Kusuoka, *A diffusion process on a fractal*, Probabilistic methods in mathematical physics (Katata/Kyoto, 1985), Academic Press, Boston, MA, 1987, pp. 251–274. MR933827

[25] Mark M. Meerschaert, Erkan Nane, and P. Vellaisamy, *Fractional Cauchy problems on bounded domains*, Ann. Probab. **37** (2009), no. 3, 979–1007, DOI 10.1214/08-AOP426. MR2537547

[26] Umberto Mosco, *Energy functionals on certain fractal structures*, J. Convex Anal. **9** (2002), no. 2, 581–600. Special issue on optimization (Montpellier, 2000). MR1970574

[27] René L. Schilling, Renming Song, and Zoran Vondraček, *Bernstein functions*, 2nd ed., De Gruyter Studies in Mathematics, vol. 37, Walter de Gruyter & Co., Berlin, 2012. Theory and applications, DOI 10.1515/9783110269338. MR2978140

[28] Bruno Toaldo, *Convolution-type derivatives, hitting-times of subordinators and time-changed C_0-semigroups*, Potential Anal. **42** (2015), no. 1, 115–140, DOI 10.1007/s11118-014-9426-5. MR3297989

[29] Robert S. Strichartz, *Differential equations on fractals: A tutorial*, Princeton University Press, Princeton, NJ, 2006. MR2246975

[30] Vladimir V. Uchaikin, *Fractional derivatives for physicists and engineers. Volume II*, Nonlinear Physical Science, Higher Education Press, Beijing; Springer, Heidelberg, 2013. Applications, DOI 10.1007/978-3-642-33911-0. MR3051755

Department of Basic and Applied Sciences for Engineering Sapienza University of Rome via A. Scarpa 10, Rome, Italy

Email address: raffaela.capitanelli@uniroma1.it

Contemporary Mathematics
Volume **828**, 2025
https://doi.org/10.1090/conm/828/16592

Existence of optimal shapes for heat diffusions across irregular interfaces

Gabriel Claret and Anna Rozanova-Pierrat

Dedicated to the memory of Bernard Sapoval

ABSTRACT. We consider a heat transmission problem across an irregular interface – that is, non-Lipschitz or fractal – between two media (a hot one and a cold one). The interface is modelled as the support of a d-upper regular measure. We introduce the properties of the interior and exterior trace operators for two-sided extension domains, which allow to prove the well-posedness (in the sense of Hadamard) of the problem on a large class of domains, which contains regular domains, but also domains with variable boundary dimension. Then, we prove the convergence in the sense of Mosco of the energy form connected to the heat content of one of the domains and the heat transfer for (ε, ∞)-domains. Finally, we prove the existence of an optimal shape maximizing the heat energy transfer in a class of (ε, ∞)-domains, allowing fractal boundaries, while that optimum can generally not be reached in the class of Lipschitz domains.

1. Introduction

Optimizing the efficiency of heat transfers is a problem which often arises in the design of industrial objects, especially in the context of new technologies (for the cooling of microprocessors for example). It is linked to the heat propagation shortly after the beginning of the cooling of an object. Intuitively, the greater the area of the exchange surface, the most efficient the heat transfer (see Figs. 1 and 2, where the volumes of the initially cold and hot media are preserved, while the lengths of the interfaces increase). The idea that the shape of the interface between media has a significant impact on the speed of the diffusive heat transfer is well known from a physical perspective; for instance, it results from notable works by de Gennes [**DG82**]. The heat propagation can be described by the increase of heat content of the initially cold medium. The matter of the efficiency of the heat transfer is then related with the asymptotic expansion of the heat content for small times t. De Gennes [**DG82**] stated that the asymptotic behaviour of the heat content in an n-dimensional domain is proportional to the measure of the interface multiplied by $t^{(n-d)/2}$, where d is the Hausdorff dimension of the interface.

2020 *Mathematics Subject Classification.* Primary 35K05, 49Q10; Secondary 35D30, 35K20, 35B38, 49J45.

Key words and phrases. Heat equation, shape optimization, trace operators, extension domains, fractals, Mosco convergence.

The heat propagation for the small times is also directly connected to the area of the interior Minkowski sausage (*i.e.* the $\sqrt{t}$-neighbourhood of the interface on the hot side). That argument was investigated experimentally and numerically in the case of prefractal configurations (see [**Man83**]) in [**RPGS12**]. It was established analytically in [**BGRP16**] in the framework of regular and d-set boundaries for the heat transmission model between two media, described by different diffusion coefficients and an interface with finite or infinite resistivity (see (1.1)).

For that matter, it is natural to consider shapes of finite volumes (due to industrial constraints) but with large or rough surfaces, which is why it appears consistent to think of objects with boundaries described by fractals [**Man67**]. Indeed, fractal shapes are self-similar (or scale invariant), which means they present irregularities at all scales, making the perimeter of the object infinite: the Hausdorff dimension of the boundary is non-integer. Therefore, the short-time heat transfer is more efficient across fractal boundaries, the Hausdorff dimension of which is superior to $n-1$, which is the dimension of regular (for instance, Lipschitz) boundaries. Figures 1 and 2 illustrate that phenomenon, representing the cooling of a hot domain into a cold one through generation 0 to 3 Minkowski pre-fractal interfaces. In Figure 1 (generations 0 and 1), the heat only spreads to the part of the cold domain sticking to the boundary. Nonetheless, the enhanced cooling near the angles of the generation 1 curve is already visible. In line with that observation, the heat transfer appears more efficient as the interface grows more irregular: in Figure 2 (generations 2 and 3), the heat spreads further into the cold domain within the same time frame as in the previous cases, progressively reaching the overall shape of the fractal boundary rather than sticking to the interface. Consequently, the hot domain cools down more efficiently (represented by the yellow areas in the hot domains), and even more so for the generation 3 interface.

There are many mathematical results on the asymptotic behaviour of the heat content in the case of one medium with a hot boundary (*i.e.*, Dirichlet boundary condition): for a C^3 boundary [**vdBLG94, vdBG94**], for a polygonal boundary [**vdBS90**] and for a triadic Von Koch snowflake [**FLV95**]. The special case of a countable disjoint union of scaled copies of a given open set in $\mathbb{R}^n$ is studied in [**LV96**]. The case of a continuous Riemannian metric (without discontinuity of the diffusion coefficients across the interface) on smooth compact n-dimensional Riemannian manifolds with a smooth boundary was considered in [**GK03**]. The case of continuous transmission boundary conditions for the expansion of the heat kernel on the diagonal was studied in [**PNB05**]. A survey of results on the asymptotic expansion of the heat kernel for different boundary conditions can be found in [**Vas03**].

Although it is not the approach followed in this work, problems regarding heat diffusion have also been widely studied from a probabilistic perspective, relying on the connection between the analytical concept of Dirichlet forms and the probabilistic notion of Feller semi-group [**CF12**]. More specifically, it is well known that solutions to the heat equation can be expressed in terms of Brownian diffusions, see for instance [**Hun56**, Paragraph 6] and [**Doo55, Ros51**]. In those papers, the solution to the Dirichlet heat problem is expressed in terms of the parabolic measure. Given a point in the domain, the measure of a time interval and a Borel subset of the boundary can be expressed as the probability for a Brownian motion starting from that point to reach the boundary subset within the time interval

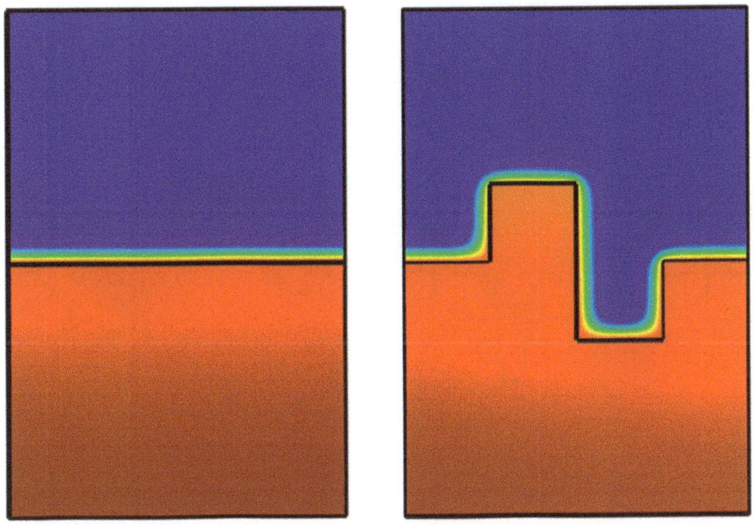

FIGURE 1. Numerical simulation (with COMSOL Multiphysics) of the cooling of a hot medium (in red) by a cold medium (in blue) across a generation 0 (on the left) and a generation 1 (on the right) Minkowski pre-fractal curve. The propagation time is the same for both figures. The heat distributions satisfy homogeneous Neumann boundary conditions on the outer boundaries (a thermally insulated cavity) and the diffusion coefficients of two media are supposed to be distinct. The volume of the media is constant, due the symmetric property of Minkowski pre-fractal curve.

(see also [**Wu79**]). With that approach, domain regularity is of little relevance: boundary points are split into regular points (*i.e.*, starting points for Brownian motions hitting the boundary arbitrarily soon almost surely) and irregular points. The latter form a null set for the parabolic measure [**Doo54**] and are connected to the thinness of the domain. Though the domain under study needs no specific regularity, the boundary conditions are seen as continuous functions. The analytical approach however demands some boundary regularity but allows less smooth conditions. For that matter, rather than strictly equivalent, the analytical and probabilistic approaches to the heat problem can be regarded as complementary.

Generally speaking, fractal geometries have been vastly studied for their potential industrial applications and their omnipresence in nature, see [**Man83**], as well as [**SRG85**] for instance. From a mathematical perspective, various types of heat models have been studied on irregular and fractal domains. In [**CD21**], the heat equation and other diffusion models are studied on random Koch snowflakes for Dirichlet, Neumann and Robin boundary conditions by approximation of the fractal shape by the associated pre-fractal sequence. In [**LV14**], the existence and uniqueness of solutions to diffusion problems on Koch domains with Wentzell boundary conditions are proved. In [**CCLRC23**] the heat exchange interface consists of a so-called Koch-mixture. See [**Cap10, Lan04, MV03**] for other problems set on domains with Koch fractal boundaries. Those boundaries are examples of d-sets [**JW84**] for $d = \ln 4/\ln 3$, which have a fixed Hausdorff dimension (which

FIGURE 2. Numerical simulation (COMSOL Multiphysics) of the cooling of a hot medium (in red) by a cold medium (in blue) across a generation 2 (on the left) and a generation 3 (on the right) Minkowski pre-fractal curve. The cooling time and boundary conditions are the same as for Figure 1.

is d). So far, from an analytical perspective, problems regarding heat diffusions on irregular shapes have been set on domains with at least d-set boundaries, see for instance [**BGRP16, CL23**]. Other boundary value problems have been studied on domains with d-set boundaries (see [**ARP19**] on the Poincaré-Steklov – or Dirichlet-to-Neumann – operator for Laplacian transport with Robin boundary conditions), but also boundaries with varying Hausdorff dimension. Here, as in [**HMRP$^+$21, DRPT22**] (see also [**LRP21**] for a more detailed discussion), we use boundaries described as the support of a finite upper-regular Borel measure, as in [**HRPT23**], where Poisson-type problems are studied on domains with boundaries of bounded Hausdorff dimension, with Dirichlet, Neumann and Robin boundary conditions. Recently, problems with Dirichlet and Neumann boundary conditions were even studied on extension domains, without a specified boundary measure [**Cla23, CHRT24**]. However, when the variational formulation of the problem considered contains a boundary integral (as for the heat transmission problem considered here), the choice of a boundary measure becomes crucial.

In this paper, we model the heat transfer problem as follows (see also [**BGRP16**]). Consider a hot domain Ω^+ with boundary $\partial\Omega$ and cold complement domain Ω^- (so that Ω^+ and Ω^- are separated by their boundary $\partial\Omega$). The domain Ω^+ is characterized by a thermal diffusivity $D^+ > 0$ and its complement by $D^- > 0$, where those coefficients are assumed to be distinct. The thermal resistance of the boundary is modelled by a continuous function $\Lambda : \partial\Omega \to [0, +\infty]$. If the heat distribution u is initially unitary on Ω^+ and null on Ω^- – modelling the cooling of Ω^+ by Ω^- – then it is solution to the following problem:

$$\begin{cases} \partial_t u^\pm - D^\pm \Delta u^\pm = 0 & \text{on }]0,+\infty[\times\Omega^\pm, \\ D^- \dfrac{\partial u^-}{\partial \nu} = \Lambda(u^- - u^+) & \text{on }]0,+\infty[\times\partial\Omega, \\ D^+ \dfrac{\partial u^+}{\partial \nu} = D^- \dfrac{\partial u^-}{\partial \nu} & \text{on }]0,+\infty[\times\partial\Omega, \\ u^+(0,\cdot) = 1 & \text{on } \Omega^+, \\ u^-(0,\cdot) = 0 & \text{on } \Omega^-. \end{cases} \tag{1.1}$$

To consider boundary value problems such as Problem (1.1) on irregular shapes (for instance, fractals), the first concern regards the definition of the boundary conditions. On the one hand, the solution we seek is defined on $\Omega^+ \cup \Omega^-$, which means it cannot simply be restricted to the boundary. To generalize that idea, we use the trace operators. In the case of Lipschitz boundaries, that notion was defined by Agmon in 1965 [**Agm10**, p. 37] and studied notably by Nečas in 1967 [**Neč12**, Subsections 2.4 - 2.6], Adams in 1975 [**Ada75**, Subsections V.3, VII.9] and Marschall in 1987 [**Mar87**], and the trace space (*i.e.*, the range of the trace operators) corresponds to the Sobolev space $H^{1/2}$ on the boundary. Then, it was generalized to d-set boundaries (see [**JW84, JW95**]) in [**Wal91**], in which case the trace space can be identified with the Besov space $B^{2,2}_\beta$, where $\beta = 1 - (n-d)/2$ and with n, the dimension of the ambient space. Later on, the trace operator was defined for d-upper regular boundaries [**Jon09, Bie09, HRPT21**], in which case it is proved the trace space is a Hilbert space, although it can no longer be identified with either one of the previous spaces. All three approaches rely on the existence of a boundary measure (Lebesgue's measure, a d-measure and a d-upper regular measure respectively) to define that operator. In 2009, Biegert [**Bie09**] relied on the extension property of H^1-Sobolev extension domains to define a trace operator which, in particular, does not depend on the choice of a boundary measure. On the other hand, fractals are nowhere differentiable, which prevents from defining the normal derivatives in the usual way. For that matter, it is necessary to consider a weak definition of normal derivatives as in [**Lan02**] (see also [**Cla23**] on extension domains) via Green's formula, as elements of the dual of the trace space. Altogether, the notions of trace operators and weak normal derivatives, as well as the functional framework for domains with irregular boundaries from [**CWHM17**] allow to consider boundary value problems such as problem (1.1) from a variational perspective on a broad class of domains (see Definition 2.1), including domains with fractal boundaries or with changing Hausdorff boundary dimensions.

In this article, we are interested in the existence of an optimal shape for the heat transfer problem between two media separated by a resistive boundary. The shape is optimal in the sense that it minimizes the total energy of the hot medium as the heat propagates during a fixed time $T > 0$. In other words, we are seeking an optimal shape for the interface in some admissible class (first, in a class of uniform Lipschitz surfaces with uniformly bounded lengths and then, in a class of surfaces with uniformly bounded dimensions) to cool the initially hot medium in the most efficient way on a time interval $[0, T]$ (the problem is described by (1.1) and the energy to minimize, by (3.1)). The main interest is for T sufficiently small, however the results obtained hold for any fixed $T > 0$.

Regarding the heat equation, the existence of a Lipschitz minimizer to the heat energy in a class of insulated Lipschitz domains with bounded volume is proved in [**BNNT22**]. The existence of an optimal heating boundary to maximize the heat diffusion is proved in [**HS92**] in the case of plane domains. Still in the plane, it is proved in [**TZZ18**] there exists an optimal shape to approximate a given heat distribution in a class of domains with a given number of holes and setting a homogeneous Dirichlet condition on the boundary.

The difficulty of dealing with questions regarding the existence of optimal shapes is highlighted in [**BB05, BG16**] for problems with Robin boundary conditions, or as soon as the variational formulation displays an integral with respect to the boundary measure (due to non-homogeneous Robin or Neumann conditions). The main obstacle arises from the non-continuity of boundary perimeters for the convergence in the sense of characteristic functions (they are only lower semi-continuous [**HP18**, Proposition 2.3.7]). As the shape of a boundary varies, so does its measure. The natural notion of convergence for those measures is the weak convergence. The complexity stems from the fact that the weak limit of a sequence of Hausdorff measures of co-dimension 1 (which are the 'natural' surface measures for Lipschitz boundaries) is not necessarily a Hausdorff measure itself. To overcome that problem in the framework of Lipschitz boundaries, relaxation methods have been developed using free-boundary problems with discontinuities in [**HP18**, Proposition 2.3.6]. However, those methods do not preserve compactness properties and therefore, do not ensure the existence of an optimal shape. Historically, the framework of compact classes for the existence of optimal shapes (adapted to boundary problems with homogeneous Neumann conditions) is set by the famous results of Chenais [**Che75**]. That class of uniformly Lipschitz domains [**Che75**] is compact for the three usual types of domain convergence: in the sense of characteristic functions, compact sets and in the sense of Hausdorff. However, as for Robin-type boundary value problems, that class is not sufficient for our heat transfer problem.

That problem was recently solved in [**MKNORP21, HRPT21, HMRP$^+$21, HRPT23**]. In [**MKNORP21**] is proved the existence of an optimal Lipschitz shape on which there exists a measure, equivalent to the perimeter measure, achieving the infimum of the acoustic energy in the framework of the Helmholtz equation with a complex-valued Robin-type boundary condition. The admissible class of domains was chosen with the perimeter measures on the Lipschitz boundaries. Since the optimal shape, albeit Lipschitz, is provided with a boundary measure equivalent but not necessarily equal to the perimeter measure, the infimum of the energy is not necessarily reached. Results of the same type are obtained in the framework of a non-homogeneous Neumann type boundary condition in [**HRPT21**] in the framework of the linear elasticity system with the aim to maximize the stability of an architectural structure. The relaxation of the assumption to work in the class of domains with the uniformly bounded measure of the perimeters is to work in the class of uniform domains with uniformly bounded d-upper and s-lower boundary measures [**HMRP$^+$21**]. This class is compact for the weak convergence of boundary measures and thus allows to provide the existence of an optimal shape minimizing the energy coming from the stationary elliptic variational problems corresponding to Robin-type or non-homogeneous Neumann boundary conditions (see [**HMRP$^+$21**] for the Helmholtz problem, [**HRPT21**] for the linear elasticity system and [**HRPT23**] for an abstract elliptic problem). The shape optimization for the time dependant transmission problems, which we consider in this paper for system (1.1), has never been previously considered. Here, we use a similar class of admissible shapes to [**HRPT21**] and adapt the method to the case of the heat diffusion problem, which is a time-dependant transmission problem.

The main results of this paper are the following: on a large class of domains, referred to as 'two-sided admissible domains' and which can have boundaries with varying Hausdorff dimension, we introduce interior and exterior trace operators. We characterize them in terms of isometries with the Sobolev space H^1, which is new in the framework of two-sided admissible domains. We also identify their kernels as the closures of the sets of C^∞ functions with compact supports, as in the regular case (Theorem 2.3). We define interior and exterior weak normal derivatives in the same spirit as what is done in [**Lan02**] on d-sets. In the framework of our two-sided admissible domains, we prove the well-posedness (in the sense of Hadamard) of a generalized version of problem (1.1), adding a source term, considering general initial conditions and without assuming the thermal diffusivity is constant inside or outside (Theorem 2.7). We introduce the energy form corresponding to the total heat content of the domain Ω^+ throughout the diffusion and the heat transfer from Ω^+ to Ω^-, and prove the convergence of that energy form in the sense of Mosco as we approximate (ε, ∞)-domains [**JW84**] (Theorem 3.2). Finally, similarly to [**MKNORP21**, Theorem 3.2] but this time in the new context of the time-depending heat transmission problem, we prove that the class of Lipschitz domains provides only the infimum of the speed heat transfer (Theorem 3.6): optimal shapes should be endowed with boundary measures which are not necessarily Hausdorff measures. For that matter, we extend the class of admissible domains to a class of (ε, ∞)-domains (or uniform domains) with measures of bounded dimensions (initially introduced in [**HMRP$^+$21**]), in which we prove the existence of an optimal shape for our heat transfer framework in Theorem 3.9.

This paper is organised as follows: in Section 2 we introduce the class of two-sided extension domains, the trace operators and the weak normal derivatives (Subsection 2.1). We prove the isometric properties of the trace. In Subsection 2.2, we state a generalized version of problem (1.1) and prove its well-posedness. In Section 3, we prove the Mosco convergence of the heat energy form (Subsection 3.1). Then, in Subsection 3.2, we consider the shape optimization problem: we start with seeking an optimal shape in a class of Lipschitz domains and then turn to a class of (ε, ∞)-domains.

Throughout this paper, $n \geq 2$ is the dimension of the ambient space. If A and B are topological spaces, $C(A, B)$ denotes the set of all continuous functions mapping A to B. All domains considered coincide with the interior of their closure.

2. Framework and well-posedness

2.1. Functional framework. We begin with defining the general class of domains on which our study is carried out.

DEFINITION 2.1 (Two-sided admissible domains). Let Ω be a bounded domain of $\mathbb{R}^n$ with connected boundary, such that the Hausdorff dimension of its boundary $\partial\Omega$ is (locally) included in $[n-1, n[$. Let μ be a positive finite Borel measure with $\operatorname{supp}\mu = \partial\Omega$. The couple (Ω, μ) is called a two-sided admissible domain if

(i) $\Omega^+ = \Omega$ and its complement domain $\Omega^- := \mathbb{R}^n \backslash \overline{\Omega}$ are H^1-extension domains of $\mathbb{R}^n$ [**HKT08**]: there exist bounded linear extension operators $\mathrm{E}^\pm : H^1(\Omega^\pm) \to H^1(\mathbb{R}^n)$ with norms depending only on Ω (and n);
(ii) the boundary measure μ is d-upper regular for a fixed $d \in [n-1, n]$: there exists $c_d > 0$ such that

$$\forall x \in \partial\Omega,\ \forall r \in]0,1], \quad \mu(B_r(x)) \leq c_d\, r^d, \tag{2.1}$$

where $B_r(x)$ is the open ball of $\mathbb{R}^n$ of center x and radius r.

Typical examples of two-sided admissible domains are Lipschitz domains [**Cal61, Ste70**], (ε, ∞)-domains (defined in Subsection 3.1) in $\mathbb{R}^2$ [**Jon81**] and non-tangentially accessible (NTA) domains [**Nys94, Nys96**]. We recall that a domain $\Omega = \Omega^+$ is an NTA domain (see [**JK82**]) if there exist $M, r_0 > 0$ such that

(i) Ω^+ satisfies the corkscrew condition:

$$\forall y \in \partial\Omega,\ \forall r < r_0,\ \exists x \in \Omega^+, \quad |x - y| < r \quad \text{and} \quad d(x, \partial\Omega) > M^{-1} r;$$

(ii) Ω^- satisfies the corkscrew condition as well;
(iii) Ω^+ satisfies the Harnack chain condition: there exists $C > 0$ such that for all $\rho > 0$, $\delta \geq 1$ and $x_1, x_2 \in \Omega^+$ with $d(x_{1,2}, \partial\Omega) \geq \rho$ and $|x_1 - x_2| \leq \delta\rho$, there exist a chain of open balls $(B_k)_{k \in [\![1,N]\!]}$ with N bounded by a constant depending only on δ, with $x_1 \in B_1$ and $x_2 \in B_N$ and such that

$$B_k \cap B_{k+1} \neq \varnothing \quad \text{and} \quad C^{-1}\, diam(B_k) \leq d(B_k, \partial\Omega) \leq C\, diam(B_k).$$

Condition (2.1) implies that the Hausdorff dimension of the boundary is bounded below by d. However, unlike for d-sets, that dimension can vary, which allows to consider domains with boundaries which are partly fractal and partly Lipschitz for example (see Figure 3). For Lipschitz boundaries, usual choices are $\mu = \lambda^{(n-1)}$, the $(n-1)$-dimensional Lebesgue measure and $\mu = \mathcal{H}^{(n-1)}$, the $(n-1)$-dimensional Hausdorff measure.

By [**HKT08**, Theorem 5], condition (i) from Definition 2.1 implies that $\Omega^\pm$ are n-sets (d-sets for $d = n$). Roughly speaking, that means the domain Ω should have neither inward nor outward cusps (see Figure 4).

In this section, whenever (Ω, μ) is a two-sided admissible domain of $\mathbb{R}^n$, we use the following notations:

- $\Omega^+ := \Omega$ is thought of as the hot domain and its complement domain $\Omega^- := \mathbb{R}^n \backslash \overline{\Omega}$ is thought of as cold;
- if $u : (t, x) \in]0, +\infty[\times \mathbb{R}^n \longmapsto u(t, x) \in \mathbb{R}$, then $u^\pm := u|_{(]0,+\infty[\times\Omega^\pm)}$ are its restrictions to Ω^+ and Ω^- respectively.

The H^1-extension property of $\Omega^\pm$ ensures the existence of bounded linear interior and exterior trace operators from $H^1(\Omega^+)$ and $H^1(\Omega^-)$ respectively (as in [**Cla23**]). The interior trace operator is defined in the same way as in [**Bie09**, Section 5] (see also [**HRPT21**]). Note that its definition does not depend on the way elements of $H^1(\Omega^+)$ are extended to $H^1(\mathbb{R}^n)$ [**Bie09**, Theorem 6.1]. Therefore, as in the d-regular case [**Wal91**, Theorem 1], we use the following equivalent definition:

DEFINITION 2.2 (Trace operators). Let (Ω, μ) be a two-sided admissible domain of $\mathbb{R}^n$.

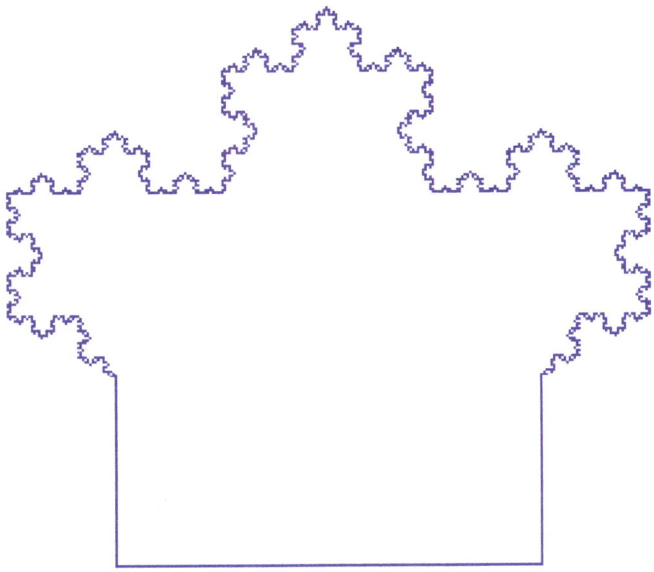

FIGURE 3. An example of a two-sided admissible domain of $\mathbb{R}^2$, lying inside the blue line. The upper part of the boundary is a Von Koch curve, which is a $\frac{\ln 4}{\ln 3}$-set (represented by a fifth generation pre-fractal [source: Wikipedia]). The lower part consists of a portion of a square, which is Lipschitz. The boundary is endowed with the 1-dimensional Lebesgue measure on the lower part and the $\frac{\ln 4}{\ln 3}$-dimensional Hausdorff measure on the upper part: condition (2.1) is satisfied for $d = 1$.

FIGURE 4. An example of a domain Ω which is not a two-sided extension domain due to the presence of an inward and an outward cusp: neither Ω^+ nor Ω^- is an H^1-extension domain.

• The interior trace operator $\mathrm{Tr}^+ : H^1(\Omega^+) \to \mathrm{Tr}^+(H^1(\Omega^+)) \subset L^2(\partial\Omega, \mu)$ is defined by

$$\mathrm{Tr}^+ u(x) = \lim_{r\to 0^+} \frac{1}{\lambda^{(n)}(\Omega^+ \cap B_r(x))} \int_{\Omega^+\cap B_r(x)} u(y)\,\mathrm{d}y, \quad x \in \partial\Omega \text{ q.e.,} \tag{2.2}$$

where $\lambda^{(n)}$ denotes Lebesgue's measure on $\mathbb{R}^n$.

• The exterior trace operator $\mathrm{Tr}^- : H^1(\Omega^-) \to \mathrm{Tr}^-(H^1(\Omega^-)) \subset L^2(\partial\Omega, \mu)$ is defined by

$$\mathrm{Tr}^- u(x) = \lim_{r\to 0^+} \frac{1}{\lambda^{(n)}(\Omega^- \cap B_r(x))} \int_{\Omega^-\cap B_r(x)} u(y)\,\mathrm{d}y, \quad x \in \partial\Omega \text{ q.e.} \tag{2.3}$$

In that definition, the term quasi-everywhere (q.e.) is understood in terms of the capacity Cap_2 from [**Bie09**, Section 2]. Unlike the space $L^2(\partial\Omega, \mu)$, the range of the trace operators does not depend on μ. As for [**HMRP$^+$21**, Theorem 5.1], it follows from [**Bie09**, Corollaries 7.3 and 7.4] and the fact that the measure on $\partial\Omega$ is finite (see [**AH96**, Theorems 7.2.2 and 7.3.2]) that the limits (2.2) and (2.3) exist μ-a.e.

The following theorem summarizes the main properties of the trace operators and a characterization of their range (see also [**Cla23**]). Among those properties are the characterization of the trace operator as an isometry and of its kernel, which are new in our framework. Results of a similar nature in a different setting can be found in [**CWHM17**] for planar screens, in which traces on subsets of hyperplanes are considered.

THEOREM 2.3 (Trace theorem). *Let (Ω, μ) be a two-sided admissible domain of $\mathbb{R}^n$. Then, the following assertions hold.*

(1) *The trace operators $\mathrm{Tr}^\pm : H^1(\Omega^\pm) \to L^2(\partial\Omega, \mu)$ are linear and compact.*

(2) *The image of the trace operators*

$$\mathcal{B}(\partial\Omega) := \mathrm{Tr}^+(H^1(\Omega^+)) = \mathrm{Tr}^-(H^1(\Omega^-))$$

is a Hilbert space endowed with either one of the equivalent norms:

$$\|f\|_{\mathrm{Tr}^\pm} := \min\{\|v\|_{H^1(\Omega^\pm)} \mid f = \mathrm{Tr}^\pm v\}. \tag{2.4}$$

The embedding $\mathcal{B}(\partial\Omega) \subset L^2(\partial\Omega, \mu)$ is dense and compact. In what follows, we denote $\|\cdot\|_{\mathcal{B}(\partial\Omega)} := \|\cdot\|_{\mathrm{Tr}^+}$.

(3) $\mathrm{Ker}(\mathrm{Tr}^\pm) = H^1_0(\Omega^\pm) := \overline{C_0^\infty(\Omega^\pm)}^{\|\cdot\|_{H^1(\Omega^\pm)}}$ *is the closure in H^1 of the set of indefinitely differentiable functions with compact support in $\Omega^\pm$.*

(4) *The trace operators $\mathrm{Tr}^\pm : H^1(\Omega^\pm) \to (\mathcal{B}(\partial\Omega), \|\cdot\|_{\mathrm{Tr}^\pm})$ are partial isometries with operator norms equal to 1.*

(5) $\mathrm{Tr}^\pm : V_1(\Omega^\pm) \to \mathcal{B}(\partial\Omega)$ *define isometries, where*

$$V_1(\Omega^\pm) = \{v \in H^1(\Omega^\pm) \mid (-\Delta + 1)v = 0 \text{ weakly}\} \tag{2.5}$$

are the spaces of 1-harmonic functions on $\Omega^\pm$ endowed with the standard H^1 norms. It holds:

$$V_1(\Omega^\pm) = \mathrm{Ker}(\mathrm{Tr}^\pm)^\perp = (H^1_0(\Omega^\pm))^\perp. \tag{2.6}$$

PROOF. Point (1) follows from [**HRPT21**], see also [**Bie09**, Corollary 7.4] and [**RP21**, Proposition 3]. Point (2) follows partially from the definition, see also [**HMRP⁺21**, Theorem 1],[**HRPT21**, Theorem 5.1], [**DRPT22**, Theorem 2], where the norm was defined as:

$$\|f\|_{\mathrm{Tr}^+} = \inf\{\|v\|_{H^1(\Omega^+)} \mid f = \mathrm{Tr}^+ v\}.$$

Here, we specify that the infimum is reached. To prove that, it is sufficient to use the continuity of the trace operator from $H^1(\Omega)$ to its image and Stampacchia's theorem [**Bre87**, Theorem V.6], ensuring the existence of $u \in H^1_f(\Omega)$, where

$$H^1_f(\Omega^+) := \left\{v \in H^1(\Omega^+) \;\middle|\; \mathrm{Tr}^+ v = f \text{ on } \partial\Omega\right\}, \tag{2.7}$$

unique minimizer of $\|\cdot\|_{H^1(\Omega^+)}$ on $H^1_f(\Omega^+)$. It is equivalent to the weak well-posedness on $H^1(\Omega^+)$ of the non-homogeneous Dirichlet boundary value problem for $(-\Delta+1)$:

$$\begin{cases} -\Delta u + u = 0 & \text{on } \Omega^+, \\ \mathrm{Tr}^+ u = f. \end{cases} \tag{2.8}$$

The weak solution to problem (2.8) is understood in the sense of its variational formulation on $H^1_f(\Omega^+)$:

$$\forall v \in H^1_f(\Omega^+), \quad \langle u, v\rangle_{H^1(\Omega^+)} = \|u\|^2_{H^1(\Omega^+)}, \tag{2.9}$$

which, in particular, implies:

$$\forall v \in H^1_f(\Omega^+), \quad \Big(\|v\|_{H^1(\Omega^+)} = \|f\|_{\mathcal{B}(\partial\Omega)} \quad \Longleftrightarrow \quad u = v\Big). \tag{2.10}$$

To prove Point (3), we use the following characterization from [**FOT10**, Corollary 2.3.1 and Example 2.3.1], holding regardless of the boundary regularity:

$$H^1_0(\mathbb{R}^n\backslash\partial\Omega) = \left\{v \in H^1(\mathbb{R}^n) \;\middle|\; \tilde{v} = 0 \text{ q.e. on } \partial\Omega\right\},$$

where $\tilde{v}$ denotes the quasi-continuous version of v. $u \in H^1(\Omega^+)$ can be extended to $H^1_0(\mathbb{R}^n\backslash\partial\Omega)$ if and only if $u \in H^1_0(\Omega^+)$, and it can be extended to the space on the right hand side of the equality if and only if $\mathrm{Tr}^+ u = 0$. Points (4) and (5) are corollaries of the previous points, of Stampacchia's theorem in particular. Let us prove (2.6). For $u \in H^1(\Omega^+)$, by Green's formula (see Definition 2.5 below), the following equivalences hold:

$$\begin{aligned} u \in V_1(\Omega^+) &\Longleftrightarrow -\Delta u + u = 0 \text{ weakly on } \Omega^+ \\ &\Longleftrightarrow \forall v \in \mathrm{Ker}(\mathrm{Tr}^+), \quad \langle u, v\rangle_{H^1(\Omega^+)} = 0 \\ &\Longleftrightarrow u \in \mathrm{Ker}(\mathrm{Tr}^+)^\perp. \end{aligned}$$

Since $\mathrm{Tr}^+ : \mathrm{Ker}(\mathrm{Tr}^+)^\perp \to \mathcal{B}(\partial\Omega)$ is bijective, by (2.10), it is an isometry. Therefore, since $\mathrm{Ker}(\mathrm{Tr}^+)^\perp \subset H^1(\Omega^+)$ is a closed Hilbert space endowed with $\|\cdot\|_{H^1}$ (as an orthogonal complement), the trace operator $\mathrm{Tr}^+ : H^1(\Omega^+) \to \mathcal{B}(\partial\Omega)$ is a partial isometry of operator norm equal to 1. The results regarding Tr^- are proved in the same way. □

REMARK 2.4. As mentioned before, the trace space $\mathcal{B}(\partial\Omega)$ does not depend on the boundary measure μ. If $\partial\Omega$ is Lipschitz, then $\mathcal{B}(\partial\Omega) = H^{1/2}(\partial\Omega)$ in the sense of equivalent norms. If $\partial\Omega$ is a d-set for some $n-2 < d < n$, then [**JW84, Wal91**] $\mathcal{B}(\partial\Omega) = B^{2,2}_\beta(\partial\Omega)$ with $\beta = 1 - \frac{n-d}{2} > 0$.

To understand problem (1.1), we also need to define normal derivatives on Ω. The class of two-sided admissible domains contains fractal shapes, which are known for being nowhere differentiable. For that matter, we cannot use the usual definition, which relies on the existence of a normal vector. For that matter, we use the weak definition of normal derivatives as elements of the dual of the trace space, similarly to what is done in [**Lan02**].

DEFINITION 2.5 (Weak normal derivatives). Consider a two-sided admissible domain (Ω, μ) of $\mathbb{R}^n$.

• Let $u \in H^1_\Delta(\Omega^+)$, where

$$H^1_\Delta(\Omega^+) := \{v \in H^1(\Omega^+) \mid \Delta v \in L^2(\Omega^+)\}. \tag{2.11}$$

The weak interior normal derivative of u is the element $\varphi \in \mathcal{B}'(\partial\Omega)$ such that:

$$\forall v \in H^1(\Omega^+), \quad \langle \varphi, \operatorname{Tr}^+ v\rangle_{\mathcal{B}', \mathcal{B}} = \int_{\Omega^+} (\Delta u) v \,\mathrm{d}x + \int_{\Omega^+} \nabla u \cdot \nabla v \,\mathrm{d}x, \tag{2.12}$$

denoted by $\dfrac{\partial^+ u}{\partial \nu} := \varphi$.

• Let $u \in H^1_\Delta(\Omega^-)$ ((2.11) with Ω^- instead of Ω^+). The weak exterior normal derivative of u is the element $\psi \in \mathcal{B}'(\partial\Omega)$ such that:

$$\forall v \in H^1(\Omega^-), \quad \langle \psi, \operatorname{Tr}^- v\rangle_{\mathcal{B}', \mathcal{B}} = -\int_{\Omega^-} (\Delta u) v \,\mathrm{d}x - \int_{\Omega^-} \nabla u \cdot \nabla v \,\mathrm{d}x, \tag{2.13}$$

denoted by $\dfrac{\partial^- u}{\partial \nu} := \psi$.

The sign convention in the definition of the weak exterior normal derivative corresponds, in the regular case, to choosing the outward normal vector to Ω^+ to define both derivatives. The boundary conditions in the heat transfer equation from problem (1.1) are boundary jumps, defined as follows in the case of two-sided admissible domains.

DEFINITION 2.6 (Boundary jump values). Consider a two-sided admissible domain (Ω, μ) of $\mathbb{R}^n$ and a thermal diffusivity $D \in \mathcal{T}(\mathbb{R}^n)$, where

$$\begin{aligned}\mathcal{T}(\mathbb{R}^n) := \Big\{ D \in L^\infty(\mathbb{R}^n) \,\Big|\, &\operatorname*{ess\,inf}_{\mathbb{R}^n} D > 0 \quad \text{and} \\ &\exists V \text{ neighbourhood of } \partial\Omega,\ D|_{V\setminus\partial\Omega} \in C(V\setminus\partial\Omega,]0, +\infty[) \Big\}.\end{aligned} \tag{2.14}$$

Let $u \in V(\mathbb{R}^n)$, where

$$V(\mathbb{R}^n) := \{v \in L^2(\mathbb{R}^n) \mid v^+ \in H^1(\Omega^+),\ v^- \in H^1(\Omega^-)\}, \tag{2.15}$$

endowed with the norm:

$$\|\cdot\|_{V(\mathbb{R}^n)} := \sqrt{\|\cdot\|^2_{L^2(\mathbb{R}^n)} + \|D\nabla\cdot\|^2_{L^2(\Omega^+\cup\Omega^-)^n}}. \tag{2.16}$$

The jump in trace of u across $\partial\Omega$ is defined by:

$$[\![\operatorname{Tr} u]\!] := \operatorname{Tr}^+ u - \operatorname{Tr}^- u.$$

In addition, if $u^+ \in H^1_\Delta(\Omega^+)$ (Eq. (2.11)) and $u^- \in H^1_\Delta(\Omega^-)$, then the jump in co-normal derivative of u is denoted by:

$$\left[\!\!\left[D\frac{\partial u}{\partial \nu} \right]\!\!\right] := D^+ \frac{\partial^+ u}{\partial \nu} - D^- \frac{\partial^- u}{\partial \nu},$$

where, for $x \in \partial\Omega$, $D^{\pm}(x) = \lim_{\substack{y \to x \\ y \in \Omega^{\pm}}} D(y)$.

Let (Ω, μ) be a two-sided admissible domain of $\mathbb{R}^n$. We consider the following heat transmission problem:

$$(2.17)\qquad \begin{cases} \partial_t u - D\Delta u = f & \text{on }]0,+\infty[\times(\mathbb{R}^n\backslash\partial\Omega),\\ D^-\dfrac{\partial^- u}{\partial\nu} + \Lambda[\![\operatorname{Tr} u]\!] = 0 & \text{on }]0,+\infty[\times\partial\Omega,\\ \left[\!\!\left[D\dfrac{\partial u}{\partial\nu}\right]\!\!\right] = 0 & \text{on }]0,+\infty[\times\partial\Omega,\\ u(0,\cdot) = u_0 & \text{on } \mathbb{R}^n\backslash\partial\Omega, \end{cases}$$

where $u_0 \in L^2(\mathbb{R}^n)$, $f \in L^2(]0,+\infty[, L^2(\mathbb{R}^n))$, $D \in \mathcal{T}(\mathbb{R}^n)$ and $\Lambda \in C(\partial\Omega, [0,+\infty])$. Those quantities can be interpreted as follows:

- u is the (time-dependent) heat distribution on $\mathbb{R}^n$;
- $u_0 \in L^2(\mathbb{R}^n)$ is the initial heat distribution. In the example of problem (1.1), $u_0 = \mathbb{1}_{\Omega^+}$: the domain Ω^+ is initially hot, while its complement is cold, and the problem models the heat diffusion from Ω^+ to Ω^-;
- $f \in L^2(]0,+\infty[, L^2(\mathbb{R}^n))$ is the (time-dependent) heat source on $\mathbb{R}^n$;
- $D \in \mathcal{T}(\mathbb{R}^n)$ is the thermal diffusivity. In the example of problem (1.1), $D = D^+\mathbb{1}_{\Omega^+} + D^-\mathbb{1}_{\Omega^-}$ with $D^+ \neq D^-$, which models the diffusion properties of two homogeneous media;
- $\Lambda \in C(\partial\Omega, [0,+\infty])$ is the resistivity of the heat conduction by the boundary. If $\Lambda = 0$, then the boundary $\partial\Omega$ acts as a thermal insulator. If $\Lambda = +\infty$, then the heat distribution u is continuous across the boundary.

In the case of a Lipschitz domain, the measure μ is chosen as the Hausdorff measure $\mathcal{H}^{(n-1)}$, which is the usual way to measure the perimeter of those domains. With that choice, the model is consistent with the physical phenomenon.

2.2. Well-posedness of the model. We begin with proving that our problem is well-posed.

Let $u_0 \in L^2(\mathbb{R}^n)$, $f \in L^2(]0,+\infty[, L^2(\mathbb{R}^n))$, $D \in \mathcal{T}(\mathbb{R}^n)$ and $\Lambda \in C(\partial\Omega, [0,+\infty])$. Problem (2.17) is understood in term of its variational formulation, given by:

$$(2.18)\qquad \begin{cases} \forall t > 0,\ \forall v \in V(\mathbb{R}^n), \quad \langle \partial_t u(t), v\rangle_{L^2(\mathbb{R}^n)} + a(u(t), v) = \langle f(t), v\rangle_{L^2(\mathbb{R}^n)},\\ u(0,\cdot) = u_0 \in L^2(\mathbb{R}^n), \end{cases}$$

where

$$(2.19)\qquad a(u,v) := \int_{\mathbb{R}^n\backslash\partial\Omega} D\nabla u\cdot\nabla v \,\mathrm{d}x + \int_{\{\Lambda<+\infty\}} \Lambda[\![\operatorname{Tr} u]\!][\![\operatorname{Tr} v]\!]\,\mathrm{d}\mu.$$

THEOREM 2.7 (Well-posedness). *Let (Ω, μ) be a two-sided admissible domain of $\mathbb{R}^n$. Let $u_0 \in L^2(\mathbb{R}^n)$, $f \in L^2(]0,+\infty[, L^2(\mathbb{R}^n))$, $D \in \mathcal{T}(\mathbb{R}^n)$ and $\Lambda \in C(\partial\Omega, [0,+\infty])$. Then, there exists a unique solution u which is in the space $H^1(\mathbb{R}_+, L^2(\mathbb{R}^n)) \cap L^2(\mathbb{R}_+, V(\mathbb{R}^n))$ to the variational formulation (Eq. (2.18)).*

For all $t \geq 0$, that solution satisfies the following energy identity:

$$(2.20)\qquad \frac{1}{2}\int_{\mathbb{R}^n} |u(t)|^2\,\mathrm{d}x + \int_0^t a(u(\tau), u(\tau))\,\mathrm{d}\tau = \frac{1}{2}\int_{\mathbb{R}^n} |u_0(x)|^2\,\mathrm{d}x + \int_0^t f(\tau)\,\mathrm{d}\tau.$$

PROOF. For all $T > 0$, the existence and uniqueness of the solution u in the space $H^1([0,T], L^2(\mathbb{R}^n)) \cap L^2(]0,T[, V(\mathbb{R}^n))$ to formulation (2.18) on the time interval $[0,T]$ for $T > 0$ given $u_0 \in L^2(\mathbb{R}^n)$ is proved as in [**Eva10**, Paragraph 7.1.2]. Here, the bilinear form $a : V(\mathbb{R}^n) \times V(\mathbb{R}^n) \to \mathbb{R}$ satisfies the following coercivity estimate:

$$\forall v \in V(\mathbb{R}^n), \quad a(v,v) \geq \|v\|^2_{V(\mathbb{R}^n)} - \|v\|^2_{L^2(\mathbb{R}^n)}.$$

The uniqueness of the solution on $[0,T]$ for all $T > 0$ (given $u_0 \in L^2(\mathbb{R}^n)$) proves the solutions on each time interval are compatible: if u_1 is a solution on $[0, T_1]$ and u_2 is a solution on $[0, T_2]$, then $u_1 = u_2$ on $[0, T_1 \wedge T_2]$. That allows to define a unique global solution $u \in H^1(\mathbb{R}_+, L^2(\mathbb{R}^n)) \cap L^2(\mathbb{R}_+, V(\mathbb{R}^n))$. □

REMARK 2.8. Since $H^1(\mathbb{R}_+, L^2(\mathbb{R}^n)) \subset C(\mathbb{R}_+, L^2(\mathbb{R}^n))$, u is also an element of $C(\mathbb{R}_+, L^2(\mathbb{R}^n))$.

The problem being well-posed, the heat distribution is unique given the parameters of the problem and it is possible to consider the properties of the energy form related to the heat content in Ω^+ and the heat transfer from Ω^+ to Ω^-.

3. Convergence and optimization

In this section, we consider the energy form associated to the heat content inside a domain Ω and its convergence as we approximate Ω. Throughout this section, the diffusion time $T > 0$ is fixed, U denotes a (large) bounded Lipschitz domain and $\Lambda \in C(U, [0, +\infty])$. In what follows, we assume $\Lambda < +\infty$ on U to simplify the expressions. Retrieving the general case can be done by replacing '$\partial\Omega$' with '$\partial\Omega \cap \{\Lambda < +\infty\}$' in the definition of the energy form (Eq. (3.1) below). All the results stated hereafter remain valid once that modification has been made. We assume the thermal diffusivity D is a constant $D^+ > 0$ on all the interior domains, and another constant $D^- > 0$ on the exterior domains, as in the case of problem (1.1). Without loss of generality, we set $D^+ = 1$.

3.1. Mosco convergence of the energy forms. If (Ω, μ) is a two-sided admissible domain of $\mathbb{R}^n$ with $\Omega \subset U$, we define the energy functional $J_T(\Omega, \mu)$ on $L^2(]0,T[, L^2(U))$ by:

$$J_T(\Omega,\mu)(u) := \int_0^T \int_\Omega |\nabla u|^2 \, \mathrm{d}x \, \mathrm{d}t + \int_0^T \int_\Omega |u|^2 \, \mathrm{d}x \, \mathrm{d}t + \int_0^T \int_{\partial\Omega} \Lambda \, |[\![\mathrm{Tr}\, u]\!]|^2 \, \mathrm{d}\mu \, \mathrm{d}t. \tag{3.1}$$

We prove the convergence of the energy form J_T defined by (3.1) in the sense of Mosco [**Mos94**, Definition 2.1.1] along an approximating sequence of (ε, ∞)-domains for some $\varepsilon > 0$. We recall that Ω is a (ε, ∞)-domain [**JW84**] if for all $x, y \in \Omega$, there exists a rectifiable arc γ drawn on Ω and joining x and y, of length $\ell(\gamma)$ and such that:

(i) $\ell(\gamma) \leq \frac{|x-y|}{\varepsilon}$;
(ii) for all $z \in \gamma$, $d(z, \partial\Omega) \geq \varepsilon \frac{|x-z||y-z|}{|x-y|}$.

Mosco convergence is defined as follows.

DEFINITION 3.1 (Mosco convergence). Let H be a Hilbert space. Let $(F_m)_{m\in\mathbb{N}}$ be a sequence of quadratic forms on H. $(F_m)_{m\in\mathbb{N}}$ is said to converge in the sense of Mosco (or M-converge) to a quadratic form F on H if the following conditions hold:

(i) if $(x_m)_{m\in\mathbb{N}} \in H^{\mathbb{N}}$ and $x \in H$ are such that $x_m \rightharpoonup x$, then
$$F(x) \leq \liminf_{m\to\infty} F_m(x_m);$$

(ii) for all $x \in H$, there exists $(x_m)_{m\in\mathbb{N}} \in H^{\mathbb{N}}$ such that $x_m \to x$ and
$$F(x) = \lim_{m\to\infty} F_m(x_m).$$

To prove such convergence, if Ω is an H^1-extension domain, we denote by $\mathrm{E}_\Omega : H^1(\Omega) \to H^1(U)$ a linear bounded H^1-extension operator. If $u \in L^2(]0,T[, H^1(\Omega))$, the extension $\mathrm{E}_\Omega u$ is understood as follows:

$$\mathrm{E}_\Omega u : t \in [0,T] \longmapsto \big(x \in U \longmapsto [\mathrm{E}_\Omega u(t)](x)\big). \tag{3.2}$$

In that sense, E_Ω is also regarded as $L^2(]0,T[, H^1(\Omega)) \to L^2(]0,T[, H^1(U))$.

The following theorem states the Mosco convergence of the energy forms (3.1) along a converging sequence of domains. Its proofs uses similar methods as those used in [**HRPT21**, Theorem 8], however the main differences are twofold: the energy functional considered here is time-dependant and the boundary term we consider is connected to the jump in trace instead of the trace value.

THEOREM 3.2 (Mosco convergence of the energy forms). *Let $\varepsilon > 0$. Let $(\Omega_m)_{m\in\mathbb{N}}$ be sequence of (ε,∞)-domains included in U and $(\mu_m)_{m\in\mathbb{N}}$ be a sequence of finite Borel measures with* $\operatorname{supp}\mu_m = \partial\Omega_m$*, all satisfying* (2.1) *for the same $d \in [n-1, n[$ and the same constant $c_d > 0$. Assume there exist a domain Ω and a measure μ with* $\operatorname{supp}\mu = \partial\Omega$ *such that:*

(i) *$(\Omega_m)_{m\in\mathbb{N}}$ converges to Ω in the sense of characteristic functions and Hausdorff (denoted by $\Omega_m \xrightarrow{\mathbb{1},H} \Omega$);*

(ii) *$(\mu_m)_{m\in\mathbb{N}}$ converges to μ weakly (denoted by $\mu_m \rightharpoonup \mu$), i.e.,*
$$\forall \varphi \in C(\overline{U}, \mathbb{R}), \quad \int_{\partial\Omega_m} \varphi \,\mathrm{d}\mu_m \xrightarrow[m\to\infty]{} \int_{\partial\Omega} \varphi \,\mathrm{d}\mu.$$

Then,

$$\lim_{m\to\infty} J_T(\Omega_m, \mu_m) = J_T(\Omega, \mu) \quad \textit{in the sense of Mosco.} \tag{3.3}$$

PROOF. Let $(u_m) \in L^2(]0,T[, L^2(U))^{\mathbb{N}}$ and $u \in L^2(]0,T[, L^2(U))$ such that $u_m \rightharpoonup u$. We aim at proving condition (i) from Definition 3.1, that is:

$$J_T(\Omega,\mu)(u) \leq \liminf_{m\to\infty} J_T(\Omega_m, \mu_m)(u_m). \tag{3.4}$$

Up to passing to a subsequence, we may assume:

$$\liminf_{m\in\mathbb{N}} J_T(\Omega_m,\mu_m)(u_m) = \lim_{m\to\infty} J_T(\Omega_m,\mu_m)(u_m),$$

and

$$\forall m \in \mathbb{N}, \quad u_m|_{\Omega_m} \in L^2(]0,T[, H^1(\Omega_m)),$$

so that the limit is finite. Then, from the definition of J_T, it holds:

$$\sup_{m\in\mathbb{N}} \|u_m|_{\Omega_m}\|_{L^2(]0,T[,H^1(\Omega_m))} < +\infty. \tag{3.5}$$

By [**HRPT21**, Proposition 4, Remark 8], there exists a sequence of linear bounded extension operators $(\mathrm{E}_{\Omega_m} : H^1(\Omega_m) \to H^1(U))_{m\in\mathbb{N}}$ and a constant $c_{\mathrm{E}} > 0$ such that:

$$\forall m \in \mathbb{N}, \quad \|\mathrm{E}_{\Omega_m}(u_m|_{\Omega_m})\|_{L^2(]0,T[,H^1(U))} \leq c_{\mathrm{E}}\|u_m|_{\Omega_m}\|_{L^2(]0,T[,H^1(\Omega_m))},$$

hence, by (3.5), $(\mathrm{E}_{\Omega_m}(u_m|_{\Omega_m}))_{m\in\mathbb{N}}$ (in the sense of (3.2)) is bounded in the Hilbert space $L^2(]0,T[,H^1(U))$. Up to passing to a subsequence, we may assume there exists $u_\infty \in L^2(]0,T[,H^1(U))$ such that $\mathrm{E}_{\Omega_m}(u_m|_{\Omega_m}) \rightharpoonup u_\infty$ in $L^2(]0,T[,H^1(\Omega_m))$. From the convergence of the domains in the sense of characteristic functions, we may also assume $\mathbb{1}_{\Omega_m} \to \mathbb{1}_\Omega$ a.e., hence:

$$\begin{aligned} \mathbb{1}_{\Omega_m}\mathrm{E}_{\Omega_m}(u_m|_{\Omega_m}) &\rightharpoonup \mathbb{1}_\Omega u_\infty && \text{in } L^2(]0,T[,L^2(U)),\\ \mathbb{1}_{\Omega_m}\nabla\mathrm{E}_{\Omega_m}(u_m|_{\Omega_m}) &\rightharpoonup \mathbb{1}_\Omega \nabla u_\infty && \text{in } L^2(]0,T[,L^2(U)^n). \end{aligned}$$

Since $\mathbb{1}_{\Omega_m}\mathrm{E}_{\Omega_m}(u_m|_{\Omega_m}) = \mathbb{1}_{\Omega_m}u_m \rightharpoonup \mathbb{1}_\Omega u$ in $L^2(]0,T[,L^2(U))$ (see also [**Eva10**, Paragraph 7.1.2, Theorem 3] for the initial condition), we can deduce:

$$u_\infty|_\Omega = u|_\Omega. \tag{3.6}$$

In the same way, up to passing to a subsequence, we can assume that for all $m \in \mathbb{N}$, $u_m|_{U\backslash\overline{\Omega}_m} \in L^2(]0,T[,H^1(U\backslash\overline{\Omega}_m))$, where $\overline{\Omega}_m$ denotes the closure of Ω_m, and obtain:

$$\mathbb{1}_{U\backslash\overline{\Omega}_m}u_m \rightharpoonup \mathbb{1}_{U\backslash\overline{\Omega}}u \quad \text{in } L^2(]0,T[,H^1(U)). \tag{3.7}$$

Using a similar decomposition as in the proof of [**HRPT21**, Theorem 5] (*i.e.*, approximating the H^1-extensions of u with smooth functions) yields:

$$\begin{aligned} \int_0^T\int_{\partial\Omega_m} \Lambda|\operatorname{Tr}^+_{\partial\Omega_m}u_m|^2\,\mathrm{d}\mu_k\,\mathrm{d}t &\longrightarrow \int_0^T\int_{\partial\Omega}\Lambda|\operatorname{Tr}^+_{\partial\Omega}u|^2\,\mathrm{d}\mu\,\mathrm{d}t,\\ \int_0^T\int_{\partial\Omega_m} \Lambda|\operatorname{Tr}^-_{\partial\Omega_m}u_m|^2\,\mathrm{d}\mu_k\,\mathrm{d}t &\longrightarrow \int_0^T\int_{\partial\Omega}\Lambda|\operatorname{Tr}^-_{\partial\Omega}u|^2\,\mathrm{d}\mu\,\mathrm{d}t,\\ \int_0^T\int_{\partial\Omega_m} \Lambda(\operatorname{Tr}^+_{\partial\Omega_m}u_m)(\operatorname{Tr}^-_{\partial\Omega_m}u_m)\,\mathrm{d}\mu_k\,\mathrm{d}t &\longrightarrow \int_0^T\int_{\partial\Omega}\Lambda(\operatorname{Tr}^+_{\partial\Omega}u)(\operatorname{Tr}^-_{\partial\Omega}u)\,\mathrm{d}\mu\,\mathrm{d}t, \end{aligned} \tag{3.8}$$

The limits from Eq. (3.8) along with the lower semi-continuity of the weak limits from Eqs. (3.6) and (3.7) yield (3.4).

To complete this proof, given $u \in L^2(]0,T[,L^2(U))$, we seek a sequence $(u_m) \in L^2(]0,T[,L^2(U))^{\mathbb{N}}$ such that $u_m \to u$ and $J_T(\Omega_m,\mu_m)(u_m) \to J_T(\Omega,\mu)(u)$. We may assume $u|_{U\backslash\partial\Omega} \in L^2(]0,T[,H^1(U\backslash\partial\Omega))$ so that the limit is finite. We define

$$u_m := (\mathrm{E}_\Omega u)|_{\Omega_m} + (\mathrm{E}_{U\backslash\overline{\Omega}}u)|_{U\backslash\overline{\Omega}_m},$$

where $\mathrm{E}_\Omega : H^1(\Omega) \to H^1(U)$ and $\mathrm{E}_{U\backslash\overline{\Omega}} : H^1(U\backslash\overline{\Omega}) \to H^1(U)$ are linear continuous extension operators. Then, by dominated convergence,

$$\int_0^T\int_{\Omega_m}\left(|u_m|^2 + |\nabla u_m|^2\right)\mathrm{d}x\,\mathrm{d}t \longrightarrow \int_0^T\int_\Omega\left(|u|^2+|\nabla u|^2\right)\mathrm{d}x\,\mathrm{d}t.$$

That limit, along with the limits of boundary integrals given by Eq. (3.8) yield $J_T(\Omega_m,\mu_m)(u_m) \to J_T(\Omega,\mu)(u)$, which is condition (ii) from Definition 3.1. □

Given the definition of convergence in the sense of Mosco and Theorem 3.2, we can deduce the Γ-convergence of the energy forms. In particular, by [**DM93**, Corollary 7.20], the convergence of the energy forms yields the following result on convergence of minimizers. Note that those minimizers depend notably on the diffusion time $T > 0$.

COROLLARY 3.3 (Convergence of minimizers). *Let $\varepsilon > 0$. Let $(\Omega_m)_{m\in\mathbb{N}}$ be sequence of (ε, ∞)-domains included in U and $(\mu_m)_{m\in\mathbb{N}}$ be a sequence of finite Borel measures with* $\operatorname{supp}\mu_m = \partial\Omega_m$*, all satisfying* (2.1) *for the same $d \in [n-1, n[$ and the same constant $c_d > 0$. Assume there exists a domain Ω and a measure μ with* $\operatorname{supp}\mu = \partial\Omega$ *such that:*

(i) *$(\Omega_m)_{m\in\mathbb{N}}$ converges to Ω in the sense of characteristic functions and Hausdorff;*
(ii) *$(\mu_m)_{m\in\mathbb{N}}$ converges to μ weakly.*

If $(u_m)_{m\in\mathbb{N}} \in L^2(]0,T[, L^2(U))^{\mathbb{N}}$ is such that for all $m \in \mathbb{N}$, u_m is a minimizer of $J_T(\Omega_m, \mu_m)$, then any accumulation point u of $(u_m)_{m\in\mathbb{N}}$ is a minimizer of $J_T(\Omega, \mu)$ and it holds:

$$J_T(\Omega, \mu)(u) = \limsup_{m\to\infty} J_T(\Omega_m, \mu_m)(u_m).$$

If $(u_m)_{m\in\mathbb{N}}$ converges to u in $L^2(]0,T[, L^2(U))$, then u is a minimizer of $J_T(\Omega, \mu)$ and it holds:

$$J_T(\Omega, \mu)(u) = \lim_{m\to\infty} J_T(\Omega_m, \mu_m)(u_m).$$

Consequently, if $(\Omega_m, \mu_m)_{m\in\mathbb{N}}$ is a sequence of (ε, ∞)-domains for some $\varepsilon > 0$ converging to (Ω, μ) in the sense of Theorem 3.2, then energy minimizers for $(\Omega_m, \mu_m)_{m\in\mathbb{N}}$ allow to determine energy minimizers for the limit domain. In the next section, we prove the existence of an optimal shape in a class of (ε, ∞)-domains. In particular, that shape can be fractal: Corollary 3.3 allows to approximate an energy minimizer u (and the associated energy) on the optimal shape with minimizers on regular shapes.

3.2. Shape optimization. We wish to prove the existence of a domain Ω along with a boundary measure which maximize the heat transfer from Ω^+ to Ω^- in a class of domains of fixed volume. To that end, we consider the following energy functional:

$$\mathcal{J}_T(\Omega, \mu) := J_T(\Omega, \mu)(u_{(\Omega,\mu)}), \tag{3.9}$$

where (Ω, μ) is a two-sided admissible domain of $\mathbb{R}^n$, J_T is defined by (3.1) and $u_{(\Omega,\mu)}$ is the unique weak solution to problem (2.17) for (Ω, μ) (in the sense of Theorem 2.7). To prove the existence of an optimal shape (Ω_*, μ_*):

$$\mathcal{J}_T(\Omega_*, \mu_*) = \min_{(\Omega,\mu)\in\mathcal{U}} \mathcal{J}_T(\Omega, \mu),$$

we consider several classes of admissible domains $\mathcal{U}$: a class of Lipschitz domains on the one hand (Paragraph 3.2.1), and a class of uniform domains on the other hand (Paragraph 3.2.2).

3.2.1. *In a class of Lipschitz domains.* We start with seeking an optimal shape in a class of Lipschitz subdomains of U (which is still a large bounded Lipschitz domain). More specifically, we wish to prove the existence of an optimal domain Ω which satisfies the ε-cone property [**HP18**, Definition 2.4.1] for a given $\varepsilon > 0$, *i.e.* for all $x \in \partial\Omega$, there exists $\xi_x \in \mathbb{R}^n$ unitary such that for all $y \in \overline{\Omega} \cap B_\varepsilon(x)$, it holds:

$$\left\{z \in \mathbb{R}^n \mid \langle z - y, \xi_x\rangle > \cos(\varepsilon)|z - y| \ \text{ and } \ 0 < |z - y| < \varepsilon\right\} \subset \Omega.$$

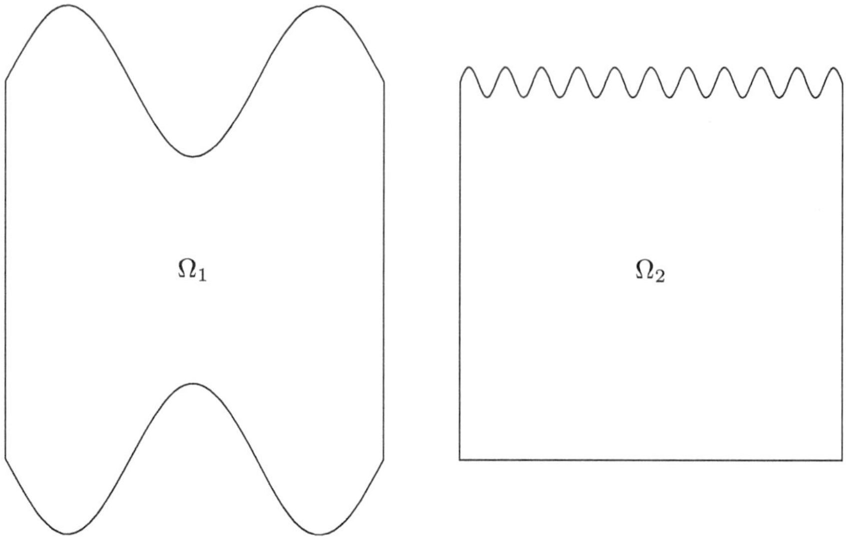

FIGURE 5. Two Lipschitz domains, Ω_1 and Ω_2. The domain Ω_1 has a greater perimeter than Ω_2, but the boundary of Ω_2 presents oscillations at a higher frequency than that of Ω_1. Altogether, there exists a constant $\hat{c} > 0$ such that (3.10) holds with $\hat{c}$ for Ω_1, but not for Ω_2.

A bounded domain Ω of $\mathbb{R}^n$ is Lipschitz if and only if it satisfies the ε-cone property for some $\varepsilon > 0$ [**HP18**, Theorem 2.4.7]. For $\varepsilon > 0$, let $\Theta(U, \varepsilon)$ be the set of subdomains of U with connected boundary satisfying the ε-cone property.

We also require the domain Ω to be of fixed volume (with respect to $\lambda^{(n)}$), its boundary to be confined in $\overline{G}$ where $G \subsetneq U$ is a domain of $\mathbb{R}^n$ and such that, for a given $\hat{c} > 0$, it holds:

$$\forall r > 0,\ \forall x \in \partial\Omega, \quad \mathcal{H}^{(n-1)}(\partial\Omega \cap B_r(x)) \leq \hat{c}\, r^{n-1}, \tag{3.10}$$

where $\mathcal{H}^{(n-1)}$ is the $(n-1)$-dimensional Hausdorff measure (here, with support $\partial\Omega$). Note that, by boundedness of U, this implies there exists M depending on $\hat{c}$ such that $\mathcal{H}^{(n-1)}(\partial\Omega) \leq M$. Condition (3.10) is a restriction on the perimeter of the domains considered both globally (for r large enough) and locally (for r small), as illustrated by Figure 5.

From those conditions, we may define the class of admissible Lipschitz domains as follows (see Figure 6).

DEFINITION 3.4 (Admissible Lipschitz domains). Let $\varepsilon > 0$, $V > 0$ and $\hat{c} > 0$. Let $G \subsetneq U$ be a bounded domain of $\mathbb{R}^n$. The class of admissible Lipschitz domains is defined by:

$$\begin{aligned} \mathcal{U}_{U,\varepsilon}(G, V, \hat{c}) := \{\Omega \in \Theta(U, \varepsilon) \mid \lambda^{(n)}(\Omega) = V, \quad \partial\Omega \subset \overline{G}, \\ \text{and} \quad \forall r > 0,\ \forall x \in \partial\Omega,\ \mathcal{H}^{(n-1)}(\partial\Omega \cap B_r(x)) \leq \hat{c}\, r^{n-1}\}. \end{aligned} \tag{3.11}$$

A key ingredient in optimization problems is the compactness of the class of admissible domains with respect to some domain topology. In our case, the class of admissible Lipschitz domains is compact with respect to the convergence in the

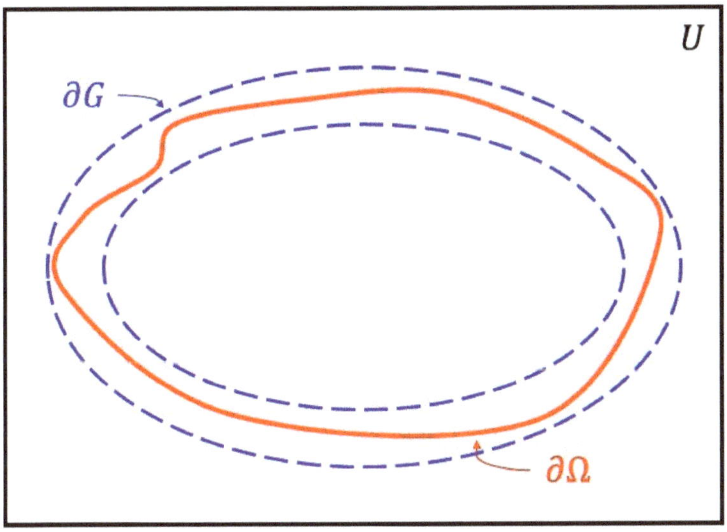

FIGURE 6. An example of an admissible Lipschitz domain, lying inside the red boundary $\partial\Omega$. U is the interior of the black rectangle. G lies between the blue dashed lines ∂G.

sense of characteristic functions, Hausdorff and compact sets [**HP18**, Definitions 2.2.3, 2.2.8 and 2.2.21]. However, it is not compact with respect to the weak convergence of boundary measures, in the sense that the weak limit of a sequence of Hausdorff measures is not necessarily a Hausdorff measure itself.

THEOREM 3.5. *Let $\varepsilon > 0$, $V > 0$ and $\hat{c} > 0$. Let $G \subsetneq U$ be a bounded domain of $\mathbb{R}^n$. Then, the class of admissible Lipschitz domains $\mathcal{U}_{U,\varepsilon}(G, V, \hat{c})$ is compact for the convergence in the sense of characteristic functions, Hausdorff and compact sets.*

PROOF. We follow the proof of [**MKNORP21**, Lemma 3.1], in dimension $n \geq 2$ and considering the Hausdorff measure $\mathcal{H}^{(n-1)}$ instead of the Lebesgue measure $\lambda^{(n-1)}$ (which coincide for $n = 2$). Such a modification can be seen as rescaling the constant $\hat{c}$. □

Since the class of admissible Lipschitz domains is compact with respect to domain convergence, it is possible to find optimal shapes in it for problems with homogeneous Dirichlet or Neumann boundary conditions. However, in our case, the energy form $\mathcal{J}_T$ also takes the boundary measure into account and we prove in the following theorem that although an optimal geometry exists in that class, its boundary is not necessarily endowed with the Hausdorff measure. Considering a boundary measure which is not the Hausdorff measure for a Lipschitz domain can be seen as measuring the perimeter of the volume in an unusual manner, in particular one which is not consistent with the physical modelling of heat diffusion problems. A similar phenomenon can be highlighted in the setting of [**MKNORP21**, Theorem 3.2], in which a Helmholtz problem in the plane with a Robin boundary condition is studied. Note that the optimal shapes discussed in the following theorem depend on the parameters which were set at the beginning of this section, notably the transfer time T.

THEOREM 3.6 (Lipschitz optimal shape). *Let $\varepsilon > 0$, $V > 0$ and $\hat{c} > 0$. Let $G \subsetneq U$ be a bounded domain of $\mathbb{R}^n$. Then, there exists $\Omega_* \in \mathcal{U}_{U,\varepsilon}(G, V, \hat{c})$ and a finite $(n-1)$-dimensional positive measure μ_* on $\partial\Omega_*$ equivalent to the Hausdorff measure $\mathcal{H}^{(n-1)}$ such that, for all Borel set $\Gamma \subset \partial\Omega_*$:*

$$\mu_*(\Gamma) \geq \mathcal{H}^{(n-1)}(\Gamma), \tag{3.12}$$

and

$$J_T(\Omega_*, \mathcal{H}^{(n-1)})(u_{(\Omega_*,\mu_*)}) \leq \inf_{\Omega \in \mathcal{U}_{U,\varepsilon}(G,V,\hat{c})} \mathcal{J}_T(\Omega, \mathcal{H}^{(n-1)}) = \mathcal{J}_T(\Omega_*, \mu_*),$$

where $\mathcal{J}_T$ is defined by (3.9) and $u_{(\Omega_*,\mu_*)}$ denotes the solution to problem (2.17) on (Ω_*, μ_*) in the sense of Theorem 2.7.

PROOF. Let $(\Omega_m)_{m\in\mathbb{N}} \in \mathcal{U}_{U,\varepsilon}(G, V, \hat{c})^{\mathbb{N}}$ be such that:

$$\mathcal{J}_T(\Omega_m, \mathcal{H}^{(n-1)}) \xrightarrow[m\to\infty]{} \inf_{\Omega \in \mathcal{U}_{U,\varepsilon}(G,V,\hat{c})} \mathcal{J}_T(\Omega, \mathcal{H}^{(n-1)}) \geq 0.$$

Then, by compactness of the class of admissible Lipschitz domains (Theorem 3.5), we may assume (up to passing to a subsequence) there exists $\Omega_* \in \mathcal{U}_{U,\varepsilon}(G, V, \hat{c})$ such that

$$\Omega_m \xrightarrow[m\to\infty]{\mathbb{1},H,K} \Omega_*,$$

where K stands for convergence in the sense of compact sets. Minor changes in the proof of [**MKNORP21**, Lemma 3.1, Point 3] (*i.e.*, considering subsets of the boundary instead of the whole boundary) yields the existence of μ_* supported by $\partial\Omega_*$ satisfying (3.12). For $m \in \mathbb{N}$, let u_m be the solution to problem (2.17) on $(\Omega_m, \mathcal{H}^{(n-1)})$ in the sense of Theorem 2.7. Since the problem is well-posed (see Eq. (2.20)) and by uniform boundedness of the extension operators $\mathrm{E}_{\Omega_m} : H^1(\Omega_m) \to H^1(U)$, $(\mathrm{E}_{\Omega_m} u_m)_{m\in\mathbb{N}}$ forms a bounded sequence of $L^2(]0,T[, H^1(U))$ and $(\mathrm{E}_{\Omega_m}(\partial_t u_m))_{m\in\mathbb{N}}$ forms a bounded sequence of $L^2(]0,T[, H^1(U)')$. Hence, there exists u_∞ such that (up to passing to a subsequence):

$$\begin{aligned} u_m &\xrightarrow[m\to\infty]{} u_\infty && \text{in } L^2(]0,T[, H^1(U)), \\ \partial_t u_m &\xrightarrow[m\to\infty]{} \partial_t u_\infty && \text{in } L^2(]0,T[, H^1(U)'). \end{aligned}$$

Proceeding as in the proof of Theorem 3.2 leading to Eq. (3.6), we prove $u_\infty|_\Omega = u_*|_\Omega$, where u_* is the weak solution to problem (2.17) on (Ω_*, μ_*). Up to passing to a subsequence, we may assume the convergence in the sense of characteristic functions in an a.e. pointwise convergence. Hence, by dominated convergence (for the integrals on the bulk) and using the boundary integral convergences from Eq. (3.8), we can deduce:

$$\mathcal{J}_T(\Omega_m, \mathcal{H}^{(n-1)}) = J_T(\Omega_m, \mathcal{H}^{(n-1)})(u_m) \xrightarrow[m\to\infty]{} J_T(\Omega_*, \mu_*)(u_*) = \mathcal{J}_T(\Omega_*, \mu_*).$$

Since $\mu_* \geq \mathcal{H}^{(n-1)}$, it follows that:

$$J_T(\Omega_*, \mathcal{H}^{(n-1)})(u_{(\Omega_*,\mu_*)}) \leq \mathcal{J}_T(\Omega_*, \mu_*).$$

□

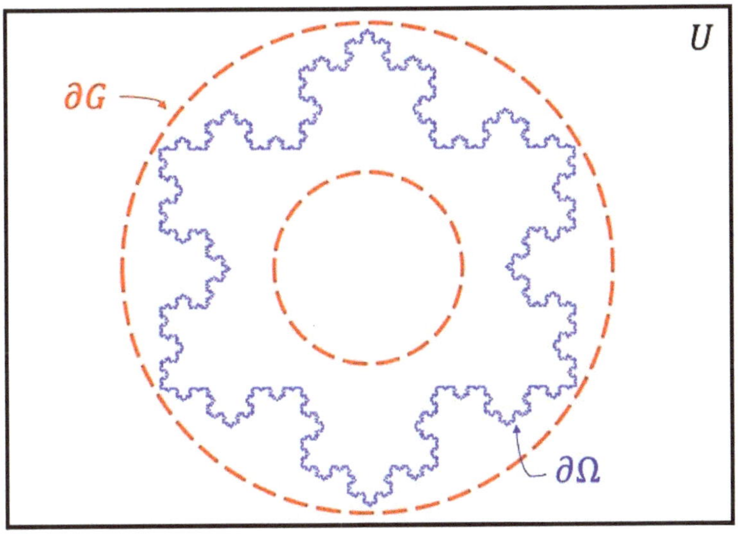

FIGURE 7. An example of a admissible uniform domain which is not an admissible Lipschitz domain. The domain lies inside the blue Von Koch curve (represented by a fifth generation pre-fractal [source: Wikipedia]) and its boundary is endowed with the $\frac{\ln 4}{\ln 3}$-dimensional Hausdorff measure. U is the interior of the black rectangle. G lies between the red dashed lines ∂G.

3.2.2. *In a class of uniform domains.* As it was proved in Theorem 3.6, the measure on the boundary of an optimal Lipschitz geometry is not necessarily the Hausdorff measure $\mathcal{H}^{(n-1)}$, which means, in some sense, that the energy infimum can be reached outside of the class of Lipschitz admissible boundaries (implicitly endowed with the Hausdorff measure as explained before). For that matter, we extend the class of admissible domains: as of now, we wish to find an optimal domain Ω which is an (ε,∞)-domain for some $\varepsilon > 0$ (defined in Subsection 3.1). For $\varepsilon > 0$, let $\mathcal{O}(U,\varepsilon)$ be the set of (ε,∞)-domains included in U and with a connected boundary. We also require the boundary to be the support of a measure μ satisfying the d-upper regularity condition (2.1) for some $d \in [n-1,n[$ and $c_d > 0$, as well as the s-lower regularity condition for some $s \in [n-1,n[$ and $c_s > 0$:

$$\forall x \in \partial\Omega, \quad \forall r \in]0,1], \quad c_s\, r^s \le \mu(\overline{B_r(x)}), \tag{3.13}$$

which means the Hausdorff dimension of the boundary cannot exceed s. Naturally, it must hold $d \le s$ for such a measure to exist. From those conditions, we may define the class of admissible uniform domains as follows (see Figure 7).

DEFINITION 3.7 (Admissible uniform domains). Let $\varepsilon > 0$, $V > 0$, $d, s \in [n-1,n[$ with $d \le s$ and $c_d, c_s > 0$. Let $G \subsetneq U$ be a bounded domain of $\mathbb{R}^n$. The class of admissible uniform domains is defined by:

$$\begin{aligned}\mathcal{U}^*_{U,\varepsilon}(G,V,d,s,c_d,c_s) := \big\{(\Omega,\mu) \mid \Omega \in \mathcal{O}(U,\varepsilon), \quad \lambda^{(n)}(\Omega) = V, \quad \partial\Omega \subset \overline{G},\\ \operatorname{supp}\mu = \partial\Omega \quad \text{and} \quad \mu \text{ satisfies (2.1) and (3.13)}\big\}.\end{aligned} \tag{3.14}$$

It follows from [**HRPT21**, Theorem 3] that the class of admissible uniform domains is compact with respect to the convergence in the sense of characteristic functions, Hausdorff and compact sets, but also with respect to the weak convergence of boundary measures.

THEOREM 3.8. *Let $\varepsilon > 0$, $V > 0$, $d, s \in [n-1, n[$ with $d \leq s$ and $c_d, c_s > 0$. Let $G \subsetneq U$ be a bounded domain of $\mathbb{R}^n$. Then,*

(i) *the class $\mathcal{U}^*_{U,\varepsilon}(G, V, d, s, c_d, c_s)$ of admissible uniform domains is compact in the sense of characteristic functions, Hausdorff and compact sets, as well as in the sense of weak convergence of boundary measures.*
(ii) *if $(\Omega_m, \mu_m)_{m\in\mathbb{N}} \in \mathcal{U}^*_{U,\varepsilon}(G, V, d, s, c_d, c_s)^{\mathbb{N}}$ is such that $(\mu_m)_{m\in\mathbb{N}}$ converges weakly, then $(\Omega_m)_{m\in\mathbb{N}}$ converges in the sense of characteristic functions, Hausdorff and compact sets.*

In light of Theorem 3.8, the proof of Theorem 3.6 can be adapted on the class of admissible uniform domains to prove the existence of an optimal shape. Note that, once again, the optimal shapes discussed in the following theorem depend on the parameters which were set at the beginning of this section, notably the transfer time T.

THEOREM 3.9 (Uniform optimal shape). *Let $\varepsilon > 0$, $V > 0$, $d, s \in [n-1, n[$ with $d \leq s$ and $c_d, c_s > 0$. Let $G \subsetneq U$ be a bounded domain of $\mathbb{R}^n$. Then, there exists $(\Omega_*, \mu_*) \in \mathcal{U}^*_{U,\varepsilon}(G, V, d, s, c_d, c_s)$ such that*

$$\mathcal{J}_T(\Omega_*, \mu_*) = \min_{(\Omega,\mu)\in \mathcal{U}^*_{U,\varepsilon}(G,V,d,s,c_d,c_s)} \mathcal{J}_T(\Omega, \mu),$$

where $\mathcal{J}_T$ is defined by (3.9).

Acknowledgments

The authors are grateful to Gabriel Singer for his preliminary work on the topic, conducted during an internship supervised by Anna Rozanova-Pierrat. The authors also thank Alexander Teplyaev and Michael Hinz for enlightening discussions on the topic. The motivation for this work originated from research conducted by Anna Rozanova-Pierrat with Bernard Sapoval and Denis Grebenkov in École Polytechnique, France.

References

[Ada75] R. A. Adams, *Sobolev spaces*, Pure and Applied Mathematics, Vol. 65, Academic Press [Harcourt Brace Jovanovich, Publishers], New York-London, 1975. MR450957

[Agm10] S. Agmon, *Lectures on elliptic boundary value problems*, AMS Chelsea Publishing, Providence, RI, 2010. Prepared for publication by B. Frank Jones, Jr. with the assistance of George W. Batten, Jr.; Revised edition of the 1965 original, DOI 10.1090/chel/369. MR2589244

[AH96] D. R. Adams and L. I. Hedberg, *Function spaces and potential theory*, Grundlehren der mathematischen Wissenschaften [Fundamental Principles of Mathematical Sciences], vol. 314, Springer-Verlag, Berlin, 1996, DOI 10.1007/978-3-662-03282-4. MR1411441

[ARP19] K. Arfi and A. Rozanova-Pierrat, *Dirichlet-to-Neumann or Poincaré-Steklov operator on fractals described by d-sets*, Discrete Contin. Dyn. Syst. Ser. S **12** (2019), no. 1, 1–26, DOI 10.3934/dcdss.2019001. MR3836589

[BB05] D. Bucur and G. Buttazzo, *Variational methods in shape optimization problems*, Progress in Nonlinear Differential Equations and their Applications, vol. 65, Birkhäuser Boston, Inc., Boston, MA, 2005. MR2150214

[BG16] D. Bucur and A. Giacomini, *Shape optimization problems with Robin conditions on the free boundary*, Ann. Inst. H. Poincaré C Anal. Non Linéaire **33** (2016), no. 6, 1539–1568, DOI 10.1016/j.anihpc.2015.07.001. MR3569242

[BGRP16] C. Bardos, D. Grebenkov, and A. Rozanova-Pierrat, *Short-time heat diffusion in compact domains with discontinuous transmission boundary conditions*, Math. Models Methods Appl. Sci. **26** (2016), no. 1, 59–110, DOI 10.1142/S0218202516500032. MR3417725

[Bie09] M. Biegert, *On traces of Sobolev functions on the boundary of extension domains*, Proc. Amer. Math. Soc. **137** (2009), no. 12, 4169–4176, DOI 10.1090/S0002-9939-09-10045-X. MR2538577

[BNNT22] D. Bucur, M. Nahon, C. Nitsch, and C. Trombetti, *Shape optimization of a thermal insulation problem*, Calc. Var. Partial Differential Equations **61** (2022), no. 5, Paper No. 186, 29, DOI 10.1007/s00526-022-02298-1. MR4457936

[Bre87] H. Brezis, *Analyse fonctionnelle* (French), Collection Mathématiques Appliquées pour la Maîtrise. [Collection of Applied Mathematics for the Master's Degree], Masson, Paris, 1983. Théorie et applications. [Theory and applications]. MR697382

[Cal61] A.-P. Calderón, *Lebesgue spaces of differentiable functions and distributions*, Proc. Sympos. Pure Math., Vol. IV, Amer. Math. Soc., Providence, RI, 1961, pp. 33–49. MR143037

[Cap10] R. Capitanelli, *Asymptotics for mixed Dirichlet-Robin problems in irregular domains*, J. Math. Anal. Appl. **362** (2010), no. 2, 450–459, DOI 10.1016/j.jmaa.2009.09.042. MR2557700

[CCLRC23] M. Cefalo, S. Creo, M. R. Lancia, and J. Rodríguez-Cuadrado, *Fractal mixtures for optimal heat draining*, Chaos Solitons Fractals **173** (2023), Paper No. 113750, 11, DOI 10.1016/j.chaos.2023.113750. MR4609962

[CD21] R. Capitanelli and M. D'Ovidio, *Fractional Cauchy problem on random snowflakes*, J. Evol. Equ. **21** (2021), no. 2, 2123–2140, DOI 10.1007/s00028-021-00673-7. MR4278424

[CF12] Z.-Q. Chen and M. Fukushima, *Symmetric Markov processes, time change, and boundary theory*, London Mathematical Society Monographs Series, vol. 35, Princeton University Press, Princeton, NJ, 2012. MR2849840

[Che75] D. Chenais, *On the existence of a solution in a domain identification problem*, J. Math. Anal. Appl. **52** (1975), no. 2, 189–219, DOI 10.1016/0022-247X(75)90091-8. MR385666

[CHRT24] G. Claret, M. Hinz, A. Rozanova-Pierrat, and A. Teplyaev, *Layer potential operators for transmission problems on extension domains*, March 2024, arXiv:2403.11601, 2024.

[CL23] S. Creo and M. R. Lancia, *Dynamic boundary conditions for time dependent fractional operators on extension domains*, Adv. Differential Equations **29** (2024), no. 9-10, 727–756. MR4740326

[Cla23] G. Claret, *Analyse fonctionnelle, spectrale et probabiliste sur domaines d'extension de Sobolev*, Master thesis, CentraleSupélec, Université Paris-Saclay, 2023.

[CWHM17] S. N. Chandler-Wilde, D. P. Hewett, and A. Moiola, *Sobolev spaces on non-Lipschitz subsets of* $\mathbb{R}^n$ *with application to boundary integral equations on fractal screens*, Integral Equations Operator Theory **87** (2017), no. 2, 179–224, DOI 10.1007/s00020-017-2342-5. MR3620760

[DG82] P.-G. De Gennes, *Physique des surfaces et des interfaces*, Compte Rendus de l'Académie des Sciences **295** (1982), 1061–1064 (fr).

[DM93] G. Dal Maso, *An introduction to* Γ*-convergence*, Progress in Nonlinear Differential Equations and their Applications, vol. 8, Birkhäuser Boston, Inc., Boston, MA, 1993, DOI 10.1007/978-1-4612-0327-8. MR1201152

[Doo54] J. L. Doob, *Semimartingales and subharmonic functions*, Trans. Amer. Math. Soc. **77** (1954), 86–121, DOI 10.2307/1990680. MR64347

[Doo55] J. L. Doob, *A probability approach to the heat equation*, Trans. Amer. Math. Soc. **80** (1955), 216–280, DOI 10.2307/1993013. MR79376

[DRPT22] A. Dekkers, A. Rozanova-Pierrat, and A. Teplyaev, *Mixed boundary valued problems for linear and nonlinear wave equations in domains with fractal boundaries*, Calc. Var. Partial Differential Equations **61** (2022), no. 2, Paper No. 75, 44, DOI 10.1007/s00526-021-02159-3. MR4393126

[Eva10] L. C. Evans, *Partial differential equations*, 2nd ed., Graduate Studies in Mathematics, vol. 19, American Mathematical Society, Providence, RI, 2010, DOI 10.1090/gsm/019. MR2597943

[FLV95] J. Fleckinger, M. Levitin, and D. Vassiliev, *Heat equation on the triadic von Koch snowflake: asymptotic and numerical analysis*, Proc. London Math. Soc. (3) **71** (1995), no. 2, 372–396, DOI 10.1112/plms/s3-71.2.372. MR1337471

[FOT10] M. Fukushima, Y. Oshima, and M. Takeda, *Dirichlet forms and symmetric Markov processes*, Second revised and extended edition, De Gruyter Studies in Mathematics, vol. 19, Walter de Gruyter & Co., Berlin, 2011. MR2778606

[GK03] P. B. Gilkey and K. Kirsten, *Heat content asymptotics with transmittal and transmission boundary conditions*, J. London Math. Soc. (2) **68** (2003), no. 2, 431–443, DOI 10.1112/S0024610703004526. MR1994692

[HKT08] P. Hajłasz, P. Koskela, and H. Tuominen, *Sobolev embeddings, extensions and measure density condition*, J. Funct. Anal. **254** (2008), no. 5, 1217–1234, DOI 10.1016/j.jfa.2007.11.020. MR2386936

[HMRP+21] M. Hinz, F. Magoulès, A. Rozanova-Pierrat, M. Rynkovskaya, and A. Teplyaev, *On the existence of optimal shapes in architecture*, Appl. Math. Model. **94** (2021), 676–687, DOI 10.1016/j.apm.2021.01.041. MR4216899

[HP18] A. Henrot and M. Pierre, *Shape variation and optimization*, EMS Tracts in Mathematics, vol. 28, European Mathematical Society (EMS), Zürich, 2018. A geometrical analysis; English version of the French publication [MR2512810] with additions and updates, DOI 10.4171/178. MR3791463

[HRPT21] M. Hinz, A. Rozanova-Pierrat, and A. Teplyaev, *Non-Lipschitz uniform domain shape optimization in linear acoustics*, SIAM J. Control Optim. **59** (2021), no. 2, 1007–1032, DOI 10.1137/20M1361687. MR4228309

[HRPT23] M. Hinz, A. Rozanova-Pierrat, and A. Teplyaev, *Boundary value problems on non-Lipschitz uniform domains: stability, compactness and the existence of optimal shapes*, Asymptot. Anal. **134** (2023), no. 1-2, 25–61, DOI 10.3233/asy-231825. MR4624813

[HS92] A. Henrot and J. Sokołowski, *Shape Optimization Problem for Heat Equation*, Rapport de recherche 3185, INRIA, December 1992.

[Hun56] G. A. Hunt, *Some theorems concerning Brownian motion*, Trans. Amer. Math. Soc. **81** (1956), 294–319, DOI 10.2307/1992918. MR79377

[JK82] D. S. Jerison and C. E. Kenig, *Boundary behavior of harmonic functions in non-tangentially accessible domains*, Adv. in Math. **46** (1982), no. 1, 80–147, DOI 10.1016/0001-8708(82)90055-X. MR676988

[Jon81] P. W. Jones, *Quasiconformal mappings and extendability of functions in Sobolev spaces*, Acta Math. **147** (1981), no. 1-2, 71–88, DOI 10.1007/BF02392869. MR631089

[Jon09] A. Jonsson, *Besov spaces on closed sets by means of atomic decomposition*, Complex Var. Elliptic Equ. **54** (2009), no. 6, 585–611, DOI 10.1080/17476930802669678. MR2537257

[JW84] A. Jonsson and H. Wallin, *Function spaces on subsets of* $\mathbf{R}^n$, Math. Rep. **2** (1984), no. 1, xiv+221. MR820626

[JW95] A. Jonsson and H. Wallin, *The dual of Besov spaces on fractals*, Studia Math. **112** (1995), no. 3, 285–300, DOI 10.4064/sm-112-3-285-300. MR1311704

[Lan02] M. R. Lancia, *A transmission problem with a fractal interface*, Z. Anal. Anwendungen **21** (2002), no. 1, 113–133, DOI 10.4171/ZAA/1067. MR1916407

[Lan04] M. R. Lancia, *On some second order transmission problems*, Arab. J. Sci. Eng. Sect. C Theme Issues **29** (2004), no. 2, 85–100. MR2126590

[LRP21] M. R. Lancia and A. Rozanova-Pierrat (eds.), *Fractals in engineering: theoretical aspects and numerical approximations*, SEMA SIMAI Springer Series, Springer, Cham, [2021] ©2021. Including selected presentations from the minisymposium

(MS256) held in Valencia, July 15–19, 2019; ICIAM 2019 SEMA SIMAI Springer Series, 8, DOI 10.1007/978-3-030-61803-2. MR4376020

[LV96] M. Levitin and D. Vassiliev, *Spectral asymptotics, renewal theorem, and the Berry conjecture for a class of fractals*, Proc. London Math. Soc. (3) **72** (1996), no. 1, 188–214, DOI 10.1112/plms/s3-72.1.188. MR1357092

[LV14] M. R. Lancia and P. Vernole, *Venttsel' problems in fractal domains*, J. Evol. Equ. **14** (2014), no. 3, 681–712, DOI 10.1007/s00028-014-0233-7. MR3250804

[Man67] B. B. Mandelbrot, *How Long Is the Coast of Britain? Statistical Self-Similarity and Fractional Dimension*, Science **156** (1967), no. 3775, 636–638 (en).

[Man83] B. B. Mandelbrot, *The fractal geometry of nature*, Schriftenreihe für den Referenten. [Series for the Referee], W. H. Freeman and Co., San Francisco, CA, 1982. MR665254

[Mar87] J. Marschall, *The trace of Sobolev-Slobodeckij spaces on Lipschitz domains*, Manuscripta Math. **58** (1987), no. 1-2, 47–65, DOI 10.1007/BF01169082. MR884984

[MKNORP21] F. Magoulès, T. P. K. Nguyen, P. Omnes, and A. Rozanova-Pierrat, *Optimal absorption of acoustic waves by a boundary*, SIAM J. Control Optim. **59** (2021), no. 1, 561–583, DOI 10.1137/20M1327239. MR4214360

[Mos94] U. Mosco, *Composite media and asymptotic Dirichlet forms*, J. Funct. Anal. **123** (1994), no. 2, 368–421, DOI 10.1006/jfan.1994.1093. MR1283033

[MV03] U. Mosco and M. A. Vivaldi, *Variational problems with fractal layers* (English, with English and Italian summaries), Rend. Accad. Naz. Sci. XL Mem. Mat. Appl. (5) **27** (2003), 237–251. MR2056421

[Neč12] J. Nečas, *Direct methods in the theory of elliptic equations*, Springer Monographs in Mathematics, Springer, Heidelberg, 2012. Translated from the 1967 French original by Gerard Tronel and Alois Kufner; Editorial coordination and preface by Šárka Nečasová and a contribution by Christian G. Simader, DOI 10.1007/978-3-642-10455-8. MR3014461

[Nys94] K. Nyström, *Smoothness properties of solutions to Dirichlet problems in domains with a fractal boundary*, Ph.D. thesis, University of Umeä, Umeä, 1994.

[Nys96] K. Nyström, *Integrability of Green potentials in fractal domains*, Ark. Mat. **34** (1996), no. 2, 335–381, DOI 10.1007/BF02559551. MR1416671

[PNB05] I. G. Pirozhenko, V. V. Nesterenko, and M. Bordag, *Integral equations for heat kernel in compound media*, J. Math. Phys. **46** (2005), no. 4, 042305, 21, DOI 10.1063/1.1870734. MR2131231

[Ros51] M. Rosenblatt, *On a class of Markov processes*, Trans. Amer. Math. Soc. **71** (1951), 120–135, DOI 10.2307/1990862. MR43406

[RP21] A. Rozanova-Pierrat, *Generalization of Rellich-Kondrachov theorem and trace compactness for fractal boundaries*, Fractals in engineering: theoretical aspects and numerical approximations, SEMA SIMAI Springer Ser., Springer, Cham, [2021] ©2021, pp. 155–173. MR4376027

[RPGS12] A. Rozanova-Pierrat, D. S. Grebenkov, and B. Sapoval, *Faster Diffusion across an Irregular Boundary*, Physical Review Letters **108** (2012), no. 24, 240602.

[SRG85] B. Sapoval, M. Rosso, and J.-F. Gouyet, *The fractal nature of a diffusion front and the relation to percolation*, Journal de Physique Lettres **46** (1985), no. 4, 149–156.

[Ste70] E. M. Stein, *Singular integrals and differentiability properties of functions*, Princeton Mathematical Series, No. 30, Princeton University Press, Princeton, NJ, 1970. MR290095

[TZZ18] E. Trélat, C. Zhang, and E. Zuazua, *Optimal shape design for 2D heat equations in large time*, Pure Appl. Funct. Anal. **3** (2018), no. 1, 255–269. MR3777007

[Vas03] D. V. Vassilevich, *Heat kernel expansion: user's manual*, Phys. Rep. **388** (2003), no. 5-6, 279–360, DOI 10.1016/j.physrep.2003.09.002. MR2017528

[vdBG94] M. van den Berg and P. B. Gilkey, *Heat content asymptotics of a Riemannian manifold with boundary*, J. Funct. Anal. **120** (1994), no. 1, 48–71, DOI 10.1006/jfan.1994.1022. MR1262245

[vdBLG94] M. van den Berg and J.-F. Le Gall, *Mean curvature and the heat equation*, Math. Z. **215** (1994), no. 3, 437–464, DOI 10.1007/BF02571723. MR1262526 [vdBS90] M. van den Berg and S. Srisatkunarajah, *Heat flow and Brownian motion for a region in* $\mathbf{R}^2$ *with a polygonal boundary*, Probab. Theory Related Fields **86** (1990), no. 1, 41–52, DOI 10.1007/BF01207512. MR1061947

[Wal91] H. Wallin, *The trace to the boundary of Sobolev spaces on a snowflake*, Manuscripta Math. **73** (1991), no. 2, 117–125, DOI 10.1007/BF02567633. MR1128682

[Wu79] J. M. G. Wu, *On parabolic measures and subparabolic functions*, Trans. Amer. Math. Soc. **251** (1979), 171–185, DOI 10.2307/1998688. MR531974

Laboratoire MICS, CentraleSupélec, Université Paris-Saclay, 91190 Gif-sur-Yvette, France

Email address: gabriel.claret@centralesupelec.fr

Laboratoire MICS, CentraleSupélec, Université Paris-Saclay, 91190 Gif-sur-Yvette, France

Email address: anna.rozanova-pierrat@centralesupelec.fr

Contemporary Mathematics
Volume **828**, 2025
https://doi.org/10.1090/conm/828/16593

Inverse problems in irregular domains: approximation via Mosco convergence

Simone Creo, Maria Rosaria Lancia, Gianluca Mola, and Silvia Romanelli

ABSTRACT. We consider inverse problems in an irregular domain Ω and their suitable approximations, respectively. Under suitable assumptions, after stating well-posedness results, we prove that the solutions of the approximating problems converge to the solution of the problem on Ω via Mosco convergence. We also present some applications.

Introduction

In this paper we focus on the approximation of the solution of inverse problems in an irregular domain Ω possibly with a fractal boundary.

We consider the so-called direct abstract Cauchy problem, which is formally stated as follows:

$$(P)\begin{cases} \frac{\partial u(t)}{\partial t} + \alpha A u(t) = 0, \quad t > 0, \\ u(0) = u_0, \end{cases}$$

where $\alpha > 0$, A is the generator of a suitable Dirichlet form E in $L^2(\Omega)$ and the initial datum $u_0 \neq 0$ belongs to $L^2(\Omega)$.

Our goal is twofold. Firstly we investigate the well-posedness of the so-called inverse problem $\boldsymbol{P}$ (defined in Section 2.1), consisting in finding the solution (α_*, u_*), where $u_*(t) \in L^2(\Omega)$ and $\alpha_* > 0$, with the additional condition

$$\|u_*(T)\|_2 = \rho,$$

where $\rho > 0$ is given, $T > 0$ is a fixed measuring time-instant and $\|\cdot\|_2$ denotes the norm on $L^2(\Omega)$.

Secondly, we address the question whether, under suitable assumptions, the solution of the inverse problem can be approximated in terms of solutions of problems on possibly smoother domains (as in the applications of Section 3).

2020 *Mathematics Subject Classification.* Primary 35R30, 35K90; Secondary 28A80, 47D06, 35B40.

Key words and phrases. Inverse problems, Mosco convergence, fractals, semigroups.

The authors have been supported by the Gruppo Nazionale per l'Analisi Matematica, la Probabilità e le loro Applicazioni (GNAMPA) of the Istituto Nazionale di Alta Matematica (IN-dAM). The first and second authors were also supported by MUR under the the project PRIN 2022–2022XZSAFN: "Anomalous Phenomena on Regular and Irregular Domains: Approximating Complexity for the Applied Sciences"–CUP B53D23009540006 (see the website https://www.sbai.uniroma1.it/~mirko.dovidio/prinSite/index.html).

To this aim, for every $n \in \mathbb{N}$, we introduce a suitable sequence of approximating problems (P_n) formally stated as

$$(P_n) \begin{cases} \frac{\partial u_n(t)}{\partial t} + \alpha_n A_n u_n(t) = 0, & t > 0, \\ u_n(0) = u_0^{(n)}, \end{cases}$$

where $\{\alpha_n\}_n$ is a sequence of positive numbers, A_n is the generator of a suitable Dirichlet form E_n in $L^2(\Omega)$ and $u_0^{(n)} \neq 0$ belongs to $L^2(\Omega)$.

After proving the well-posedness results for problems (P) and (P_n) via semigroup theory [**6**], we consider the sequence of inverse problems $\boldsymbol{P_n}$ (defined in Section 2.1) associated to (P_n), for every $n \in \mathbb{N}$. We prove that, for every $n \in \mathbb{N}$, there exists a unique solution (α_n, u_n) of $\boldsymbol{P_n}$ under the additional condition

$$\|u_n(T_n)\|_2 = \rho_n,$$

where $\rho_n > 0$ is given, and $T_n > 0$ is a fixed measuring time-instant. This corresponds to measure the energy of the solution-to-be u_n at a given time, leading to a quadratic condition. We remark that, in most of the physical and engineering applications, the above quantity is a feasible measurement. The well-posedness results of the inverse problems are proved in Section 2, where we extend and deepen the results proved in [**44**] (see also [**45**]).

Hence, we focus on the convergence of the sequence of solutions (α_n, u_n), with the idea that, under suitable assumptions, there exists a non trivial limit which is indeed the solution of $\boldsymbol{P}$.

The convergence of the solutions is achieved via Mosco Convergence, in the following denoted by M-Convergence. More precisely, under the assumptions on the M-convergence in $L^2(\Omega)$ of the energies E_n to E, we prove that the approximating solutions (α_n, u_n) strongly converge to (α_*, u_*), see Theorem 2.5. We remark that the M-convergence of E_n to E in $L^2(\Omega)$ yields also the convergence of the solutions of the direct problems (P_n) to the solution of problem (P), by following the approach in [**39**] and [**20**].

We point out that the results presented in Section 2 are general and can be applied in different frameworks and to BVPs with different boundary conditions. For the sake of simplicity, in Section 3 we show some applications of our abstract results to some model BVPs in fractal domains with homogeneous Dirichlet boundary conditions.

The literature on boundary value problems (BVPs) on irregular domains (possibly of fractal type), concerning well-posedness results, approximation via M-Convergence and also numerical approximation, goes back to the beginning of the 21st century. The first results are contained in [**35, 36**], where the heat equation across highly conductive thin layers of fractal type has been investigated. The approximation of BVPs in irregular domains via M-Convergence has been addressed for the first time in [**40**], see also [**38, 39**], while the numerical approximation is firstly investigated in [**9–12**]. Starting from these mentioned papers, in this direction many other results concerning linear, quasilinear and/or nonlocal operators follow; among the others, we mention e.g. [**8, 15, 18–22, 29, 30, 37, 55**].

We remark that the interest on inverse problems rises from the need to identify one or more parameters which, in a certain mathematical model, are not directly accessible to a measurement. This requires the introduction of overdeterminating conditions on the solution of the associated direct problem. In the framework of parabolic equations, the literature concerning the reconstruction of functions such as coefficients and source terms is vast; among the others, we refer to the seminal books [**31**, **50**] (see also [**52**, **53**]). Concerning the problem presented in this paper, the identification of time-dependent diffusivity has been achieved in [**24**], under the validity of time-dependent overdeterminating condition of the same type as the one considered in this paper. Nevertheless, there are only a few papers (e.g. [**3**, **7**, **41**–**43**, **58**]) where a constant rather than a function is identified. However, none of the contributions therein quoted covers the application to a domain with irregular boundary. We point out that, in the paper [**42**] (where the operator was fixed), the dependence on the data has been already tackled. The main result of this paper is the study of the approximation of the operator A by a suitable sequence A_n through Mosco convergence. This is motivated by the application on irregular sets. Nevertheless, to our knowledge, this result seems new also in the regular setting. This analysis is crucial in view of numerical approximation results which will be the object of a forthcoming paper [**17**].

The plan of the paper is the following.
In Section 1 we recall some preliminaries on the irregular domains that we are going to consider, on energy forms, operators, semigroups and on M-convergence.
In Section 2 we consider the direct and inverse problems. We prove existence and uniqueness results for the inverse problems in Theorems 2.2 and 2.3 respectively. In Theorem 2.5 we prove the main result of this paper, that is the convergence of the solutions of the inverse problems $\boldsymbol{P_n}$ to the solution of the inverse problem $\boldsymbol{P}$. Section 3 is devoted to some applications.

1. Preliminaries

1.1. (ε, δ) domains and functional spaces. We set the short notations $\mathbb{R}_+ = (0, \infty)$ and $\overline{\mathbb{R}}_+ = [0, \infty)$.

Let $\mathcal{G}$ (resp. $\mathcal{S}$) be an open (resp. closed) set of $\mathbb{R}^N$. By $L^p(\mathcal{G})$, for $p \geq 1$, we denote the Lebesgue space with respect to the Lebesgue measure $d\mathcal{L}_N$, which will be left to the context whenever that does not create ambiguity. By $L^p(\partial\mathcal{G})$ we denote the Lebesgue space on $\partial\mathcal{G}$ with respect to a positive Borel measure μ supported on $\partial\mathcal{G}$. By $\mathcal{D}(\mathcal{G})$ we denote the space of infinitely differentiable functions with compact support in $\mathcal{G}$. By $W^{s,p}(\mathcal{G})$, where $s \in \mathbb{R}_+$, we denote the usual (possibly fractional) Sobolev spaces (see [**1**], [**48**]); $W_0^{s,p}(\mathcal{G})$ is the closure of $\mathcal{D}(\mathcal{G})$ with respect to the $\|\cdot\|_{W^{s,p}}$-norm. By $C(\mathcal{S})$ we denote the space of continuous functions on $\mathcal{S}$. We write $B(P, r) = \{P' \in \mathbb{R}^N : |P' - P| < r\}$, $P \in \mathbb{R}^N, r > 0$, for the euclidean ball of radius r centered at P, and we denote by $|A|$ the Lebesgue measure of a subset $A \subset \mathbb{R}^N$.

In the following, we consider a wide class of possibly very irregular domains, the so-called (ε, δ) domains. These domains have been introduced by Jones in [**32**]. For the sake of clarity, we recall the definition.

DEFINITION 1.1. Let $Q \subset \mathbb{R}^N$ be open and connected and let $\varepsilon, \delta > 0$. For $x \in Q$, let $\mathbf{d}(x) := \inf_{y \in Q^c} |x - y|$. We say that Q is an (ε, δ) domain if, whenever $x, y \in Q$ with $|x - y| < \delta$, there exists a rectifiable arc $\gamma \subset Q$ joining x to y such that

$$\ell(\gamma) \leq \frac{1}{\varepsilon}|x - y| \quad \text{and} \quad \mathbf{d}(z) \geq \frac{\varepsilon |x - z||y - z|}{|x - y|} \text{ for every } z \in \gamma.$$

Examples of (ε, δ) domains are Lipschitz domains and the Koch snowflake domain. As pointed out by Jones [**32**], (ε, δ) domains can have a highly non-rectifiable boundary. In particular, we will consider bounded (ε, δ) domains having as boundary a d-set, see Definition 1.2 below.

DEFINITION 1.2. A closed nonempty set $\mathcal{S} \subset \mathbb{R}^N$ is a d-set (for $0 < d \leq N$) if there exist a Borel measure μ with $\operatorname{supp} \mu = \mathcal{S}$ and two positive constants c_1 and c_2 such that

$$c_1 r^d \leq \mu(\overline{B(x,r)} \cap \mathcal{S}) \leq c_2 r^d \quad \forall x \in \mathcal{S}, \quad 0 < r \leq 1. \tag{1.1}$$

The measure μ is called d-measure.

From now on, unless otherwise specified, $\Omega \subset \mathbb{R}^N$ will denote a bounded (ε, δ) domain having as boundary a d-set. We introduce the following assumption $(\mathcal{H})$, which is not needed for the abstract results presented in Section 2, but will play a crucial role in view of the applications of Section 3. We suppose that Ω can be approximated by a sequence $\{\Omega_n\}$ of domains such that, for every $n \in \mathbb{N}$,

$$(\mathcal{H}) \begin{cases} \Omega_n \text{ is bounded and Lipschitz;} \\ \Omega_n \subseteq \Omega_{n+1}; \\ \Omega = \bigcup\limits_{n=1}^{\infty} \Omega_n. \end{cases}$$

Examples of irregular domains enjoying the properties above are Koch-type fractal domains, where the approximating sequence $\{\Omega_n\}$ consists of the so-called *pre-fractal domains*, see Section 3.1.
We remark that the results of the following Sections can be achieved also for more general bounded (ε, δ) domains in $\mathbb{R}^N$ with boundary a d-set which can be approximated with suitable Lipschitz domains, following the approach of [**16**] and [**14**] for the case $N = 2$ (see also [**22**] for other possible approximations of the domain Ω in the framework of direct problems).

Throughout the paper, C denotes possibly different constants. We give the dependence of constants on some parameters in parentheses.

1.2. Energy forms, operators and semigroups. Let $H := L^2(\Omega)$, where Ω is as in Section 1.1. We denote its scalar product and its associated norm by $(\cdot,\cdot)_2$ and $\|\cdot\|_2$ respectively. We denote by $\mathfrak{L}(H)$ the space of linear and continuous functionals on H, along with its norm $\|\cdot\|_{\mathfrak{L}(H)}$.

We consider an energy form $E[\cdot]$ with effective domain $D(E) \subset L^2(\Omega)$ dense in $L^2(\Omega)$. We consider also a sequence of energy forms $E_n[\cdot]$ with effective domain $D(E_n) \subset L^2(\Omega)$ dense in $L^2(\Omega)$, for every $n \in \mathbb{N}$. We extend the forms E and E_n on the whole space $L^2(\Omega)$ by defining

$$E[u] = +\infty \text{ for every } u \in L^2(\Omega) \setminus D(E)$$

and

$$E_n[u] = +\infty \text{ for every } u \in L^2(\Omega) \setminus D(E_n).$$

We equip $D(E)$ and $D(E_n)$ with the induced norms

$$\|u\|_{D(E)} := E[u] + \|u\|_2 \tag{1.2}$$

and

$$\|u\|_{D(E_n)} := E_n[u] + \|u\|_2 \tag{1.3}$$

respectively. By $E(u,v)$ and $E_n(u,v)$ we denote the corresponding bilinear forms defined by polarization.

Following [**25**, Section 1.1], we call E a *symmetric form* if it is a non-negative definite symmetric bilinear form densely defined on H.
We now recall the notion of Dirichlet form. Following [**25**], we recall that a symmetric form E is *closed* if for every Cauchy sequence $\{u_k\} \subseteq D(E)$ there exists $u \in D(E)$ such that

$$\|u_k - u\|_{D(E)} = E[u_k - u] + \|u_k - u\|_2 \xrightarrow[k\to+\infty]{} 0.$$

Moreover, a closed symmetric form E is *Markovian* if for every $u \in D(E)$ it holds that

$$v := (u \vee 0) \wedge 1 \in D(E) \text{ and } E[v] \le E[u].$$

DEFINITION 1.3. A symmetric form E on $L^2(\Omega)$ is a Dirichlet form if it is closed and Markovian.

For the main properties of Dirichlet forms, we refer to [**25**].
From now on, we suppose that E and E_n are coercive and continuous Dirichlet forms on $L^2(\Omega)$.

As $(E, D(E))$ is a closed bilinear form on $L^2(\Omega)$ with domain $D(E)$ dense in $L^2(\Omega)$, from [**34**, Theorem 2.1, Chap. 6] there exists a unique self-adjoint non negative operator A on $L^2(\Omega)$, with domain $D(A) \subseteq D(E)$ dense in $L^2(\Omega)$, such that

$$E(u,v) = (Au,v)_2\,,\ u \in D(A), v \in D(E). \tag{1.4}$$

Moreover, in Theorem 1.3.1 of [**25**] it is proved that to each closed symmetric form E we can associate a family of linear operators $\{G_\beta, \beta > 0\}$ with the following property:

$$E(G_\beta u, v) + \beta(G_\beta u, v)_2 = (u,v)_2,\ u \in L^2(\Omega) \text{ and } v \in D(E),$$

and this family is a strongly continuous resolvent with generator A, which also generates a strongly continuous semigroup $\{e^{-tA}\}_{t\ge0}$.

Proceeding as above, for every $n \in \mathbb{N}$, we denote by A_n the unique self-adjoint non negative operator on $L^2(\Omega)$ associated to E_n, with domain $D(A_n) \subseteq D(E_n)$ dense in $L^2(\Omega)$.
Moreover, we denote by $\{G^n_\beta, \beta > 0\}$ and $\{e^{-tA_n}(t)\}_{t\ge0}$ the resolvents and the semigroups associated to E_n, for every $n \in \mathbb{N}$, respectively.

We recall here the main properties of the semigroups $\{e^{-tA}\}_{t\ge0}$ and $\{e^{-tA_n}\}_{t\ge0}$.

PROPOSITION 1.4. *Let $\{e^{-tA}\}_{t\ge0}$ and $\{e^{-tA_n}\}_{t\ge0}$ be the strongly continuous semigroups generated by the operators $-A$ and $-A_n$ respectively. Then $\{e^{-tA}\}_{t\ge0}$ and $\{e^{-tA_n}\}_{t\ge0}$ are analytic contraction semigroups in $L^2(\Omega)$.*

The proof follows from [**54**, Theorem 6.2, Chapter 4], Lumer-Phillips Theorem (see Theorem 4.3, Chapter 1 in [**49**]) and Theorem 1.3.1 in [**25**].

We now recall the notion of M-convergence of energy forms adapted to our case. This definition was first introduced by Mosco in [**46**]; here we recall the definition given in [**47**, Definition 2.1.1], which better fits our aims.

DEFINITION 1.5. Let H be a Hilbert space. A sequence of forms $\{E_n\}$ defined in H M-converges to a form E in H if the following conditions hold:

(a) for every $\{v_n\} \in H$ weakly converging to $u \in H$

$$\underline{\lim}_{n\to\infty} E_n[v_n] \geq E[u].$$

(b) for every $u \in H$ there exists a sequence $\{u_n\} \in H$ strongly converging to u in H, such that

$$\overline{\lim}_{n\to\infty} E_n[u_n] \leq E[u].$$

We now recall some important consequences of the M-convergence of energy forms.

THEOREM 1.6 (see Corollary 2.6.1 in [**47**]). *If E_n M-converges to E, then the sequence of semigroups $\{e^{-tA_n}\}$ associated with the form E_n converges to the semigroup e^{-tA} associated with the form E in the operator norm uniformly on every interval $0 < t \leq T$.*

THEOREM 1.7 (see Theorem 2.4.1 in [**47**]). *If E_n M-converges to E, then the sequence of resolvents $\left\{G^n_\beta\right\}$ associated with the form E_n converges to the resolvent G_β associated with the form E in the strong operator norm on $L^2(\Omega)$.*

From Theorem 1.7 and Theorem VIII.24 in [**51**] we deduce the following result.

THEOREM 1.8. *If E_n M-converges to E, then if λ belongs to the spectrum of A there exists an element λ_n of the spectrum of A_n such that $\lambda_n \to \lambda$ as $n \to +\infty$.*

In Proposition 1.9 we recall important properties of analytic semigroups of operators according to [**6**, Theorem 7.7].

PROPOSITION 1.9. *Let $\{e^{-tA}\}_{t\geq 0}$ be the semigroup generated by $-A$ on H. Then*

(1) *for every $u_0 \in H$, the function $u(t) = e^{-tA}u_0$, $t > 0$, is the unique solution to the Cauchy problem*

$$\begin{cases} \frac{\partial u(t)}{\partial t} + Au(t) = 0, \quad t > 0, \\ u(0) = u_0, \end{cases}$$

with regularity

$$u \in C(\overline{\mathbb{R}}_+; H) \cap C^1(\mathbb{R}_+; H) \cap C(\mathbb{R}_+; D(A));$$

(2) *the identity*

$$\frac{\partial}{\partial t}(e^{-tA}u_0) = -Ae^{-tA}u_0$$

holds for every $t > 0$ and every $u_0 \in H$;

(3) *if $\lambda > 0$ denotes the smallest eigenvalue of A, the decay estimate*

$$\|e^{-tA}\|_{\mathfrak{L}(H)} \leq e^{-\lambda t}$$

holds for every $t \geq 0$.

2. Direct and inverse problems and their approximations

In this section we state and prove the main abstract results in the paper. The first result concerns the existence and uniqueness for the solution to the inverse conduction problem for one single constant in a family of parabolic homogeneous equations in $L^2(\Omega)$. Subsequently, we prove a convergence result for the solutions of the inverse problems we shall introduce.

2.1. Direct and inverse problems. Let $\alpha > 0$ and $\{\alpha_n\}_n$ be a sequence of positive numbers. Under the hypotheses and notations of Section 1.2, we consider the following abstract Cauchy problems:

$$(P)\begin{cases} \frac{\partial u(t)}{\partial t} + \alpha A u(t) = 0, \quad t \geq 0, \\ u(0) = u_0, \end{cases}$$

and, for every $n \in \mathbb{N}$

$$(P_n)\begin{cases} \frac{\partial u_n(t)}{\partial t} + \alpha_n A_n\, u_n(t) = 0, \quad t \geq 0, \\ u_n(0) = u_0^{(n)}. \end{cases}$$

From Proposition 1.9-(1), given $u_0, u_0^{(n)} \in L^2(\Omega)$, problems (P) and (P_n), for every $n \in \mathbb{N}$, admit unique solutions given by the formulas

$$u(t) = e^{-t\alpha A} u_0, \quad t \geq 0, \tag{2.1}$$

and, for every $n \in \mathbb{N}$,

$$u_n(t) = e^{-t\alpha_n A_n} u_0^{(n)}, \quad t \geq 0, \tag{2.2}$$

respectively, where $\left\{e^{-t\alpha A}\right\}_{t\geq 0}$ is the analytic semigroup generated by the operator $-\alpha A$ and $\left\{e^{-t\alpha_n A_n}\right\}_{t\geq 0}$ is the analytic semigroup generated by the operator $-\alpha_n A_n$.

We now present the rigorous abstract formulation of the inverse problem we aim at investigating.

PROBLEM $\boldsymbol{P}$. Find a function $u_* : \overline{\mathbb{R}}_+ \to L^2(\Omega)$ and a real constant $\alpha_* > 0$ fulfilling the parabolic Cauchy problem

$$\begin{cases} \frac{\partial u(t)}{\partial t} + \alpha A u(t) = 0, \quad t > 0, \\ u(0) = u_0, \end{cases} \tag{2.3}$$

and the additional condition

$$\|u(T)\|_2 = \rho, \tag{2.4}$$

where $0 \neq u_0 \in L^2(\Omega)$ and $\rho > 0$ are given, and $T > 0$ is a fixed measuring time-instant.

Replacing (2.1) in the overdeterminating condition (2.4) at the instant $t = T$, we see that the unknown constant α must fulfill the following nonlinear equation:

$$f(\alpha) := \|e^{-T\alpha A} u_0\|_2^2 = \rho^2, \quad \alpha > 0. \tag{2.5}$$

As it is clear, the well-posedness for $\boldsymbol{P}$ is a consequence of the one for (2.5). This requires the investigation of the basic properties of function f. Indeed, the following results hold.

PROPOSITION 2.1. *Let* $f\colon \mathbb{R}_+ \to \mathbb{R}_+$ *be defined as in* (2.5), *Then:*

(i) $f \in C^1(\mathbb{R}_+)$ *with* $f'(\alpha) < 0$ *for all* $\alpha > 0$;

(ii) $\mathrm{Im} f = \big(0, \|u_0\|_2^2\big)$.

PROOF. The regularity of f follows immediately by Proposition 1.9-(1). Furthermore, Proposition 1.9-(2) implies

$$\frac{\partial}{\partial\alpha}(e^{-T\alpha A}u_0) = -TAe^{-T\alpha A}u_0$$

for all $\alpha > 0$. Then, by the chain rule, we simply get

$$f'(\alpha) = \frac{\partial}{\partial\alpha}\|e^{-T\alpha A}u_0\|_2^2 = \frac{\partial}{\partial\alpha}\left(e^{-T\alpha A}u_0, e^{-T\alpha A}u_0\right)_2 = -2T\|A^{\frac{1}{2}}e^{-T\alpha A}u_0\|_2^2 < 0.$$

This proves (i). In order to show (ii), we notice that it is a direct consequence of the continuity of f and Proposition 1.9-(3). □

After Proposition 2.1, we learn that for every $0 < \rho < \|u_0\|_2$ equation (2.5) admits a unique solution α_*. As a byproduct, the $L^2(\Omega)$-valued function

$$u_*(t) = e^{-t\alpha_* A}u_0, \quad t > 0,$$

is solving Cauchy problem (2.3) as well as the additional condition (2.4).
Summing up, we proved the next result.

THEOREM 2.2. *For every* $0 \neq u_0 \in L^2(\Omega)$ *and every* $0 < \rho < \|u_0\|_2$, *the inverse problem* $\boldsymbol{P}$ *admits a unique solution*

$$(\alpha_*, u_*) \in \mathbb{R}_+ \times \left(C(\overline{\mathbb{R}}_+; L^2(\Omega)) \cap C^1(\mathbb{R}_+; L^2(\Omega)) \cap C\left(\mathbb{R}_+; D(A)\right)\right).$$

We now investigate the following sequence of inverse problems:

PROBLEM $\boldsymbol{P_n}$. For every $n \in \mathbb{N}$, find a function $u_n : \overline{\mathbb{R}}_+ \to L^2(\Omega)$ and a real constant $\alpha_n > 0$ fulfilling the parabolic Cauchy problem

$$\begin{cases} \frac{\partial u_n(t)}{\partial t} + \alpha_n A_n u_n(t) = 0, & t > 0, \\ u_n(0) = u_0^{(n)}, \end{cases} \tag{2.6}$$

and the additional condition

$$\|u_n(T_n)\|_2 = \rho_n, \tag{2.7}$$

where $0 \neq u_0^{(n)} \in L^2(\Omega)$ and $\rho_n > 0$ are given, and $T_n > 0$ is a sequence of measuring time-instants.

The same argument displayed for solving Problem $\boldsymbol{P}$ immediately yields the following existence and uniqueness result for Problem $\boldsymbol{P_n}$.

THEOREM 2.3. *Let* $n \in \mathbb{N}$ *be fixed. For every* $0 \neq u_0^{(n)} \in L^2(\Omega)$ *and every* $0 < \rho_n < \|u_0^{(n)}\|_2$, *problem* $\boldsymbol{P_n}$ *admits a unique solution*

$$(\alpha_n, u_n) \in \mathbb{R}_+ \times \left(C(\overline{\mathbb{R}}_+; L^2(\Omega)) \cap C^1(\mathbb{R}_+; L^2(\Omega)) \cap C\left(\mathbb{R}_+; D(A_n)\right)\right),$$

where α_n *fulfills the equation*

$$f_n(\alpha) := \|e^{-T_n\alpha A_n}u_0^{(n)}\|_2^2 = \rho_n^2, \quad \alpha > 0 \tag{2.8}$$

and

$$u_n(t) = e^{-t\alpha_n A_n}u_0^{(n)}, \quad t > 0.$$

2.2. Convergence results. In this subsection, we investigate the convergence of the solutions of the approximating problems (P_n) and $\boldsymbol{P_n}$ to the solutions of problems (P) and $\boldsymbol{P}$ respectively.

From now on, we suppose that the sequence of energy forms $(E_n, D(E_n))$, associated with the sequence of operators A_n appearing in Problem (P_n), M-converges to the form $(E, D(E))$, associated with the operator A appearing in Problem (P) considered in Section 2.1.

We start from the direct problems.

THEOREM 2.4. *If* $u_0^{(n)} \xrightarrow[n\to+\infty]{} u_0$ *strongly in* $L^2(\Omega)$ *and* $\alpha_n \xrightarrow[n\to+\infty]{} \alpha$, *then for every* $t \in [0,T]$ *the solution* $u_n(t)$ *defined by* (2.2) *of problem* (P_n) *converges in* $L^2(\Omega)$ *as* $n \to +\infty$ *to the solution* $u(t)$ *defined by* (2.1) *of problem* (P). *Moreover, it also holds that* $u_n \xrightarrow[n\to+\infty]{} u$ *in* $L^2(0,T; L^2(\Omega))$.

The proof follows from Theorem 1.6.

The next theorem, which is the main result of this paper, states the convergence of the sequence of the solutions of the inverse problems $\boldsymbol{P_n}$ to the one of the inverse problem $\boldsymbol{P}$, provided the sequence of the data converges.

THEOREM 2.5. *Under the assumptions of Theorems* 2.2 *and* 2.3, *suppose that*

$$\begin{cases} \rho_n \to \rho > 0, \\ T_n \to T > 0, \\ u_0^{(n)} \to u_0 \text{ in } L^2(\Omega), \\ \inf_n \lambda_n^k > 0 \text{ for every } k \in \mathbb{N}, \end{cases}$$

where we denote by $\lambda_n^k \in \mathbb{R}_+$ *the* k*-th eigenvalue of* A_n.
Then

$$\begin{cases} \alpha_n \to \alpha_*, \\ u_n(t) \to u_*(t) \text{ in } L^2(\Omega) \text{ uniformly on every bounded subset } \mathcal{B} \subset \overline{\mathbb{R}}_+, \end{cases}$$

where the couple (α_*, u_*), *with* $u_* : \overline{\mathbb{R}}_+ \to L^2(\Omega)$ *and* $\alpha_* > 0$, *is the solution of the inverse Problem* $\boldsymbol{P}$.

In order to prove Theorem 2.5 we shall need the following result, see [**4**, Theorem 1].

LEMMA 2.6. *Let* $f_n : \operatorname{Dom} f_n \to \operatorname{Im} f_n$ *be a sequence of continuous injective real functions defined on a common bounded interval* $I \subseteq \bigcap_n \operatorname{Dom} f_n$, *uniformly converging to a function* f *on* I, *as* $n \to +\infty$. *If* f *is injective, then the sequence* f_n^{-1} *converges uniformly to* f^{-1} *on any bounded interval* $J \subseteq \bigcap_n \operatorname{Im} f_n$.

PROOF OF THEOREM 2.5. The proof consists of six steps.

STEP 1. First, we prove that the sequence of α_n is bounded. By contradiction, assume that, up to a subsequence, $\alpha_n \to +\infty$ as $n \to +\infty$. Then, exploiting the decay estimate in Proposition 1.9-(3), we see that

$$\rho_n = \left\| e^{-T_n\alpha_n A_n} u_0^{(n)} \right\|_2 \leq \left\| e^{-T_n\alpha_n A_n} (u_0^{(n)} - u_0) \right\|_2 + \left\| e^{-T_n\alpha_n A_n} u_0 \right\|_2$$

$$\leq \|u_0^{(n)} - u_0\|_2 + \left\| e^{-T_n\alpha_n A_n} \right\|_{\mathfrak{L}(H)} \|u_0\|_2$$

$$\leq \|u_0^{(n)} - u_0\|_2 + e^{-T_n\lambda_n\alpha_n} \|u_0\|_2.$$

Then, since $u_0^{(n)}$ tends to u_0 in $L^2(\Omega)$, and the convergence stated in Theorem 1.8 implies that $T_n\lambda_n\alpha_n$ diverges, we deduce the right-hand-side of the above inequality is vanishing as $n \to +\infty$. That is, $\rho_n \to \rho = 0$, which contradicts the assumption $\rho > 0$.

STEP 2. Thanks to the boundedness of the sequence $\{\alpha_n\}_n$ proved in *Step 1*, it is possible to choose a bounded interval $I \subset \mathbb{R}_+$ large enough so that $\alpha_n \in I$ for every $n \in \mathbb{N}$.

STEP 3. Define the function $f_n(\alpha)$ as in (2.8). Then, the assumptions of Lemma 2.6 are satisfied. Indeed, concerning the co-domains of functions f_n, we define the bounded interval

$$J = \bigcap_n \left(0, \|u_0^{(n)}\|_2^2\right) = \left(0, \inf_n \|u_0^{(n)}\|_2^2\right).$$

STEP 4. As an immediate consequence of Theorem 1.6, we learn that

$$f_n(\alpha) \to f(\alpha) \text{ uniformly on } I \text{ as } n \to +\infty,$$

where I is the interval constructed in *Step 2*. Since the limiting function f is injective, we can apply Lemma 2.6 to deduce

$$f_n^{-1}(\rho) \to f^{-1}(\rho) \text{ uniformly on } J \text{ as } n \to +\infty,$$

where J is the interval constructed in *Step 3*.

STEP 5. By uniform convergence, we immediately get

$$\alpha_n = f_n^{-1}(\rho_n) \xrightarrow[n\to+\infty]{} f^{-1}(\rho) = \alpha_*.$$

STEP 6. Since α_n tends to α_*, again by Theorem 1.6, we deduce that

$$e^{-t\alpha_n A_n} u_0^{(n)} \to e^{-t\alpha_* A} u_0 \text{ in } L^2(\Omega) \text{ uniformly on every bounded subset } \mathcal{B} \subset \overline{\mathbb{R}}_+.$$

Therefore, we have

$$\sup_{t\in\mathcal{B}} \|u_n(t) - u_*(t)\|_2 = \sup_{t\in\mathcal{B}} \left\| e^{-t\alpha_n A_n} u_0^{(n)} - e^{-t\alpha_* A} u_0 \right\|_2$$

$$\leq \sup_{t\in\mathcal{B}} \left\| e^{-t\alpha_n A_n} (u_0^{(n)} - u_0) \right\|_2 + \sup_{t\in\mathcal{B}} \left\| \left(e^{-t\alpha_n A_n} - e^{-t\alpha_* A}\right) u_0 \right\|_2$$

$$\leq \left\| u_0^{(n)} - u_0 \right\|_2 + \sup_{t\in\mathcal{B}} \left\| \left(e^{-t\alpha_n A_n} - e^{-t\alpha_* A}\right) u_0 \right\|_2 \to 0,$$

as $n \to +\infty$.

The proof of Theorem 2.5 is complete. □

3. Applications

In this section, we present some applications of the abstract results obtained in the previous section. Throughout this section, we will consider the case of domains $\Omega \subset \mathbb{R}^2$ with fractal boundary, see the next Subsection 3.1. Key points in the M-Convergence of the approximating energy forms E_n associated to the problems at hand are assumption $(\mathcal{H})$ and the choice of the effective domains $D(E_n)$. This latter one in turn will implicitly imply that the solutions of the approximating problems will vanish outside of the approximating domains Ω_n introduced in assumption $(\mathcal{H})$ (see (3.7) and (3.12) below).

3.1. The Koch snowflake. By the Koch snowflake K, we will denote the union of three com-planar Koch curves (see [**23**]) K_1, K_2 and K_3 as shown in Figure 1. We assume that the junction points A_1, A_3 and A_5 are the vertices of a regular triangle with unit side length, i.e. $|A_1 - A_3| = |A_1 - A_5| = |A_3 - A_5| = 1$. From now on we assume that a clockwise orientation is given on K.

The Hausdorff dimension of the Koch snowflake is given by $d_f = \frac{\ln 4}{\ln 3}$. One can define, in a natural way, a finite Borel measure μ supported on K by

$$\mu := \mu_1 + \mu_2 + \mu_3, \tag{3.1}$$

where μ_i denotes the normalized d_f-dimensional Hausdorff measure, restricted to K_i, $i = 1, 2, 3$.

In the following we denote by

$$K_{n+1} = \bigcup_{i=1}^{3} K_i^{(n+1)} \tag{3.2}$$

the closed polygonal curve approximating K at the $(n+1)$-th step. Here we denote by $K_i^{(n+1)}$ the so-called pre-fractal (polygonal) curve approximating K_i.

The measure μ has the property that there exist two positive constants c_1 and c_2 such that

$$c_1 r^{d_f} \le \mu(B(P,r) \cap K) \le c_2 r^{d_f}, \quad \forall\, P \in K. \tag{3.3}$$

In the terminology of Section 1.1, we say that K is a d_f-set.

Let $\Omega \subset \mathbb{R}^2$ denote the (open) snowflake domain (see Figure 1) and, for every integer $n \in \mathbb{N}$, let $\Omega_n \subset \mathbb{R}^2$ be the pre-fractal polygonal domains approximating Ω and let $K_n = \partial\Omega_n$ be the corresponding closed polygonal curves approximating K (see Figure 2). We remark that the sequence $\{\Omega_n\}_{n\in\mathbb{N}}$ is an increasing sequence of sets exhausting Ω, hence it satisfies assumption $(\mathcal{H})$ of Section 1.1.

We now recall the definition of Besov space specialized to our case. For generalities on Besov spaces, we refer to [**33**].

DEFINITION 3.1. Let $\mathcal{S}$ be a d-set, $0 < \gamma < 1$ and $1 \le p < \infty$. $B^{p,p}_\gamma(\mathcal{S})$ is the space of functions for which the following norm is finite:

$$\|u\|^p_{B^{p,p}_\gamma(\mathcal{S})} = \|u\|^p_{L^p(\mathcal{S})} + \iint_{|x-y|<1} \frac{|u(x) - u(y)|^p}{|x-y|^{d+p\gamma}}\, \mathrm{d}\mu(x)\, \mathrm{d}\mu(y).$$

We now state the trace theorem for functions in $W^{s,p}(Q)$ specialized to our case.

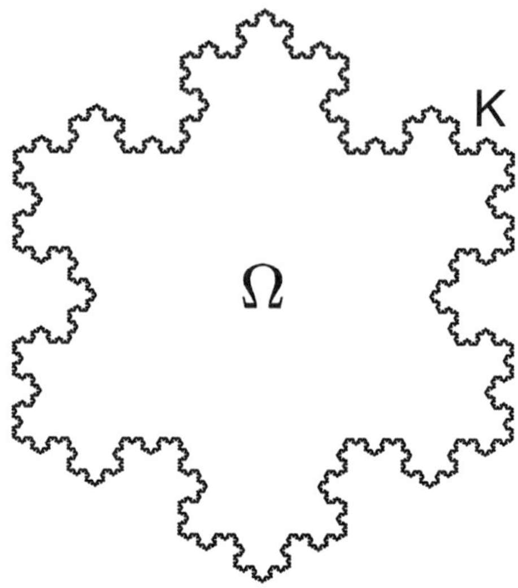

FIGURE 1. The fractal domain Ω.

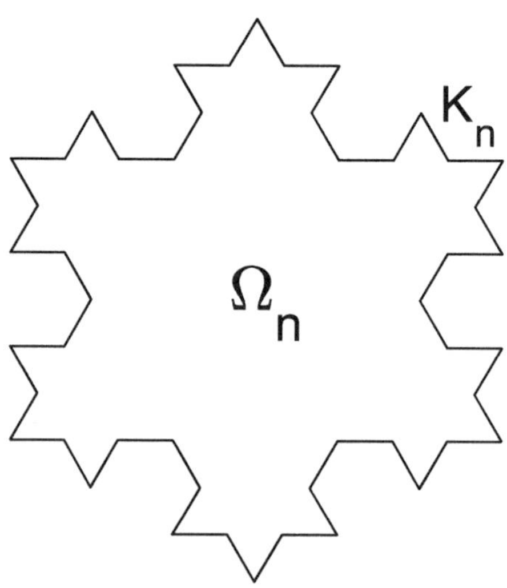

FIGURE 2. The pre-fractal domain Ω_n.

From now on, if $0 < d < 2$, we assume that

$$\frac{2-d}{p} < s < 1 \tag{3.4}$$

and we set

$$\gamma := s - \frac{2-d}{p}. \tag{3.5}$$

PROPOSITION 3.2. *Let $0 < d < 2$, $1 < p < \infty$, s be as in (3.4) and γ be as in (3.5). If $Q \subset \mathbb{R}^2$ is a (ε, δ) domain with boundary a d-set, $B^{p,p}_{\gamma}(\partial Q)$ is the trace space of $W^{s,p}(Q)$ in the following sense:*

(i) *there exists a continuous and linear operator* γ_0 *from* $W^{s,p}(Q)$ *to* $B^{p,p}_{\gamma}(\partial Q)$;
(ii) *there exists a continuous and linear operator* Ext *from* $B^{p,p}_{\gamma}(\partial Q)$ *to* $W^{s,p}(Q)$ *such that* $\gamma_0 \circ \mathrm{Ext}$ *is the identity operator in* $B^{p,p}_{\gamma}(\partial Q)$.

For the proof, we refer to [**33**, Theorem 1, Chapter VII], see also [**56**, Theorem 1]. We point out that, if $Q = \Omega$, then $\partial Q = K$ is a d_f-set. Instead, if $Q = \Omega_n$, its boundary $\partial Q = K_n$ is a 1-set. Hence, the trace space of $W^{s,p}(\Omega_n)$ is $B^{p,p}_{s-\frac{1}{p}}(K_n)$, and the latter space coincides with $W^{s-\frac{1}{p},p}(K_n)$.

3.2. Heat equation. We now introduce the energy form

$$E[u] = \begin{cases} \displaystyle\int_{\Omega} |\nabla u|^2 \mathrm{d}\mathcal{L}_2 & \text{if } u \in D(E), \\ +\infty & \text{if } u \in L^2(\Omega) \setminus D(E), \end{cases} \tag{3.6}$$

defined on $D(E) = H^1_0(\Omega)$. By $E(u,v)$ we will denote the corresponding bilinear form.

PROPOSITION 3.3. *The form* E *defined in* (3.6) *is continuous and coercive on* $H^1_0(\Omega)$. *Moreover, it is a Dirichlet form in* $L^2(\Omega)$.

By proceeding as in Section 1.2, we denote by A the non negative operator associated to the energy form E and by $D(A) \subseteq H^1_0(\Omega)$ its domain, which is dense in $L^2(\Omega)$. We denote by $\{e^{-tA}\}_{t\geq 0}$ the strongly continuous analytic contraction semigroup generated by $-A$ as in Proposition 1.4, and by $\{G_\beta, \beta > 0\}$ the strongly continuous resolvent family with generator A.

We now introduce the approximating energy forms on the pre-fractal domains Ω_n. We identify the space $H^1_0(\Omega_n)$, for every $n \in \mathbb{N}$, as

$$H^1_0(\Omega_n) = \overline{\{\phi \in \mathcal{D}(\Omega) \, : \, \phi \equiv 0 \text{ on } \Omega \setminus \Omega_n\}}^{\|\cdot\|_{H^1(\Omega)}}. \tag{3.7}$$

For every $n \in \mathbb{N}$, we define the energy forms

$$E_n[u] = \begin{cases} \displaystyle\int_{\Omega_n} |\nabla u|^2 \mathrm{d}\mathcal{L}_2 & \text{if } u \in D(E_n), \\ +\infty & \text{if } u \in L^2(\Omega) \setminus D(E_n), \end{cases} \tag{3.8}$$

defined on the domain $D(E_n) = H^1_0(\Omega_n)$. By $E_n(u,v)$ we will denote the corresponding bilinear form.

THEOREM 3.4. *For every* $n \in \mathbb{N}$, *the form* E_n *defined in* (3.8) *is continuous and coercive on* $H^1_0(\Omega_n)$. *Moreover, it is a Dirichlet form in* $L^2(\Omega)$.

Proceeding as above, we denote by A_n, $\{e^{-tA_n}\}_{t\geq 0}$ and $\{G^n_\beta, \beta > 0\}$ the generators, the semigroups and the resolvents associated to E_n, for every $n \in \mathbb{N}$, respectively. We point out that, since the coercivity constant of E_n does not depend on n, it follows that the angle of the analytic semigroup e^{-tA_n} does not depend on n too.

We now recall a M-convergence result.

THEOREM 3.5. *The sequence of forms* E_n *defined in* (3.8) *converges in the sense of Definition* 1.5 *to the form* E *defined in* (3.6).

For the proof, we refer to [**46**].

We now consider the Cauchy problems:

$$(P_L)\begin{cases}\frac{\partial u(t)}{\partial t}+\alpha Au(t)=0, & t>0,\\ u(0)=u_0,\end{cases}$$

and, for every $n\in\mathbb{N}$,

$$(P_L^{(n)})\begin{cases}\frac{\partial u_n(t)}{\partial t}+\alpha_n A_n\, u_n(t)=0, & t>0,\\ u_n(0)=u_0^{(n)},\end{cases}$$

where $A\colon D(A)\subset L^2(\Omega)\to L^2(\Omega)$ and $A_n\colon D(A_n)\subset L^2(\Omega)\to L^2(\Omega)$ are the generators associated to the energy form E and the energy forms E_n introduced in (3.6) and (3.8) respectively.

If u_0 and $u_0^{(n)}$ belong to $L^2(\Omega)$, existence and uniqueness of solutions of problems (P_L) and $(P_L^{(n)})$ in terms of the analytic semigroups generated by $-A$ and $-A_n$ respectively follow from Proposition 1.9.

We now give the strong formulation of problems (P_L) and $(P_L^{(n)})$.

THEOREM 3.6. *Let u be the unique solution of problem (P_L) and let $\gamma=\frac{d_f}{2}$. Then we have for every fixed $t>0$*

$$(\overline{P_L})\begin{cases}\frac{\partial u}{\partial t}(t,P)-\alpha\Delta u(t,P)=0 & \text{for a.e. } P\in\Omega,\\ u(t,P)=0 & \text{in } B^{2,2}_{\gamma}(K),\\ u(0,P)=u_0(P) & \text{for } P\in\overline{\Omega}.\end{cases}$$

THEOREM 3.7. *For every $n\in\mathbb{N}$, let u_n be the unique solution of problem $(P_L^{(n)})$. Then we have for every fixed $t>0$*

$$(\overline{P_L^{(n)}})\begin{cases}\frac{\partial u_n}{\partial t}(t,P)-\alpha_n\Delta u_n(t,P)=0 & \text{for a.e. } P\in\Omega_n,\\ u_n(t,P)=0 & \text{in } H^{\frac{1}{2}}(K_n),\\ u_n(0,P)=u_0^{(n)}(P) & \text{for } P\in\overline{\Omega}_n.\end{cases}$$

The proofs of these theorems can be achieved as in [**39**].

We now come to the convergence of the solutions of the direct problems. Actually, we can prove a stronger convergence of the solutions than the one stated by Theorem 2.4. For more details, we refer to [**46**, Corollary 2, p. 557].

THEOREM 3.8. *Let u and u_n be the solutions of problems $(\overline{P_L})$ and $(\overline{P_L^{(n)}})$ respectively. If $u_0^{(n)}\xrightarrow[n\to+\infty]{}u_0$ strongly in $L^2(\Omega)$ and $\alpha_n\xrightarrow[n\to+\infty]{}\alpha$, then $u_n(t)$ strongly converges to $u(t)$ in $H^1_0(\Omega)$ as $n\to+\infty$ for every $t>0$.*

We now consider also the corresponding inverse problems $\boldsymbol{P_L^{(n)}}$ and $\boldsymbol{P_L}$, which can be stated as in Subsection 2.1. Then, if we denote by (α_n,u_n) the unique solution of problem $\boldsymbol{P_L^{(n)}}$ given by Theorem 2.3 and by (α_*,u_*) the unique solution of problem $\boldsymbol{P_L}$ given by Theorem 2.2, then under the hypotheses of Theorem 2.5 we have that $\alpha_n \xrightarrow[n\to+\infty]{} \alpha_*$ and $u_n(t) \xrightarrow[n\to+\infty]{} u_*(t)$ strongly in $L^2(\Omega)$ and uniformly on every bounded subset $\mathcal{B} \subset \overline{\mathbb{R}}_+$.

3.3. The Dirichlet problem for the regional fractional Laplacian. In this section we investigate the Dirichlet problem for the regional fractional Laplacian on snowflake domains, and its approximation via pre-fractal problems.
We introduce the so-called "regional" fractional Laplacian, for the definition we refer to [**5**, **13**, **26**–**28**].
Let $s \in (0,1)$. For $\mathcal{G} \subseteq \mathbb{R}^N$ open, we define the space

$$\mathcal{L}^1_s(\mathcal{G}) := \left\{ u \colon \mathcal{G} \to \mathbb{R} \text{ measurable } : \int_{\mathcal{G}} \frac{|u(x)|}{(1+|x|)^{N+2s}} \,\mathrm{d}\mathcal{L}_N(x) < \infty \right\}.$$

The regional fractional Laplacian $(-\Delta)^s_{\mathcal{G}}$ is defined as follows, for $x \in \mathcal{G}$:

$$\begin{aligned} (-\Delta)^s_{\mathcal{G}} u(x) &= C_{N,s}\mathrm{P.V.} \int_{\mathcal{G}} \frac{u(x)-u(y)}{|x-y|^{N+2s}} \,\mathrm{d}\mathcal{L}_N(y) \\ &= C_{N,s} \lim_{\varepsilon\to 0^+} \int_{\{y\in\mathcal{G}\,:\,|x-y|>\varepsilon\}} \frac{u(x)-u(y)}{|x-y|^{N+2s}} \,\mathrm{d}\mathcal{L}_N(y), \end{aligned} \tag{3.9}$$

provided that the limit exists for every function $u \in \mathcal{L}^1_s(\mathcal{G})$. The positive constant $C_{N,s}$ is defined as follows:

$$C_{N,s} = \frac{s 2^{2s} \Gamma(\frac{N+2s}{2})}{\pi^{\frac{N}{2}} \Gamma(1-s)},$$

where Γ is the Euler function.

From now on, we suppose that

$$\frac{1}{2} < s < 1.$$

We now introduce the fractional energy forms E_s and $E^{(n)}_s$ respectively. We define

$$E_s[u] := \begin{cases} \dfrac{C_{2,s}}{2} \displaystyle\iint_{\Omega\times\Omega} \frac{(u(x)-u(y))^2}{|x-y|^{2s+2}} \,\mathrm{d}\mathcal{L}_2(x)\mathrm{d}\mathcal{L}_2(y) & \text{if } u \in D(E_s), \\ +\infty & \text{if } u \in L^2(\Omega) \setminus D(E_s), \end{cases} \tag{3.10}$$

and, for every $n \in \mathbb{N}$,

$$E^{(n)}_s[u] := \begin{cases} \dfrac{C_{2,s}}{2} \displaystyle\iint_{\Omega_n\times\Omega_n} \frac{(u(x)-u(y))^2}{|x-y|^{2s+2}} \,\mathrm{d}\mathcal{L}_2(x)\mathrm{d}\mathcal{L}_2(y) & \text{if } u \in D(E^{(n)}_s), \\ +\infty & \text{if } u \in L^2(\Omega) \setminus D(E^{(n)}_s), \end{cases} \tag{3.11}$$

where the effective domains are respectively $D(E_s) = H^s_0(\Omega)$ and $D(E^{(n)}_s) = H^s_0(\Omega_n)$, which for every $n \in \mathbb{N}$ can be identifed as in (3.7) by

$$H^s_0(\Omega_n) = \overline{\{\phi \in \mathcal{D}(\Omega) \,:\, \phi \equiv 0 \text{ on } \Omega \setminus \Omega_n\}}^{\|\cdot\|_{H^s(\Omega)}}. \tag{3.12}$$

By proceeding as in [**20**], it can be proved that E_s and $E^{(n)}_s$, for every $n \in \mathbb{N}$, are continuous and coercive Dirichlet forms on $L^2(\Omega)$. Hence, as in Section 1.2, there exist unique self-adjoint non negative operators A_s and $A^{(n)}_s$, for every $n \in \mathbb{N}$, on $L^2(\Omega)$ satisfying (1.4) and having domains $D(A_s) \subset D(E_s)$ and $D(A_s^{(n)}) \subset D(E_s^{(n)})$ respectively, which are dense in $L^2(\Omega)$. Moreover, these operators are the generator of strongly continuous analytic contraction semigroups $\{e^{-tA_s}\}_{t\geq 0}$ and $\{e^{-tA_s^{(n)}}\}_{t\geq 0}$, for every $n \in \mathbb{N}$, respectively, on $L^2(\Omega)$.

We now recall a M-convergence result.

THEOREM 3.9. *The sequence of forms $E_s^{(n)}$ defined in* (3.11) *converges in the sense of Definition* 1.5 *to the form E_s defined in* (3.10).

The proof can be achieved as in [**20**].

We now consider the following abstract Cauchy problems:

$$(P_F)\begin{cases} \frac{\partial u(t)}{\partial t} + \alpha A_s u(t) = 0, & t > 0, \\ u(0) = u_0, \end{cases}$$

and, for every $n \in \mathbb{N}$,

$$(P_F^{(n)})\begin{cases} \frac{\partial u_n(t)}{\partial t} + \alpha_n A_s^{(n)} u_n(t) = 0, & t > 0, \\ u_n(0) = u_0^{(n)}, \end{cases}$$

where $A_s^{(n)}$ and A_s are as above. If u_0 and $u_0^{(n)}$ belong to $L^2(\Omega)$, existence and uniqueness of the solutions of problems (P_F) and $(P_F^{(n)})$ for every $n \in \mathbb{N}$ follow as in Proposition 1.9.

We now consider the "strong" formulations of the above abstract Cauchy problems.

THEOREM 3.10. *Let u be the unique solution of problem (P_F) and let $\gamma = s - 1 + \frac{d_f}{2}$. Then we have for every fixed $t > 0$*

$$(\overline{P_F})\begin{cases} \frac{\partial u}{\partial t}(t,P) + \alpha(-\Delta)^s_\Omega u(t,P) = 0 & \text{for a.e. } P \in \Omega, \\ u(t,P) = 0 & \text{in } B^{2,2}_\gamma(K), \\ u(0,P) = u_0(P) & \text{for } P \in \overline{\Omega}. \end{cases}$$

THEOREM 3.11. *For every $n \in \mathbb{N}$, let u_n be the unique solution of problem $(P_F^{(n)})$. Then we have for every fixed $t > 0$*

$$(\overline{P_F^{(n)}})\begin{cases} \frac{\partial u_n}{\partial t}(t,P) + \alpha_n(-\Delta)^s_{\Omega_n} u_n(t,P) = 0 & \text{for a.e. } P \in \Omega_n, \\ u_n(t,P) = 0 & \text{in } H^{s-\frac{1}{2}}(K_n), \\ u_n(0,P) = u_0^{(n)}(P) & \text{for } P \in \overline{\Omega}_n. \end{cases}$$

The proofs of these theorems can be achieved as in [**20**]

We now come to the convergence of the solutions of the direct problems. As in Section 3.2 (Theorem 3.8), we can prove a stronger convergence of the solutions than the one stated by Theorem 2.4.

THEOREM 3.12. *Let u and u_n be the solutions of problems $(\overline{P_F})$ and $(\overline{P_F^{(n)}})$ respectively. If $u_0^{(n)} \xrightarrow[n\to+\infty]{} u_0$ strongly in $L^2(\Omega)$ and $\alpha_n \xrightarrow[n\to+\infty]{} \alpha$, then $u_n(t)$ strongly converges to $u(t)$ in $H^s_0(\Omega)$ as $n \to +\infty$ for every $t > 0$.*

PROOF. The result follows by adapting Lemma 1.8, p. 531 and Corollary 2, p. 557 of [46], since Ω_n is an increasing sequence of sets invading Ω and $\mathrm{cap}_{(s,2),\Omega}(E \setminus \Omega_n) \to 0$ when $n \to +\infty$ for any compact subset E of Ω, where, for any compact subset $E \subset \Omega$, its relative (s,p)-capacity with respect to Ω is defined by

$$\mathrm{cap}_{(s,p),\Omega}(E) = \inf\left\{\|\varphi\|^p_{W^{s,p}(\Omega)} : \varphi \in \mathcal{D}(\Omega) \text{ and } \varphi \geq 1 \text{ on } E\right\},$$

see [2] and [57]. □

Finally, we consider also the corresponding inverse problems $\boldsymbol{P_F^{(n)}}$ and $\boldsymbol{P_F}$, which can be stated as in Subsection 2.1. Then, if we denote by (α_n, u_n) the unique solution of problem $\boldsymbol{P_F^{(n)}}$ given by Theorem 2.3 and by (α_*, u_*) the unique solution of problem $\boldsymbol{P_F}$ given by Theorem 2.2, then under the hypotheses of Theorem 2.5 we have that $\alpha_n \xrightarrow[n\to+\infty]{} \alpha_*$ and $u_n(t) \xrightarrow[n\to+\infty]{} u_*(t)$ strongly in $L^2(\Omega)$ and uniformly on every bounded subset $\mathcal{B} \subset \overline{\mathbb{R}}_+$.

References

[1] D. R. Adams, *Sobolev Spaces.* (1975) Pure and Applied Mathematics, Vol. 65, Academic Press, New York-London.

[2] D. R. Adams, L. I. Hedberg, *Function Spaces and Potential Theory.* (1996) Springer-Verlag, Berlin.

[3] E. A. Artyukhin, A. S. Okhapkin, *Determination of the parameters in the generalized heat-conduction equation from transient experimental data*, J. Eng. Phys. Thermophys., 42, (1982) pp. 693–698.

[4] Erich Barvínek, Ivan Daler, and Jan Franců, *Convergence of sequences of inverse functions*, Arch. Math. (Brno) **27B** (1991), 201–204. MR1189216

[5] Krzysztof Bogdan, Krzysztof Burdzy, and Zhen-Qing Chen, *Censored stable processes*, Probab. Theory Related Fields **127** (2003), no. 1, 89–152, DOI 10.1007/s00440-003-0275-1. MR2006232

[6] H. Brezis, *Analyse Fonctionnelle, Théorie et Applications.* (1983) Masson, Paris.

[7] J. R. Cannon, *Determination of certain parameters in heat conduction problems*, J. Math. Anal. Appl. **8** (1964), 188–201, DOI 10.1016/0022-247X(64)90061-7. MR160047

[8] Raffaela Capitanelli and Maria Agostina Vivaldi, *Insulating layers and Robin problems on Koch mixtures*, J. Differential Equations **251** (2011), no. 4-5, 1332–1353, DOI 10.1016/j.jde.2011.02.003. MR2812592

[9] Massimo Cefalo, Simone Creo, Maria Rosaria Lancia, and Paola Vernole, *Nonlocal Venttsel' diffusion in fractal-type domains: regularity results and numerical approximation*, Math. Methods Appl. Sci. **42** (2019), no. 14, 4712–4733, DOI 10.1002/mma.5686. MR3992935

[10] Maria Rosaria Lancia, Massimo Cefalo, and Guido Dell'Acqua, *Numerical approximation of transmission problems across Koch-type highly conductive layers*, Appl. Math. Comput. **218** (2012), no. 9, 5453–5473, DOI 10.1016/j.amc.2011.11.033. MR2870066

[11] Massimo Cefalo and Maria Rosaria Lancia, *An optimal mesh generation algorithm for domains with Koch type boundaries*, Math. Comput. Simulation **106** (2014), 133–162, DOI 10.1016/j.matcom.2014.04.009. MR3274752

[12] Massimo Cefalo, Maria Rosaria Lancia, and Haodong Liang, *Heat-flow problems across fractal mixtures: regularity results of the solutions and numerical approximation*, Differential Integral Equations **26** (2013), no. 9-10, 1027–1054. MR3100075

[13] Zhen-Qing Chen and Takashi Kumagai, *Heat kernel estimates for stable-like processes on d-sets*, Stochastic Process. Appl. **108** (2003), no. 1, 27–62, DOI 10.1016/S0304-4149(03)00105-4. MR2008600

[14] Simone Creo, Michael Hinz, and Maria Rosaria Lancia, *Non-local boundary energy forms for quasidiscs: Codimension gap and approximation*, Under review, `arXiv:2310.04549`, 2025.

[15] Simone Creo and Maria Rosaria Lancia, *Fractional (s,p)-Robin-Venttsel' problems on extension domains*, NoDEA Nonlinear Differential Equations Appl. **28** (2021), no. 3, Paper No. 31, 33, DOI 10.1007/s00030-021-00692-w. MR4244930

[16] Simone Creo and Maria Rosaria Lancia, *The p-curl system in extension domains*, Discrete Contin. Dyn. Syst. Ser. S **17** (2024), no. 5-6, 2208–2223, DOI 10.3934/dcdss.2023113. MR4762582

[17] Simone Creo, Maria Rosaria Lancia, Gianluca Mola, Silvia Romanelli, *Approximation of inverse problems in irregular domains on varying Hilbert spaces*, In preparation.

[18] Simone Creo, Maria Rosaria Lancia, and Alexander I. Nazarov, *Regularity results for nonlocal evolution Venttsel' problems*, Fract. Calc. Appl. Anal. **23** (2020), no. 5, 1416–1430, DOI 10.1515/fca-2020-0070. MR4173830

[19] Simone Creo, Maria Rosaria Lancia, Alejandro Vélez-Santiago, and Paola Vernole, *Approximation of a nonlinear fractal energy functional on varying Hilbert spaces*, Commun. Pure Appl. Anal. **17** (2018), no. 2, 647–669, DOI 10.3934/cpaa.2018035. MR3743158

[20] Simone Creo, Maria Rosaria Lancia, and Paola Vernole, *Convergence of fractional diffusion processes in extension domains*, J. Evol. Equ. **20** (2020), no. 1, 109–139, DOI 10.1007/s00028-019-00517-5. MR4072651

[21] Adrien Dekkers and Anna Rozanova-Pierrat, *Dirichlet boundary valued problems for linear and nonlinear wave equations on arbitrary and fractal domains*, J. Math. Anal. Appl. **512** (2022), no. 1, Paper No. 126089, 45, DOI 10.1016/j.jmaa.2022.126089. MR4389010

[22] Adrien Dekkers, Anna Rozanova-Pierrat, and Alexander Teplyaev, *Mixed boundary valued problems for linear and nonlinear wave equations in domains with fractal boundaries*, Calc. Var. Partial Differential Equations **61** (2022), no. 2, Paper No. 75, 44, DOI 10.1007/s00526-021-02159-3. MR4393126

[23] K. Falconer, *The Geometry of Fractal Sets.* (1990) Cambridge Univ. Press, Cambridge, 2nd ed.

[24] Angelo Favini, Gianluca Mola, and Silvia Romanelli, *Recovering time-dependent diffusion coefficients in a nonautonomous parabolic equation from energy measurements*, Discrete Contin. Dyn. Syst. Ser. S **15** (2022), no. 6, 1439–1454, DOI 10.3934/dcdss.2022017. MR4421542

[25] Masatoshi Fukushima, Yōichi Ōshima, and Masayoshi Takeda, *Dirichlet forms and symmetric Markov processes*, De Gruyter Studies in Mathematics, vol. 19, Walter de Gruyter & Co., Berlin, 1994, DOI 10.1515/9783110889741. MR1303354

[26] Qing-Yang Guan, *Integration by parts formula for regional fractional Laplacian*, Comm. Math. Phys. **266** (2006), no. 2, 289–329, DOI 10.1007/s00220-006-0054-9. MR2238879

[27] Qing-Yang Guan and Zhi-Ming Ma, *Boundary problems for fractional Laplacians*, Stoch. Dyn. **5** (2005), no. 3, 385–424, DOI 10.1142/S021949370500150X. MR2167307

[28] Qing-Yang Guan and Zhi-Ming Ma, *Reflected symmetric α-stable processes and regional fractional Laplacian*, Probab. Theory Related Fields **134** (2006), no. 4, 649–694, DOI 10.1007/s00440-005-0438-3. MR2214908

[29] Michael Hinz, Maria Rosaria Lancia, Alexander Teplyaev, and Paola Vernole, *Fractal snowflake domain diffusion with boundary and interior drifts*, J. Math. Anal. Appl. **457** (2018), no. 1, 672–693, DOI 10.1016/j.jmaa.2017.07.065. MR3702726

[30] Michael Hinz, Anna Rozanova-Pierrat, and Alexander Teplyaev, *Non-Lipschitz uniform domain shape optimization in linear acoustics*, SIAM J. Control Optim. **59** (2021), no. 2, 1007–1032, DOI 10.1137/20M1361687. MR4228309

[31] V. Isakov, *Inverse source problems.* American Mathematical Society Providence, Rhode Island (1990)

[32] Peter W. Jones, *Quasiconformal mappings and extendability of functions in Sobolev spaces*, Acta Math. **147** (1981), no. 1-2, 71–88, DOI 10.1007/BF02392869. MR631089

[33] Alf Jonsson and Hans Wallin, *Function spaces on subsets of* $\mathbf{R}^n$, Math. Rep. **2** (1984), no. 1, xiv+221. MR820626

[34] T. Kato, *Perturbation Theory for Linear Operators*, Springer Verlag, New York (1966).

[35] Maria Rosaria Lancia, *A transmission problem with a fractal interface*, Z. Anal. Anwendungen **21** (2002), no. 1, 113–133, DOI 10.4171/ZAA/1067. MR1916407

[36] Maria Rosaria Lancia, *Second order transmission problems across a fractal surface* (English, with English and Italian summaries), Rend. Accad. Naz. Sci. XL Mem. Mat. Appl. (5) **27** (2003), 191–213. MR2056419

[37] Maria Rosaria Lancia, Alejandro Vélez-Santiago, and Paola Vernole, *Quasi-linear Venttsel' problems with nonlocal boundary conditions on fractal domains*, Nonlinear Anal. Real World Appl. **35** (2017), 265–291, DOI 10.1016/j.nonrwa.2016.11.002. MR3595327

[38] Maria Rosaria Lancia and Paola Vernole, *Irregular heat flow problems*, SIAM J. Math. Anal. **42** (2010), no. 4, 1539–1567, DOI 10.1137/090761173. MR2679587
[39] Maria Rosaria Lancia and Paola Vernole, *Venttsel' problems in fractal domains*, J. Evol. Equ. **14** (2014), no. 3, 681–712, DOI 10.1007/s00028-014-0233-7. MR3250804
[40] Maria Rosaria Lancia and Maria Agostina Vivaldi, *Asymptotic convergence of transmission energy forms*, Adv. Math. Sci. Appl. **13** (2003), no. 1, 315–341. MR2002727
[41] Alfredo Lorenzi, *Recovering two constants in a parabolic linear equation*, Journal of Physics: Conference Series, 73, paper no. 012014 (2007).
[42] Alfredo Lorenzi and Gianluca Mola, *Identification of a real constant in linear evolution equations in Hilbert spaces*, Inverse Probl. Imaging **5** (2011), no. 3, 695–714, DOI 10.3934/ipi.2011.5.695. MR2825734
[43] Luca Lorenzi, *An identification problem for the Ornstein-Uhlenbeck operator*, J. Inverse Ill-Posed Probl. **19** (2011), no. 2, 293–326, DOI 10.1515/JIIP.2011.032. MR2805857
[44] Gianluca Mola, *Identification of the diffusion coefficient in linear evolution equations in Hilbert spaces*, J. Abstr. Differ. Equ. Appl. **2** (2011), no. 1, 14–28. MR2806287
[45] Gianluca Mola, Noboru Okazawa, Jan Prüss, and Tomomi Yokota, *Semigroup-theoretic approach to identification of linear diffusion coefficients*, Discrete Contin. Dyn. Syst. Ser. S **9** (2016), no. 3, 777–790, DOI 10.3934/dcdss.2016028. MR3503640
[46] Umberto Mosco, *Convergence of convex sets and of solutions of variational inequalities*, Advances in Math. **3** (1969), 510–585, DOI 10.1016/0001-8708(69)90009-7. MR298508
[47] Umberto Mosco, *Composite media and asymptotic Dirichlet forms*, J. Funct. Anal. **123** (1994), no. 2, 368–421, DOI 10.1006/jfan.1994.1093. MR1283033
[48] J. Nečas, *Les Mèthodes Directes en Thèorie des Èquationes Elliptiques*, Masson, Paris (1967).
[49] A. Pazy, *Semigroups of linear operators and applications to partial differential equations*, Applied Mathematical Sciences, vol. 44, Springer-Verlag, New York, 1983, DOI 10.1007/978-1-4612-5561-1. MR710486
[50] A. I. Prilepko, D. G. Orlovsky, I. A. Vasin, *Methods for Solving Inverse Problems in Mathematical Physics*, CRC Press. (2000)
[51] Michael Reed and Barry Simon, *Methods of modern mathematical physics. I. Functional analysis*, Academic Press, New York-London, 1972. MR0493419
[52] A. V. Rozanova, *Controllability in a nonlinear parabolic problem with integral overedetermination*, Differ. Eq., 40 (2004) (6), pp. 853–872.
[53] A. V. Rozanova, *Controllability for a nonlinear abstract evolution equation*, Math. Notes, 76 (2004) (4), pp. 511–524.
[54] R. E. Showalter, *Hilbert Space Methods for Partial Differential Equations*, Monographs and Studies in Mathematics, Pitman, London-San Francisco (1977).
[55] Alejandro Vélez-Santiago, *A quasi-linear Neumann problem of Ambrosetti-Prodi type on extension domains*, Nonlinear Anal. **160** (2017), 191–210, DOI 10.1016/j.na.2017.05.012. MR3667682
[56] H. Wallin, *The trace to the boundary of Sobolev spaces on a snowflake*, Manuscripta Math., 73 (1981), pp. 117–126.
[57] Mahamadi Warma, *The fractional relative capacity and the fractional Laplacian with Neumann and Robin boundary conditions on open sets*, Potential Anal. **42** (2015), no. 2, 499–547, DOI 10.1007/s11118-014-9443-4. MR3306694
[58] M. Yamamoto, *Determination of constant parameters in some semilinear parabolic equations*, Ill-posed problems in natural sciences (Moscow, 1991), VSP, Utrecht (1992), pp. 439–445.

Dipartimento di Scienze di Base e Applicate per l'Ingegneria, Sapienza Università di Roma, Via Antonio Scarpa 10, 00161 Roma, Italy
Email address: simone.creo@uniroma1.it

Dipartimento di Scienze di Base e Applicate per l'Ingegneria, Sapienza Università di Roma, Via Antonio Scarpa 10, 00161 Roma, Italy
Email address: mariarosaria.lancia@uniroma1.it

Department of Science and Engeneering, Sorbonne University Abu Dhabi, Al Reem Island, Abu Dhabi, United Arab Emirates
Email address: gianluca.mola@sorbonne.ae

Dipartimento di Matematica, Università degli Studi di Bari Aldo Moro, Via Edoardo Orabona 4, 70125 Bari, Italy
Email address: silvia.romanelli@uniba.it

Contemporary Mathematics
Volume **828**, 2025
https://doi.org/10.1090/conm/828/16594

Iterated fractal drums ~ some new perspectives: Polyhedral measures, atomic decompositions and Morse theory

Claire David and Michel L. Lapidus

ABSTRACT. In this research paper, we carry on our exploration of the connections between the Complex Fractal Dimensions of an iterated fractal drum (IFD) and the intrinsic properties of the fractal involved – in our present case, the Weierstrass Curve.

In order to gain a better understanding of the differential operators associated to this everywhere singular object, we use atomic decompositions, which enable us to characterize Besov spaces on the Weierstrass Curve $\Gamma_{\mathcal{W}}$ and then, by means of trace theorems which extend the results of Alf Jonsson and Hans Wallin obtained in the case of a d-set, to identify the trace of the classical Sobolev spaces on $\Gamma_{\mathcal{W}}$. For this purpose, we construct a specific polyhedral measure, which is done by means of a sequence of polygonal neighborhoods of the Curve. We then determine the order of a Laplace-type operator on $\Gamma_{\mathcal{W}}$, thanks to the connections between Sobolev spaces and the usual Laplacian.

Moreover, we lay out some of the foundations of an extension of Morse theory dedicated to fractals, where the Complex Fractal Dimensions appear to play a major role, by means of level sets connected to the successive prefractal approximations.

Towards the end, we envision the Weierstrass Curve as the projection of a 3-dimensional vertical comb, where each horizontal row is associated to the kth cohomological infinitesimal, the *fractal signature* of the kth prefractal approximation, according to our previous results on fractal cohomology.

Weierstrass Curve, iterated fractal drum (IFD), Complex Dimensions of an IFD, box-counting (or Minkowski) dimension, cohomology infinitesimal, polyhedral measure, atomic decomposition, trace theorems, order of a Laplace-type operator, fractal Morse theory, fractal Morse indexes.

CONTENTS

2020 *Mathematics Subject Classification.* Primary 28A75, 28A80, 35R02, 57Q70.

The research of the second author was supported by the Burton Jones Endowed Chair in Pure Mathematics, as well as by grants from the U. S. National Science Foundation. This project received financial support from the CNRS via the MITI interdisciplinary programs through its exploratory research program. The research of both authors was supported by the MITI Centre National de la Recherche Scientifique (CNRS) 80PRIME, Conditions Extrêmes and by the Agence Nationale de la Recherche (ANR) FRACTALS (ANR-24-CE45-3362).

1. Introduction

In [**DL24a**], [**DL24c**], we introduced the concept of *Weierstrass Iterated Fractal Drums* (in short, Weierstrass IFDs), by analogy with the *relative fractal drums* (RFDs) involved, for instance, in the case of the Cantor Staircase, in [**LRŽ17a**], Section 5.5.4, as well as in [**LRŽ17b**] and in [**LRŽ18**]. Those iterated fractal drums simply consist in a sequence of appropriate tubular neighborhoods of prefractal polygonal approximations of the Curve.

By exploring the connections between the Complex Dimensions of those IFDs and the cohomological properties of a fractal object, we showed, in [**DL24c**], that the functions belonging to the cohomology groups associated to the Curve are obtained, by induction, as (finite or infinite) sums indexed by the underlying Complex Dimensions. In particular, to each prefractal approximation of the given iterated fractal drum, we associate the *minimal real part* of the Complex Dimensions involved. Contrary to fractal tube formulas, which are obtained for small values of a positive parameter ε, the aforementioned expansions are only valid for the value of the (multi-scales) *cohomology infinitesimal* ε associated to the scaling relationship obeyed by the Weierstrass Curve (or else, for a smaller positive infinitesimal). (See also [**DL24a**] for an exposition of these results, as well as of the computation of the Complex Dimensions via a fractal tube formula obtained in [**DL25c**].)

Our differentials δ and $\bar{\delta}$ (see [**DL24c**]) enable us to define the associated Laplacian $\delta\,\bar{\delta} + \bar{\delta}\,\delta$. This naturally raises questions as to the possible connections with the usual Laplacian (i.e., the one of classical analysis). Based on the seminal works of Alf Jonsson and Hans Wallin in [**Wal89**], [**JW84**], a hint was that it could involve the Complex Dimensions, insofar as, in the case of a d-set $\mathcal{F} \subset \mathbb{R}^n$, for $n \in \mathbb{N}^\star$ and $d > 0$, characterizations of the restrictions of classical Sobolev spaces to $\mathcal{F}$ can be obtained by means of trace theorems. More precisely, the restrictions to $\mathcal{F}$ of those Sobolev spaces are obtained as Besov spaces $B_\beta^{p,q}(\mathcal{F})$ on $\mathcal{F}$, where the index β depends explicitly on d. (Recall that $\mathcal{F} \subset \mathbb{R}^n$ is a *d-set* if it has finite and positive d dimensional Hausdorff measure.)

In the present research paper, we hereafter extend the results of Alf Jonsson and Hans Wallin in the case of our Weierstrass IFDs. This requires, in particular, the construction of a specific measure (called the polyhedral measure) on the Weierstrass IFD, which is done by means of a sequence of polygons – a polygonal neighborhood of the Curve. This enables us to define the Besov spaces on the Weierstrass Curve, by means of an atomic decomposition.

We next recall a few important facts about atomic decompositions. They came into play in the 1970s, in order to obtain a better characterization of the smoothness, in a more general context of the "measure of smoothness", which is the expression used by Hans Triebel in the Preface of his book [**Tri06**], page xi. Atomic decompositions were studied by several authors. We may cite, for instance, Ronald R. Coifman in [**Coi74**], where the author provides an explicit representation theorem for functions belonging to the Hardy space $\mathcal{H}^p(\mathbb{R}) = \{ f \in L^p(\mathbb{R}),\, \mathcal{R}_i(f) \in L^p(\mathbb{R}) \}$, for $0 < p \leqslant 1$, (where $\mathcal{R}_i(f)$ denotes the ith Riesz transform of the function f) by means of p-atoms, i.e., functions a_p with compact support and vanishing k^{th} order moments $\displaystyle\int_{-\infty}^{\infty} x^k\, a_p(x)\, dx$, for all integers $k \in \left\{0, \dots, \left[\frac{1}{p}\right] - 1\right\}$, where $[.]$ denotes the integer part.

The n-dimensional case of distributions in $\mathcal{H}^p(\mathbb{R}^n)$, for $n \geqslant 1$, was then developed by Robert H. Latter in [**Lat78**]. The latter results were extended by Alberto P. Calderon to distributions in parabolic $\mathcal{H}^p$ spaces, for $0 < p \leqslant \infty$, in [**Cal77**]. Extensions to the case of homogeneous Besov spaces were given by Michael Frazier and Björn Jawerth in [**FJ85**].

However, one had to wait until the work of Hans Triebel for the extension to d-sets of $\mathbb{R}^n$, with $0 < d < n$; see [**Tri06**], Chapter 8, Section 8. 1. 3, on pages 350–356. Along these lines, further extensions were obtained by Maryia Kabanava in the case of nested fractals; see [**Kab12**].

In our present context, we follow and extend Maryia Kabanava's approach, which enables us to characterize Besov spaces on the Weierstrass Curve. Thanks to the connections between Sobolev spaces and the usual Laplacian, we are then able to connect explicitly the order of the restriction of the usual Laplacian on the Curve – or, rather, on the IFD – and the maximal real part of the Complex Dimensions, namely, the Minkowski Dimension of the Curve.

Note that whereas atomic decompositions may appear as *magical* tools, which, in a sense, convert the study of function spaces into the study of sequence spaces, they have drawbacks, insofar as all the information relative to the involved function space is not contained in the associated sequence space. Indeed, the coefficients of the optimal atomic decomposition of a function f generally only depends on f. One can think of *local*, vs *global* decompositions. In order to overcome the resulting loss of information, a slightly different approach was presented by Hans Triebel in [**Tri97**]; see, especially, in the case of general Besov spaces, Chapter 4, Section 20, building on the spectral theory for fractal pseudodifferential operators. More precisely, the function spaces on the involved fractal (viewed as a d-set) are directly adapted from the case of the boundary $\partial\Omega \subset \mathbb{R}^{n-1}$ of a C^∞-domain in $\mathbb{R}^n$, where $\partial\Omega$ is reduced "via an atlas of finitely many C^∞ charts to corresponding spaces on $\mathbb{R}^{n-1}$" (the quote is from Section 20. 1, on page 151, in. *loc. cit.*).

Still to circumvent the aforementioned drawbacks of atomic decompositions, significant advances were then made by Ingrid Daubechies, who introduced localized, compactly supported wavelets; see the expositary paper [**Dau92**]. In our present context, we cannot, for the moment, apply such elegant methods. However, we retain the important idea that an essential – and unavoidable step – is to question the quality of the sequence involved in the atomic decomposition. In the specific case of a closed subset $\mathcal{F}$ of $\mathbb{R}^n$, the above disadvantages were also highlighted by Alf Jonsson in [**Jon09**], where the author explains that, contrary to the (more general) case studied by Michael Frazier and Björn Jawerth in [**FJ85**], "the definition of $B_\alpha^{p,q}(\mathcal{F})$ is not satisfactory from the point of view that it is not constructive" (the quote is from page 586, in. *loc. cit.*). It is for this reason that Alf Jonsson considered the (even) more specific case of sets preserving Markov's Inequality. In its original form, that goes back to 1889 (see [**Mar48**]), Markov's Inequality states that, for any nonnegative integer n, and any complex polynomial P of degree d_P:

$$\|P'\|_{[-1,1]} \leqslant d_P^2 \, \|P\|_{[-1,1]}.$$

By extension, a closed nonempty subset $\mathcal{F}$ of $\mathbb{R}^n$ *preserves Markov's Inequality* if, for all positive integers N, any real polynomial P with d_P variables, of degree at most equal to N, any point $X \in \mathcal{F}$, and any real number $r \in]0,1]$:

$$\max_{\mathcal{F} \cap \mathcal{B}(X,r)} |\nabla P| \leqslant \frac{c(P,n,N,\mathcal{F})}{r} \max_{\mathcal{F} \cap \mathcal{B}(X,r)} |P|,$$

where $c(P,n,N,\mathcal{F})$ denotes a positive constant that depend on n, P, N and $\mathcal{F}$.

Such an assumption enabled Alf Jonsson to build polynomial approximations, by means of orthogonal polynomials. Such an approach therefore ensures "more satisfactory trace theorems" (the quote is from page 595, in *loc. cit.*). Indeed, all is finally about describing the smoothness properties of functions, where the degree of the polynomials involved in the approximation (defined on the intersection of $\mathcal{F}$ with squares or balls) provides the informations related to the level of smoothness. In this light, we clearly understand that Markov's Inequality ensures the *quality* of the approximation.

Going further, in the case when $p \geqslant 1$, Alf Jonsson introduces a space $\widetilde{B}_\alpha^{p,q}(\mathcal{F})$, obtained via local polynomial approximation, and proves that it is equal to the Besov space $B_\alpha^{p,q}(\mathcal{F})$ obtained thanks to atomic decomposition.

Now, a very interesting point is that fractal curves generally preserve Markov's Inequality. For instance, it is the case of the von Koch's snowflake, as is studied in the work by Hans Wallin in [**Wal91**], where the author provides the trace to the (fractal) boundary for functions in a Sobolev space.

In our present context, the Weierstrass Curve $\Gamma_{\mathcal{W}}$, also preserves Markov's Inequality, as in proved in Property 4.5. This ensures that our characterization of Besov spaces on $\Gamma_{\mathcal{W}}$ is obtained in a constructive manner.

Switching to more geometric topics, following our quest for the connections between the Complex Dimensions and the intrinsic properties of a fractal, we showed, in our previous work [**DL24c**], that, in the case of junction vertices, i.e., points belonging to consecutive prefractal graphs, the aforementioned maximal real part of the Complex Dimensions, associated to the cohomology group involved, changes. Hence, a change of shape – when one switches from a prefractal approximation, to the consecutive polygonal approximation, also corresponds to the occurrence of new polygons – is closely connected to a change of fractal dimensions. In this light, it was natural to explore further connections between the Complex Dimensions of a fractal object – the Weierstrass Curve, or the Weierstrass IFDs – and a suitable analog of Morse theory: given the Complex Dimensions and the fractal Morse indexes, can we, in some sense, reconstruct the fractal? Towards the end of the paper, we begin to lay the foundations for addressing this challenging and very interesting inverse topological and geometric problem.

Our main results in the present setting can be found in the following places:

i. In Section 3, Property 3.12, where we introduce the polyhedral measure on the Weierstrass IFD. In particular, we prove that this polyhedral measure is well defined, as well as nontrivial, and is a bounded and singular Borel measure on the Weierstrass Curve, whose total mass is positive and given by a suitable renormalization procedure (see Theorem 3.14.

ii. In Section 4, where our polyhedral measure enables us to extend the aforementioned results of Alf Jonsson and Hans Wallin to the case of the Weierstrass IFD. More precisely, we define *the atomic decomposition* of a function defined on the IFD; see Definition 4.15. We also give a direct proof of the fact that the Weierstrass Curve satisfies Markov's Inequality; see Property 4.5. We can then define and characterize the Besov spaces on the IFD; see Definition 4.22, along with Property 4.27. In the end, we establish a trace theorem in this context (Theorem 4.34), and (in Corollary 4.36) deduce from it the order of a Laplace-type operator on the IFD, which is in agreement with previous results of Robert S. Strichartz in the case of the Sierpiński Gasket $\mathcal{SG}$ in [**Str03**]; namely, our Laplace-type operator is not of order 2, unlike for the classical Laplacian. We also note that our trace theorem 4.34, connects, in particular, the Sobolev space on a suitable open neighborhood of the Weierstrass Curve $\Gamma_{\mathcal{W}}$ with the Besov space on $\Gamma_{\mathcal{W}}$, for the appropriate values of the exponents involved.

iii. In Section 5, where we use the Complex Dimensions, along with the fractal cohomology, of the Weierstrass IFD, obtained and developed by the authors in [**DL25c**], [**DL24c**] (see also [**DL24a**]), in order to propose an extension of Morse theory devoted to fractals. In particular, a *minimal real Complex Dimension* is associated to each prefractal approximation (in Definition 5.6), along with *a cohomological vertex integer*, associated, this time, to each vertex of the prefractal approximation; see Definition 5.8. We also define new Morse indexes – applicable to fractal curves such as the Weierstrass Curve; see Definition 5.21. Furthermore, we introduce the notion of a *fractal Morse function* (in Definition 5.18), and show, in Property 5.20, that the Weierstrass function is a fractal Morse function.

iv. In Section 6, we explain how the Weierstrass Curve can be viewed as the projection of a 3-dimensional vertical comb, the teeth of which are directly connected with the cohomological vertex integers. This provides us with new research directions for future work, in connection with our fractal Morse theory.

For a thorough discussion of the theory of Complex Dimensions, we refer the interested reader to [**LvF13**] and to [**LRŽ17a**], in the case of fractal strings and of (relative) higher-dimensional fractal drums, respectively; see also the survey article on the subject, [**Lap19**].

In closing this introduction, we mention that, for clarity and by necessity of concision, we work throughout this paper with the important example of the Weierstrass Curve and the associated Weierstrass IFD. We stress, however, that we expect that our main results will extend to a large class of fractal curves and their associated IFDs – as well as, eventually, to a large class of higher-dimensional fractal manifolds obtained by means of polyhedral prefractal approximations.

2. Geometry of the Weierstrass Curve

We begin by reviewing the main geometric properties of the Weierstrass Curve (and of the associated IFD), which will be key to our work in the rest of this paper.

This section is an excerpt of the corresponding one in [**DL25c**]. Also, all of the material in this section is essential to the rest of the paper.

Henceforth, we place ourselves in the Euclidean plane, equipped with a direct orthonormal frame. The usual Cartesian coordinates are denoted by (x, y). The horizontal and vertical axes will be respectively referred to as $(x'x)$ and $(y'y)$.

NOTATION 1 (**Set of all nonnegative integers, and intervals**). As in Bourbaki [**Bou04**] (Appendix E. 143), we denote by $\mathbb{N} = \{0, 1, 2, \ldots\}$ the set of all nonnegative integers, and set $\mathbb{N}^\star = \mathbb{N} \setminus \{0\}$.

Given a, b with $-\infty \leqslant a \leqslant b \leqslant \infty$, $]a, b[= (a, b)$ denotes an open interval, while, for example, $]a, b] = (a, b]$ denotes a half-open, half-closed interval.

NOTATION 2 (**Wave inequality symbol**). Given two positive-valued functions f and g, defined on a subset $\mathcal{I}$ of $\mathbb{R}$, we use the following notation, for all $x \in \mathcal{I}$: $f(x) \lesssim g(x)$ when there exists a strictly positive constant C such that, for all $x \in \mathcal{I}$, $f(x) \leqslant C\, g(x)$.

NOTATION 3 (**Weierstrass parameters**). In the sequel, λ and N_b are two real numbers such that

$$0 < \lambda < 1 \quad , \quad N_b \in \mathbb{N}^\star \quad \text{and} \quad \lambda\, N_b > 1. \qquad (\clubsuit)$$

Note that this implies that $N_b > 1$; i.e., $N_b \geqslant 2$, if $N_b \in \mathbb{N}^\star$, as will be the case in this paper.

As is explained in [**Dav19**], we deliberately made the choice to introduce the notation N_b which replaces the initial number b, in so far as, in Hardy's paper [**Har16**] (in contrast to Weierstrass' original article [**Wei75**]), b is any positive real number satisfying $\lambda\, b > 1$, whereas we deal here with the specific case of a positive integer, which accounts for the natural notation N_b.

DEFINITION 2.1 (**Weierstrass function, Weierstrass Curve**). We consider the *Weierstrass function* $\mathcal{W}$ (also called, in short, the $\mathcal{W}$*-function*) defined, for any real number x, by

$$\mathcal{W}(x) = \sum_{n=0}^{\infty} \lambda^n \cos\left(2\,\pi\, N_b^n\, x\right).$$

We call the associated graph the *Weierstrass Curve*, and denote it by $\Gamma_{\mathcal{W}}$.

Due to the one-periodicity of the $\mathcal{W}$-function (since $N_b \in \mathbb{N}^\star$), from now on, and without loss of generality, we restrict our study to the interval $[0, 1[= [0, 1)$. Note that $\mathcal{W}$ is continuous, and hence, bounded on all of $\mathbb{R}$. In particular, $\Gamma_{\mathcal{W}}$ is a (nonempty) compact subset of $\mathbb{R}^2$.

PROPERTY 2.2 (**Scaling properties of the Weierstrass function, and consequences** [**DL25c**]). *Since* $\mathcal{W}(x) = \sum_{n=0}^{\infty} \lambda^n \cos\left(2\,\pi\, N_b^n\, x\right)$ *for any real number* x, *one also has*

$$\begin{aligned}\mathcal{W}(N_b\, x) &= \sum_{n=0}^{\infty} \lambda^n \cos\left(2\,\pi\, N_b^{n+1}\, x\right) = \frac{1}{\lambda} \sum_{n=1}^{\infty} \lambda^n \cos\left(2\,\pi\, N_b^n\, x\right) \\ &= \frac{1}{\lambda} \left(\mathcal{W}(x) - \cos\left(2\,\pi\, x\right)\right),\end{aligned}$$

which yields, for any strictly positive integer m *and any* j *in* $\{0, \ldots, \# V_m\}$,

$$\mathcal{W}\left(\frac{j}{(N_b - 1)\, N_b^m}\right) = \lambda\, \mathcal{W}\left(\frac{j}{(N_b - 1)\, N_b^{m-1}}\right) + \cos\left(\frac{2\,\pi\, j}{(N_b - 1)\, N_b^m}\right).$$

By induction, one then obtains that

$$\mathcal{W}\left(\frac{j}{(N_b-1)\,N_b^m}\right)=\lambda^m\,\mathcal{W}\left(\frac{j}{(N_b-1)}\right)+\sum_{k=0}^{m-1}\lambda^k\,\cos\left(\frac{2\,\pi\,N_b^k\,j}{(N_b-1)\,N_b^m}\right).$$

DEFINITION 2.3 (**Weierstrass Complexified function**). We introduce the *Weierstrass Complexified function* $\mathcal{W}_{comp}$, defined, for any real number x, by

$$\mathcal{W}_{comp}(x)=\sum_{n=0}^{\infty}\lambda^n\,e^{2\,i\,\pi\,N_b^n\,x}. \tag{1}$$

Clearly, $\mathcal{W}_{comp}$ is also a continuous and 1-periodic function on $\mathbb{R}$.

NOTATION 4 (**Logarithm**). Given $y>0$, $\ln y$ denotes the natural logarithm of y, while, given $a>1$, $\ln_a y=\dfrac{\ln y}{\ln a}$ denotes the logarithm of y in base a; so that, in particular, $\ln=\ln_e$.

NOTATION 5 (**Minkowski dimension and Hölder exponent**). For the parameters λ and N_b satisfying condition (♣) (see Notation 3), we denote by

$$D_{\mathcal{W}}=2+\frac{\ln\lambda}{\ln N_b}=2-\ln_{N_b}\frac{1}{\lambda}\in\,]1,2[\tag{2}$$

the box-counting dimension (or Minkowski dimension) of the Weierstrass Curve $\Gamma_{\mathcal{W}}$, which happens to be equal to its Hausdorff dimension [**KMPY84**], [**BBR14**], [**She18**], [**Kel17**]. We point out that the results in our previous papers([**DL25a**] and [**DL23a**] respectively), also provide a direct geometric proof of the fact that $D_{\mathcal{W}}$, the Minkowski dimension (or box-counting dimension) of $\Gamma_{\mathcal{W}}$, exists and takes the above values, as well as of the fact that $\mathcal{W}$ is Hölder continuous with *optimal* Hölder exponent

$$2-D_{\mathcal{W}}=-\frac{\ln\lambda}{\ln N_b}=\ln_{N_b}\frac{1}{\lambda}.$$

CONVENTION (**The Weierstrass Curve as a cyclic curve**). In the sequel, we identify the points $(0,\mathcal{W}(0))$ and $(1,\mathcal{W}(1))=(1,\mathcal{W}(0))$. This is justified by the fact that the Weierstrass function $\mathcal{W}$ is 1-periodic, since N_b is an integer.

REMARK 2.4. The above convention makes sense, because, in addition to the periodicity property of the $\mathcal{W}$–function, the points $(0,\mathcal{W}(0))$ and $(1,\mathcal{W}(1))$ have the same vertical coordinate.

PROPERTY 2.5 (**Symmetry with respect to the vertical line $x=\frac{1}{2}$**). *Since, for any $x\in[0,1]$,*

$$\mathcal{W}(1-x)=\sum_{n=0}^{\infty}\lambda^n\,\cos\left(2\,\pi\,N_b^n-2\,\pi\,N_b^n\,x\right)=\mathcal{W}(x),$$

the Weierstrass Curve is symmetric with respect to the vertical straight line $x=\dfrac{1}{2}$.

PROPOSITION 2.6 (**Nonlinear and noncontractive iterated function system (IFS)**). *Following our previous work* [**Dav18**], *we approximate the restriction $\Gamma_{\mathcal{W}}$ to $[0,1[\times\mathbb{R}$, of the Weierstrass Curve, by a sequence of graphs, built via* *an iterative process. For this purpose, we use the nonlinear iterated function system (IFS) consisting of a finite family of* C^∞ *maps from* $\mathbb{R}^2$ *to* $\mathbb{R}^2$ *and denoted by*

$$\mathcal{T}_{\mathcal{W}} = \{T_0, \ldots, T_{N_b-1}\},$$

where, for any integer i *belonging to* $\{0, \ldots, N_b - 1\}$ *and any point* (x, y) *of* $\mathbb{R}^2$,

$$T_i(x, y) = \left(\frac{x+i}{N_b}, \lambda\, y + \cos\left(2\,\pi\left(\frac{x+i}{N_b}\right)\right)\right).$$

Note that unlike in the classical situation, these maps are not contractions. Nevertheless, $\Gamma_{\mathcal{W}}$ *can be recovered from this IFS in the usual way, as we next explain.*

PROPERTY 2.7 (**Attractor of the IFS [Dav18], [Dav19]**). *The Weierstrass Curve* $\Gamma_{\mathcal{W}}$ *is the attractor of the IFS* $\mathcal{T}_{\mathcal{W}}$*, and hence, is the unique nonempty compact subset* $\mathcal{K}$ *of* $\mathbb{R}^2$ *satisfying* $\mathcal{K} = \bigcup_{i=0}^{N_b-1} T_i(\mathcal{K})$*; in particular,* $\Gamma_{\mathcal{W}} = \bigcup_{i=0}^{N_b-1} T_i(\Gamma_{\mathcal{W}})$.

NOTATION 6 (**Fixed points**). For any integer i belonging to $\{0, \ldots, N_b - 1\}$, we denote by

$$P_i = (x_i, y_i) = \left(\frac{i}{N_b-1}, \frac{1}{1-\lambda}\cos\left(\frac{2\,\pi\, i}{N_b-1}\right)\right)$$

the unique fixed point of the map T_i; see [**Dav19**].

DEFINITION 2.8 (**Sets of vertices, prefractals**). We denote by V_0 the ordered set (according to increasing abscissae) of the points

$$\{P_0, \ldots, P_{N_b-1}\}.$$

The set of points V_0 – where, for any integer i in $\{0, \ldots, N_b - 2\}$, the point P_i is linked to the point P_{i+1} – constitutes an oriented finite graph, ordered according to increasing abscissae, which we will denote by $\Gamma_{\mathcal{W}_0}$. Then, V_0 is called *the set of vertices* of the graph $\Gamma_{\mathcal{W}_0}$.

For any positive integer m, i.e., for $m \in \mathbb{N}^\star$, we set $V_m = \bigcup_{i=0}^{N_b-1} T_i(V_{m-1})$.

The set of points V_m, where two consecutive points are linked, is an oriented finite graph, ordered according to increasing abscissae, called the $\boldsymbol{m^{th}}$ **order** $\boldsymbol{\mathcal{W}}$**-prefractal**. Then, V_m is called *the set of vertices* of the prefractal $\Gamma_{\mathcal{W}_m}$; see Figure 2.

PROPERTY 2.9 (**Density of the set** $V^\star = \bigcup_{n\in\mathbb{N}} V_n$ **in the Weierstrass Curve [DL25c]**). *The set* $V^\star = \bigcup_{n\in\mathbb{N}} V_n$ *is dense in the Weierstrass Curve* $\Gamma_{\mathcal{W}}$.

DEFINITION 2.10 (**Adjacent vertices, edge relation**). For any $m \in \mathbb{N}$, the prefractal graph $\Gamma_{\mathcal{W}_m}$ is equipped with an edge relation $\underset{m}{\sim}$, as follows: two vertices X and Y of $\Gamma_{\mathcal{W}_m}$ (i.e., two points belonging to V_m) will be said to be *adjacent* (i.e., *neighboring* or *junction points*) if and only if the line segment $[X, Y]$ is an edge of $\Gamma_{\mathcal{W}_m}$; we then write $X \underset{m}{\sim} Y$. Note that this edge relation depends on m, which means that points adjacent in V_m might not remain adjacent in V_{m+1}.

We refer to part *iv.* of Property 2.11, along with Figure 1, for the definition of the polygons $\mathcal{P}_{m,k}$ and $\mathcal{Q}_{m,k}$ associated with the Weierstrass Curve.

PROPERTY 2.11 ([**Dav18**]). *For any* $m \in \mathbb{N}$*, the following statements hold:*

i. $V_m \subset V_{m+1}$.

ii. $\#V_m = (N_b - 1)\, N_b^m + 1$, *where* $\#V_m$ *denotes the number of elements in the finite set* V_m.

iii. *The prefractal graph* $\Gamma_{\mathcal{W}_m}$ *has exactly* $(N_b - 1)\, N_b^m$ *edges.*

iv. *The consecutive vertices of the prefractal graph* $\Gamma_{\mathcal{W}_m}$ *are the vertices of* N_b^m *simple nonregular polygons* $\mathcal{P}_{m,k}$. *For any strictly positive integer* m, *the junction point between two consecutive polygons* $\mathcal{P}_{m,k}$ *and* $\mathcal{P}_{m,k+1}$ *is the point*

$$\left(\frac{(N_b-1)\,k}{(N_b-1)\,N_b^m}, \mathcal{W}\left(\frac{(N_b-1)\,k}{(N_b-1)\,N_b^m}\right)\right), \quad 1 \leqslant k \leqslant N_b^m - 1.$$

Hence, the total number of junction points is $N_b^m - 1$. *For instance, in the case* $N_b = 3$, *the polygons are all triangles; see Figure* 1.

We call extreme vertices *of the polygon* $\mathcal{P}_{m,k}$ *the junction points*

$$\mathcal{V}_{initial}\left(\mathcal{P}_{m,k}\right) = \left(\frac{(N_b-1)\,k}{(N_b-1)\,N_b^m}, \mathcal{W}\left(\frac{(N_b-1)\,k}{(N_b-1)\,N_b^m}\right)\right),$$

for $0 \leqslant k \leqslant N_b^m - 1$, *and*

$$\mathcal{V}_{end}\left(\mathcal{P}_{m,k}\right) = \left(\frac{(N_b-1)\,(k+1)}{(N_b-1)\,N_b^m}, \mathcal{W}\left(\frac{(N_b-1)\,(k+1)}{(N_b-1)\,N_b^m}\right)\right),$$

for $0 \leqslant k \leqslant N_b^m - 2$.

In the sequel, we will denote by $\mathcal{P}_0$ ***the initial polygon***, *whose vertices are the fixed points of the maps* T_i, $0 \leqslant i \leqslant N_b - 1$, *introduced in Notation* 6 *and Definition* 2.8, *i.e.,* $\{P_0, \ldots, P_{N_b-1}\}$; *see, again, Figure* 1.

In the same way, the consecutive vertices of the prefractal graph $\Gamma_{\mathcal{W}_m}$, *distinct from the fixed points* P_0 *and* P_{N_b-1} *(see Notation* 6*), are also the vertices of* $N_b^m - 1$ *simple nonregular polygons* $\mathcal{Q}_{m,j}$, *for* $1 \leqslant j \leqslant N_b^m - 1$, *this time with at most* N_b *sides. For any integer* j *such that* $1 \leqslant j \leqslant N_b^m - 2$, *one obtains each polygon* $\mathcal{Q}_{m,j}$ *by connecting the point number* j *to the point number* $j+1$ *if* $j \equiv i \bmod N_b$, *for* $1 \leqslant i \leqslant N_b - 1$, *and the point number* j *to the point number* $j - N_b + 1$ *if* $j \equiv 0 \bmod N_b$.

As previously, we call extreme vertices *of the polygon* $\mathcal{Q}_{m,k}$ *the points*

$$\mathcal{V}_{initial}\left(\mathcal{Q}_{m,k}\right) = \left(\frac{(N_b-1)\,k}{(N_b-1)\,N_b^m}, \mathcal{W}\left(\frac{(N_b-1)\,k}{(N_b-1)\,N_b^m}\right)\right),$$

for $1 \leqslant k \leqslant N_b^m - 1$, *and*

$$\mathcal{V}_{end}\left(\mathcal{Q}_{m,k}\right) = \left(\frac{(N_b-1)\,(k+1)}{(N_b-1)\,N_b^m}, \mathcal{W}\left(\frac{(N_b-1)\,(k+1)}{(N_b-1)\,N_b^m}\right)\right),$$

for $1 \leqslant k \leqslant N_b^m - 2$.

DEFINITION 2.12 (**Polygonal sets**). For any $m \in \mathbb{N}$, we introduce the following polygonal sets

$$\mathcal{P}_m = \left\{\mathcal{P}_{m,k},\, 0 \leqslant k \leqslant N_b^m - 1\right\} \quad \text{and} \quad \mathcal{Q}_m = \left\{\mathcal{Q}_{m,k},\, 0 \leqslant k \leqslant N_b^m - 2\right\}.$$

NOTATION 7. For any $m \in \mathbb{N}$, we denote by:

i. $X \in \mathcal{P}_m$ (resp., $X \in \mathcal{Q}_m$) a vertex of a polygon $\mathcal{P}_{m,k}$, with $0 \leqslant k \leqslant N_b^m - 1$ (resp., a vertex of a polygon $\mathcal{Q}_{m,k}$, with $1 \leqslant k \leqslant N_b^m - 2$).

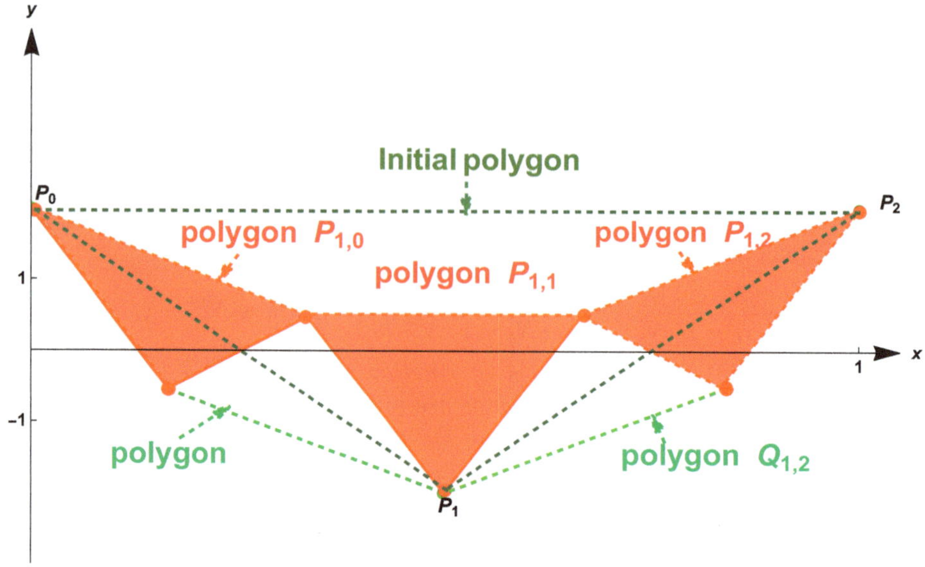

FIGURE 1. The initial polygon $\boldsymbol{\mathcal{P}_0}$, and the respective polygons $\boldsymbol{\mathcal{P}_{1,0}}$, $\boldsymbol{\mathcal{P}_{1,1}}$, $\boldsymbol{\mathcal{P}_{1,2}}$, $\boldsymbol{\mathcal{Q}_{1,1}}$, $\boldsymbol{\mathcal{Q}_{1,2}}$, in the case when $\boldsymbol{\lambda = \frac{1}{2}}$ and $\boldsymbol{N_b = 3}$. (See also Figure 2.)

ii. $\mathcal{P}_m \bigcup \mathcal{Q}_m$ the reunion of the polygonal sets $\mathcal{P}_m$ and $\mathcal{Q}_m$, which consists in the set of all the vertices of the polygons $\mathcal{P}_{m,k}$, with $0 \leqslant k \leqslant N_b^m - 1$, along with the vertices of the polygons $\mathcal{Q}_{m,k}$, with $1 \leqslant k \leqslant N_b^m - 2$. In particular, $X \in \mathcal{P}_m \bigcup \mathcal{Q}_m$ simply denotes a vertex in $\mathcal{P}_m$ or $\mathcal{Q}_m$.

iii. $\mathcal{P}_m \bigcap \mathcal{Q}_m$ the intersection of the polygonal sets $\mathcal{P}_m$ and $\mathcal{Q}_m$, which consists in the set of all the vertices of both a polygon $\mathcal{P}_{m,k}$, with $0 \leqslant k \leqslant N_b^m - 1$, and a polygon $\mathcal{Q}_{m,k'}$, with $1 \leqslant' k \leqslant N_b^m - 2$.

DEFINITION 2.13 (**Vertices of the prefractals, elementary lengths, heights and angles**). Given a strictly positive integer m, we denote by $\left(M_{j,m}\right)_{0 \leqslant j \leqslant (N_b-1)\, N_b^m}$ *the set of vertices* of the prefractal graph $\Gamma_{\mathcal{W}_m}$. One thus has, for any integer j in $\left\{0, \ldots, (N_b - 1)\, N_b^m\right\}$:

$$M_{j,m} = \left(\frac{j}{(N_b - 1)\, N_b^m}, \mathcal{W}\left(\frac{j}{(N_b - 1)\, N_b^m}\right)\right).$$

We also introduce, for any integer j in $\left\{0, \ldots, (N_b - 1)\, N_b^m - 1\right\}$:

i. the elementary horizontal lengths:

$$L_m = \frac{1}{(N_b - 1)\, N_b^m};$$

ii. the elementary lengths:

$$l_{j,j+1,m} = d\left(M_{j,m}, M_{j+1,m}\right) = \sqrt{L_m^2 + h_{j,j+1,m}^2},$$

where $h_{j,j+1,m}$ is defined in *iii.* just below.

iii. the elementary heights:

$$h_{j,j+1,m} = \left|\mathcal{W}\left(\frac{j+1}{(N_b-1)\,N_b^m}\right) - \mathcal{W}\left(\frac{j}{(N_b-1)\,N_b^m}\right)\right| ;$$

iv. the minimal height:

(3) $$h_m^{inf} = \inf_{0\leqslant j\leqslant (N_b-1)\,N_b^m-1} h_{j,j+1,m},$$

along with the maximal height:

(4) $$h_m = \sup_{0\leqslant j\leqslant (N_b-1)\,N_b^m-1} h_{j,j+1,m},$$

v. the geometric angles:

$$\theta_{j-1,j,m} = \left((y'y), (\widehat{M_{j-1,m}M_{j,m}})\right), \quad \theta_{j,j+1,m} = \left((y'y), (\widehat{M_{j,m}M_{j+1,m}})\right),$$

where $(y'y)$ denotes the vertical axis, which yield **the following value of the geometric angle between consecutive edges**, namely, $[M_{j-1,m}\,M_{j,m}, M_{j,m}\,M_{j+1,m}]$, with $\arctan = \tan^{-1}$:

$$\theta_{j-1,j,m} + \theta_{j,j+1,m} = \arctan\frac{L_m}{h_{j-1,j,m}} + \arctan\frac{L_m}{h_{j,j+1,m}}.$$

(Note that, of course, $\theta_{j-1,j,m} = \arctan\frac{L_m}{h_{j-1,j,m}}$ and $\theta_{j,j+1,m} = \arctan\frac{L_m}{h_{j,j+1,m}}$.)

PROPERTY 2.14. *For the geometric angle* $\theta_{j-1,j,m}$, $0\leqslant j\leqslant (N_b-1)\,N_b^m$, $m\in\mathbb{N}$, *we have the following relation:*

$$\tan\theta_{j-1,j,m} = \frac{L_m}{h_{j-1,j,m}}.$$

PROPERTY 2.15 (**A consequence of the symmetry with respect to the vertical line** $x=\frac{1}{2}$). *For any strictly positive integer* m *and any* j *in* $\{0,\ldots,\#V_m\}$, *we have that*

$$\mathcal{W}\left(\frac{j}{(N_b-1)\,N_b^m}\right) = \mathcal{W}\left(\frac{(N_b-1)\,N_b^m-j}{(N_b-1)\,N_b^m}\right),$$

which means that the points

$$\left(\frac{(N_b-1)\,N_b^m-j}{(N_b-1)\,N_b^m}, \mathcal{W}\left(\frac{(N_b-1)\,N_b^m-j}{(N_b-1)\,N_b^m}\right)\right)$$

and

$$\left(\frac{j}{(N_b-1)\,N_b^m}, \mathcal{W}\left(\frac{j}{(N_b-1)\,N_b^m}\right)\right)$$

are symmetric with respect to the vertical line $x=\frac{1}{2}$*; see Figure* 3.

DEFINITION 2.16 (**Left-side and right-side vertices**). Given nonnegative integers m, k such that $0\leqslant k\leqslant N_b^m-1$, and a polygon $\mathcal{P}_{m,k}$, we define:

i. The set of its *left-side vertices* as the set of the first $\left[\frac{N_b-1}{2}\right]$ vertices, where $[y]$ denotes the integer part of the real number y.

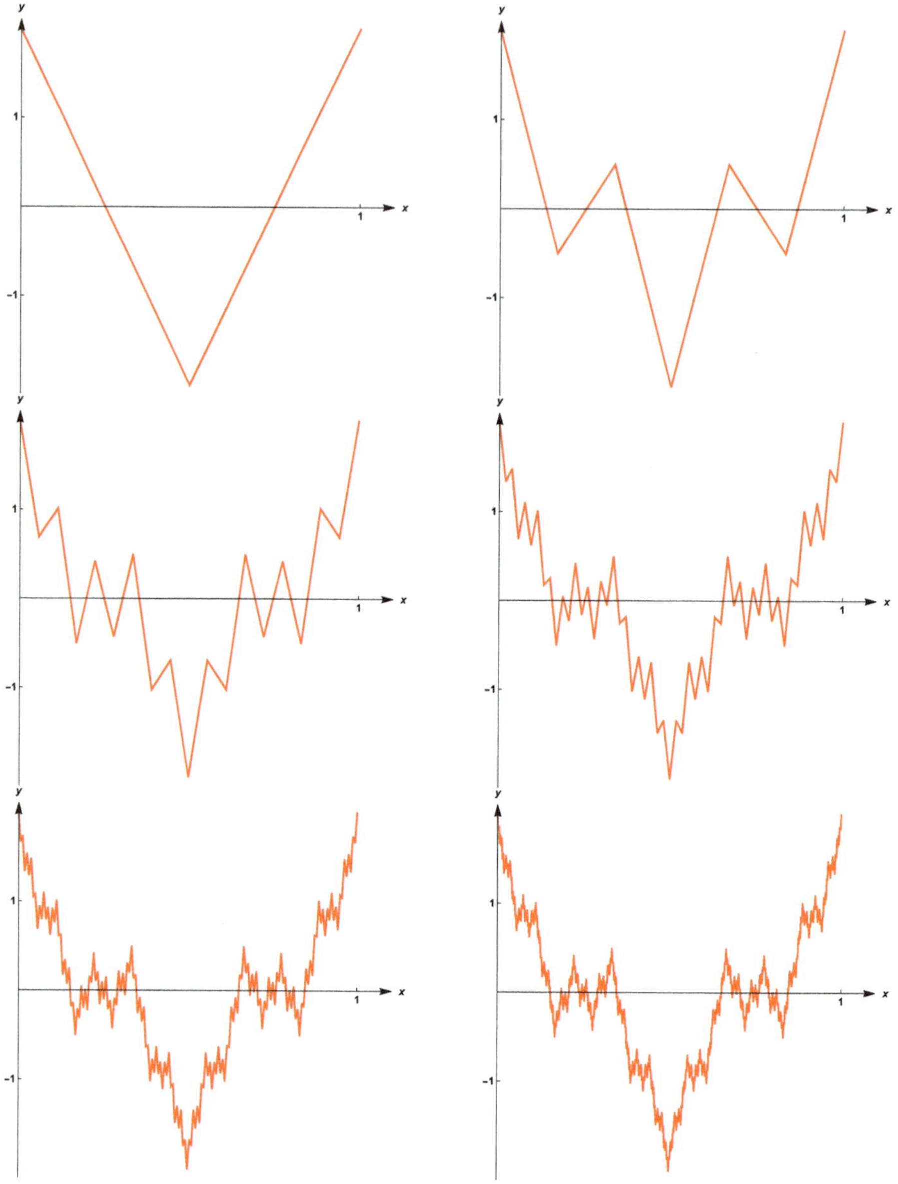

FIGURE 2. The prefractal graphs $\boldsymbol{\Gamma_{\mathcal{W}_0}}$, $\boldsymbol{\Gamma_{\mathcal{W}_1}}$, $\boldsymbol{\Gamma_{\mathcal{W}_2}}$, $\boldsymbol{\Gamma_{\mathcal{W}_3}}$, $\boldsymbol{\Gamma_{\mathcal{W}_4}}$, $\boldsymbol{\Gamma_{\mathcal{W}_5}}$, in the case when $\boldsymbol{\lambda = \frac{1}{2}}$ and $\boldsymbol{N_b = 3}$. For example, $\boldsymbol{\Gamma_{\mathcal{W}_1}}$ is on the right side of the top row, while $\boldsymbol{\Gamma_{\mathcal{W}_4}}$ is on the left side of the bottom row.

ii. The set of its *right-side vertices* as the set of the last $\left[\frac{N_b - 1}{2}\right]$ vertices.

When the integer N_b is odd, we define the bottom vertex as the $\left(\frac{N_b - 1}{2}\right)^{th}$ one; see Figure 4.

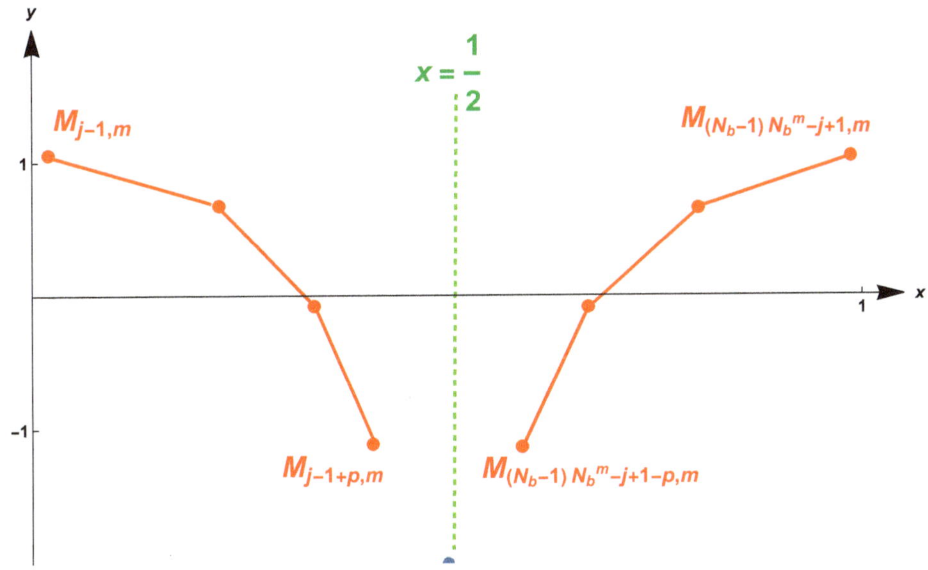

FIGURE 3. Symmetric points with respect to the vertical line $x = \frac{1}{2}$.

PROPERTY 2.17 ([**DL25c**]). *For any integer j in $\{0, \ldots, N_b - 1\}$:*

$$\begin{aligned}\mathcal{W}\left(\frac{j}{N_b-1}\right) &= \sum_{n=0}^{\infty} \lambda^n \cos\left(2\pi N_b^n \frac{j}{(N_b-1)}\right)\\ &= \sum_{n=0}^{\infty} \lambda^n \cos\left(\frac{2\pi j}{N_b-1}\right)\\ &= \frac{1}{1-\lambda} \cos\left(\frac{2\pi j}{N_b-1}\right).\end{aligned}$$

PROPERTY 2.18 ([**DL25c**]). *For $0 \leqslant j \leqslant \frac{N_b-1}{2}$ (resp., for $\frac{N_b-1}{2} \leqslant j \leqslant N_b - 1$), we have that*

$$\mathcal{W}\left(\frac{j+1}{N_b-1}\right) - \mathcal{W}\left(\frac{j}{N_b-1}\right) \leqslant 0 \quad \left(\textit{resp.}, \ \mathcal{W}\left(\frac{j+1}{N_b-1}\right) - \mathcal{W}\left(\frac{j}{N_b-1}\right) \geqslant 0\right).$$

NOTATION 8 (**Signum function**). The *signum function* of a real number x is defined by

$$\operatorname{sgn}(x) = \begin{cases} -1, & \text{if } x < 0,\\ 0, & \text{if } x = 0,\\ +1, & \text{if } x > 0.\end{cases}$$

PROPERTY 2.19 ([**DL25c**]). *Given any strictly positive integer m, we have the following properties:*

i. For any j in $\{0, \ldots, \#V_m\}$, the point

$$\left(\frac{j}{(N_b-1)\,N_b^m}, \mathcal{W}\left(\frac{j}{(N_b-1)\,N_b^m}\right)\right)$$

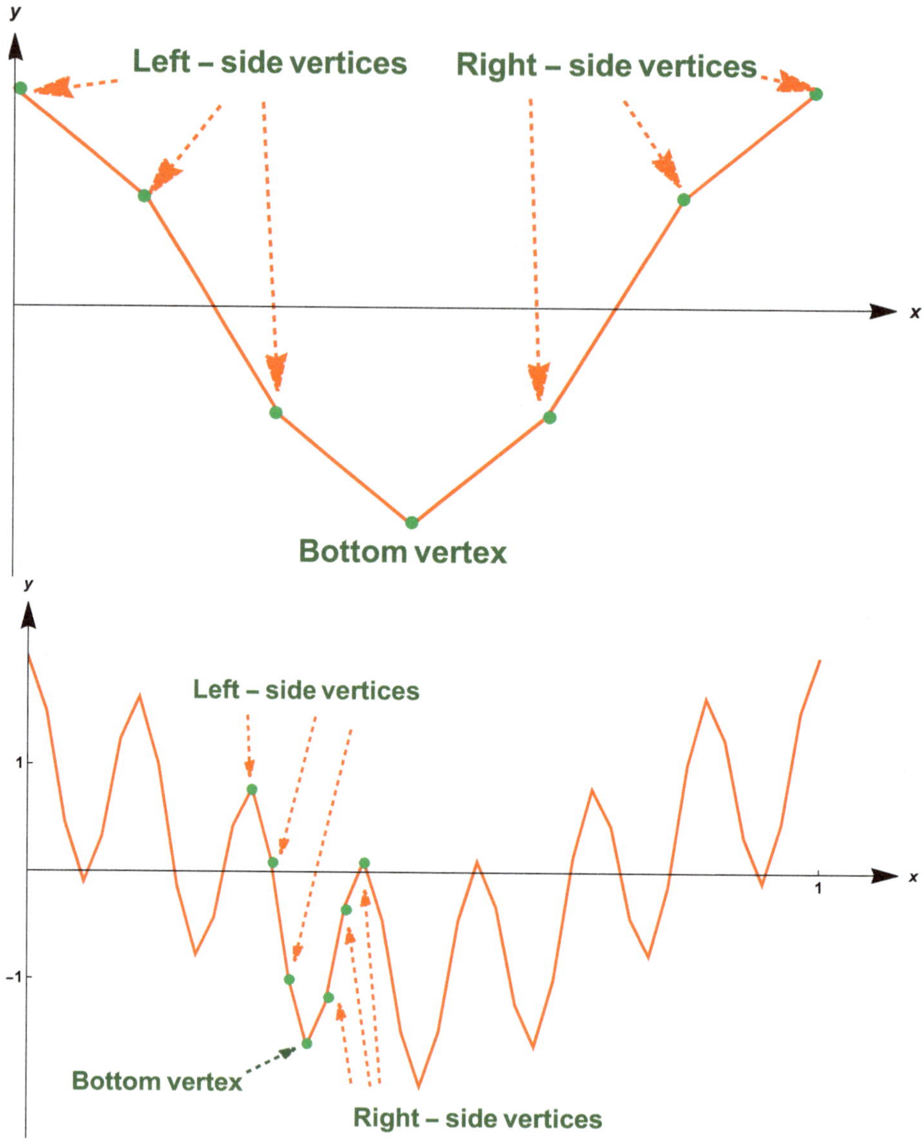

FIGURE 4. The Left and Right-Side Vertices.

is the image of the point

$$\left(\frac{j}{(N_b-1)\,N_b^{m-1}}-i,\mathcal{W}\left(\frac{j}{(N_b-1)\,N_b^{m-1}}-i\right)\right),$$

which can also be written as

$$\left(\frac{j-i\,(N_b-1)\,N_b^{m-1}}{(N_b-1)\,N_b^{m-1}},\mathcal{W}\left(\frac{j-i\,(N_b-1)\,N_b^{m-1}}{(N_b-1)\,N_b^{m-1}}\right)\right)$$

under the map T_i, where $i \in \{0,\ldots,N_b-1\}$ is arbitrary.

Consequently, for $0 \leqslant j \leqslant N_b - 1$, ***the*** j^{th} ***vertex of the polygon*** $\mathcal{P}_{m,k}$, $0 \leqslant k \leqslant N_b^m - 1$, *i.e., the point*

$$\left(\frac{(N_b-1)\,k+j}{(N_b-1)\,N_b^m}, \mathcal{W}\left(\frac{(N_b-1)\,k+j}{(N_b-1)\,N_b^m} \right) \right),$$

is the image of the point

$$\left(\frac{(N_b-1)\left(k-i\,(N_b-1)\,N_b^{m-1}\right)+j}{(N_b-1)\,N_b^{m-1}}, \mathcal{W}\left(\frac{(N_b-1)\left(k-i\,(N_b-1)\,N_b^{m-1}\right)+j}{(N_b-1)\,N_b^{m-1}} \right) \right);$$

under the map T_i, *where* $i \in \{0,\ldots,N_b-1\}$ *is again arbitrary. This latter point is also* ***the*** j^{th} ***vertex of the polygon*** $\mathcal{P}_{m-1,k-i\,(N_b-1)\,N_b^{m-1}}$. *Therefore, there is an exact correspondance between vertices of the polygons at consecutive steps* $m-1$, m.

ii. Given j *in* $\{0,\ldots,N_b-2\}$ *and* k *in* $\{0,\ldots,N_b^m-1\}$, *we have that*

$$\begin{aligned} &\operatorname{sgn}\left(\mathcal{W}\left(\frac{k\,(N_b-1)+j+1}{(N_b-1)\,N_b^m} \right) - \mathcal{W}\left(\frac{k\,(N_b-1)+j}{(N_b-1)\,N_b^m} \right) \right) \\ &\qquad = \operatorname{sgn}\left(\mathcal{W}\left(\frac{j+1}{N_b-1} \right) - \mathcal{W}\left(\frac{j}{N_b-1} \right) \right). \end{aligned}$$

PROOF.

i. Given $m \in \mathbb{N}^\star$, let us consider $i \in \{0,\ldots,N_b-1\}$. The image of the point

$$\left(\frac{j}{(N_b-1)\,N_b^{m-1}} - i, \mathcal{W}\left(\frac{j}{(N_b-1)\,N_b^{m-1}} - i \right) \right)$$

under the map T_i is obtained by applying the analytic expression given in Property 2.6, to the coordinates of this point, which, thanks to Property 2.2 above, yields the expected result, namely,

$$\left(\frac{j}{(N_b-1)\,N_b^m}, \lambda\, \underbrace{\mathcal{W}\left(\frac{j}{(N_b-1)\,N_b^{m-1}} - i \right)}_{\substack{\mathcal{W}\left(\frac{j}{(N_b-1)\,N_b^{m-1}} \right) \\ \text{(by 1-periodicity)}}} + \cos \frac{2\pi\, j}{(N_b-1)\,N_b^m}, \right)$$

i.e., also,

$$\left(\frac{j}{(N_b-1)\,N_b^m}, \mathcal{W}\left(\frac{j}{(N_b-1)\,N_b^m} \right) \right).$$

ii. See [**DL25c**].

□

PROPERTY 2.20 (**Lower bound and upper bound for the elementary heights** [**DL25c**]). *For any strictly positive integer* m *and any* j *in* $\{0,\ldots,(N_b-1)\,N_b^m\}$, *we have the following estimates, where* L_m *is the elementary horizontal length introduced in part i. of Definition* 2.13*:*

$$C_{inf}\, L_m^{2-D_\mathcal{W}} \leqslant \underbrace{\left| \mathcal{W}\left((j+1)\,L_m\right) - \mathcal{W}\left(j\,L_m\right) \right|}_{h_{j,j+1,m}} \leqslant C_{sup}\, L_m^{2-D_\mathcal{W}}, \tag{5}$$

where the finite and positive constants C_{inf} and C_{sup} are given by

$$C_{inf} = (N_b-1)^{2-D_{\mathcal{W}}} \min_{0\leqslant j\leqslant N_b-1,\, \mathcal{W}\left(\frac{j+1}{N_b-1}\right)\neq \mathcal{W}\left(\frac{j}{N_b-1}\right)} \left|\mathcal{W}\left(\frac{j+1}{N_b-1}\right)-\mathcal{W}\left(\frac{j}{N_b-1}\right)\right|$$

and

$$C_{sup} = (N_b-1)^{2-D_{\mathcal{W}}} \left(\max_{0\leqslant j\leqslant N_b-1} \left|\mathcal{W}\left(\frac{j+1}{N_b-1}\right)-\mathcal{W}\left(\frac{j}{N_b-1}\right)\right| + \frac{2\pi}{(N_b-1)\,(\lambda\, N_b-1)}\right).$$

One should note, in addition, that these constants C_{inf} and C_{sup} depend on the initial polygon $\mathcal{P}_0$.

As a consequence, we also have that

$$C_{inf}\, L_m^{2-D_{\mathcal{W}}} \leqslant h_m^{inf} \leqslant C_{sup}\, L_m^{2-D_{\mathcal{W}}} \quad \text{and} \quad C_{inf}\, L_m^{2-D_{\mathcal{W}}} \leqslant h_m \leqslant C_{sup}\, L_m^{2-D_{\mathcal{W}}},$$

where h_m^{inf} and h_m respectively denote the minimal and maximal heights introduced in part iv. of Definition 2.13.

THEOREM 2.21 (**Sharp local discrete reverse-Hölder properties of the Weierstrass function** [**DL25c**]). *For any $m \in \mathbb{N}$, let us consider a pair of real numbers (x,x') such that*

$$x = \frac{(N_b-1)\,k+j}{(N_b-1)\,N_b^m} = ((N_b-1)\,k+j)\; L_m,$$

and

$$x' = \frac{(N_b-1)\,k+j+\ell}{(N_b-1)\,N_b^m} = ((N_b-1)\,k+j+\ell)\; L_m,$$

where $0 \leqslant k \leqslant N_b^m - 1$. We then have the following (discrete, local) reverse-Hölder inequality, *with sharp Hölder exponent $-\dfrac{\ln\lambda}{\ln N_b} = 2 - D_{\mathcal{W}}$:*

$$C_{inf}\, |x'-x|^{2-D_{\mathcal{W}}} \leqslant \left|\mathcal{W}(x')-\mathcal{W}(x)\right|,$$

where $(x,\mathcal{W}(x))$ and $\left(x',\mathcal{W}(x')\right)$ are adjacent vertices of the same m^{th} prefractal approximation, $\Gamma_{\mathcal{W}_m}$, with $m \in \mathbb{N}$ arbitrary. Here, C_{inf} is given as in Property 2.20 *just above.*

COROLLARY 2.22 (**Optimal Hölder exponent for the Weierstrass function (see** [**DL25c**])). *The local reverse Hölder property of Theorem* 2.21 *just above – in conjunction with the Hölder condition satisfied by the Weierstrass function (see also* [**Zyg02**], *Chapter II, Theorem 4.9, on page 47) – shows that the Codimension*

$$2 - D_{\mathcal{W}} = -\frac{\ln\lambda}{\ln N_b} \in \,]0,1[$$

is the best (i.e., optimal) Hölder exponent for the Weierstrass function (as was originally shown, by a completely different method, by G. H. Hardy in [**Har16**]*).*

Note that, as a consequence, since the Hölder exponent is strictly smaller than one, it follows that the Weierstrass function $\mathcal{W}$ is nowhere differentiable.

COROLLARY 2.23 (**Coming from Property** 2.19). *Thanks to Property* 2.20, *one may now write, for any strictly positive integer m and any integer j in $\{0,\ldots,(N_b-1)\,N_b^m-1\}$:*

i. *for the elementary heights:*

$$h_{j-1,j,m} = L_m^{2-D_{\mathcal{W}}}\,\mathcal{O}(1)\,; \tag{6}$$

ii. *for the elementary quotients:*

$$\frac{h_{j-1,j,m}}{L_m} = L_m^{1-D_{\mathcal{W}}}\,\mathcal{O}(1)\,, \tag{7}$$

and where

$$0 < C_{inf} \leqslant \mathcal{O}(1) \leqslant C_{sup} < \infty.$$

Recall from Property 2.20, *that* C_{inf} *and* C_{sup} *are independent of* $m \in \mathbb{N}^\star$.

COROLLARY 2.24 (**Nonincreasing sequence of geometric angles (coming from Property** 2.19; see [**DL25c**])). *For the* ***geometric angles*** $\theta_{j-1,j,m}$, $0 \leqslant j \leqslant (N_b-1)\,N_b^m$, $m \in \mathbb{N}$, *introduced in part v. of Definition* 2.13, *we have the following result:*

$$\tan\theta_{j-1,j,m} = \frac{L_m}{h_{j-1,j,m}}(N_b-1) > \tan\theta_{j-1,j,m+1},$$

which yields

$$\theta_{j-1,j,m} > \theta_{j-1,j,m+1} \quad \text{and} \quad \theta_{j-1,j,m+1} \lesssim L_m^{D_{\mathcal{W}}-1}.$$

COROLLARY 2.25 (**Local extrema (coming from Property 2.19)**).

i. *The set of local maxima of the Weierstrass function on the interval* $[0,1]$ *is given by*

$$\left\{\left(\frac{(N_b-1)\,k}{N_b^m},\mathcal{W}\left(\frac{(N_b-1)\,k}{N_b^m}\right)\right) : 0 \leqslant k \leqslant N_b^m-1,\ m \in \mathbb{N}\right\},$$

and corresponds to the extreme vertices of the polygons $\mathcal{P}_{m,k}$ *and* $\mathcal{Q}_{m,k}$ *(see Property* 2.11*) at a given step* m *(i.e., they are the vertices connecting consecutive polygons; see part iv. of Property* 2.11*).*

ii. *For odd values of* N_b, *the set of local minima of the Weierstrass function on the interval* $[0,1]$ *is given by*

$$\left\{\left(\frac{(N_b-1)\,k+\frac{N_b-1}{2}}{(N_b-1)\,N_b^m},\mathcal{W}\left(\frac{(N_b-1)\,k+\frac{N_b-1}{2}}{(N_b-1)\,N_b^m}\right)\right) : 0 \leqslant k \leqslant N_b^m-1,\ m \in \mathbb{N}\right\},$$

and corresponds to the bottom vertices of the polygons $\mathcal{P}_{m,k}$ *and* $\mathcal{Q}_{m,k}$ *at a given step* m*; see also part iv. of Property* 2.11.

PROPERTY 2.26 (**Existence of reentrant angles** [**DL25c**]).

i. *The initial polygon* $\mathcal{P}_0$, *admits* ***reentrant interior angles***, *at a vertex* P_j, *with* $0 < j \leqslant N_b-1$, *in the sense that, with the right-hand rule (according to which angles are measured in a counterclockwise direction), we have that*

$$\left(\widehat{(P_jP_{j+1}),(P_jP_{j-1})}\right) > \pi,$$

for

$$0 < j \leqslant \frac{N_b-3}{4} \quad \text{or} \quad \frac{3\,N_b-1}{4} \leqslant j < N_b-1$$

(see Figure 5*), which does not occur for values of* $N_b < 7$.

The number of reentrant angles is then equal to $2\left[\frac{N_b-3}{4}\right]$.

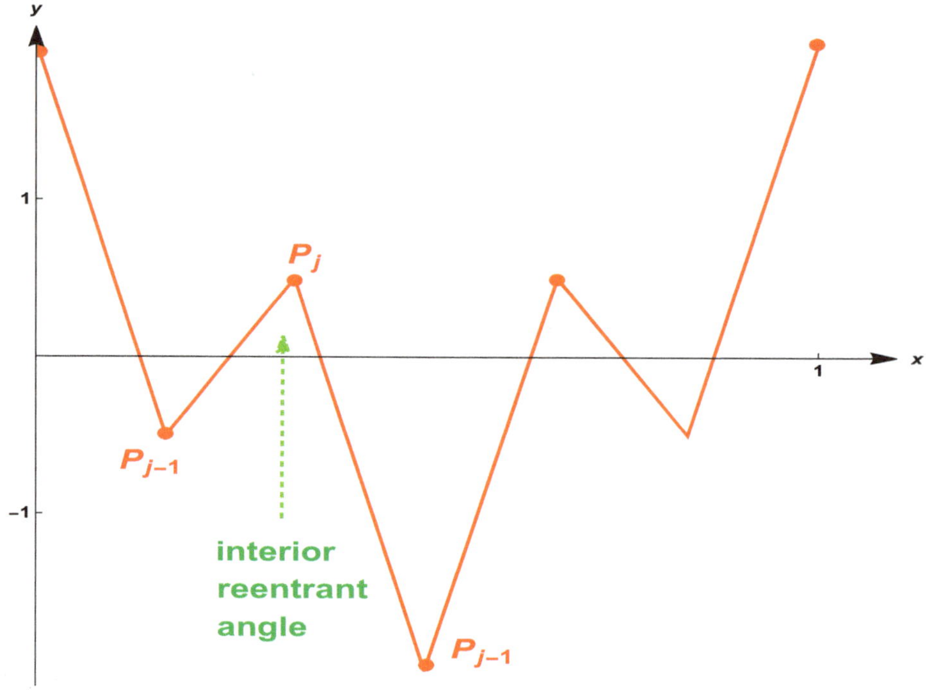

FIGURE 5. An interior reentrant angle. Here, $\boldsymbol{N_b = 7}$ and $\boldsymbol{\lambda = \frac{1}{2}}$.

ii. *At a given step* $m \in \mathbb{N}^\star$*, with the above convention, a polygon* $\mathcal{P}_{m,k}$ *admits reentrant interior angles in the sole cases when* $N_b \geqslant 7$*, at vertices* M_{k+j}*,* $1 \leqslant k \leqslant N_b^m$*,* $0 < j \leqslant N_b - 1$*, as well as in the case when*

$$0 < j \leqslant \frac{N_b - 3}{4} \quad or \quad \frac{3\,N_b - 1}{4} \leqslant j < N_b - 1.$$

The number of reentrant angles is then equal to $2\,N_b^m \left[\dfrac{N_b - 3}{4}\right]$.

REMARK 2.27. Note that due to the respective definitions of the polygons $\mathcal{P}_{m,k}$ and $\mathcal{Q}_{m,k}$, the existence of reentrant interior angles for $\mathcal{P}_{m,k}$ at a vertex M_{k+j}, for $1 \leqslant k \leqslant N_b^m$, $0 < j \leqslant N_b - 1$, also results in the existence of reentrant interior angles for $\mathcal{Q}_{m,k}$ at the vertices M_{k+j-1}, $1 \leqslant k \leqslant N_b^m$, $1 < j \leqslant N_b - 1$ and M_{k+j+1}, $1 \leqslant k \leqslant N_b^m$, $0 < j \leqslant N_b - 2$.

3. Polyhedral measure on the Weierstrass IFD

Our results on fractal cohomology obtained in [**DL24c**] (see also [**DL25c**], [**DL24c**]) have highlighted the role played by specific threshold values for the number $\varepsilon_m^m > 0$ at any step $m \in \mathbb{N}$ of the prefractal graph approximation; namely, the mth *cohomology infinitesimal* introduced in Definition 3.1 just below.

DEFINITION 3.1 (**m^{th} cohomology infinitesimal** [**DL25c**], [**DL24c**] **and m^{th} intrinsic cohomology infinitesimal** [**DL25a**], [**DL23a**]). From now on, given any $m \in \mathbb{N}$, we will call m^{th} *cohomology infinitesimal* the number $\varepsilon_m^m > 0$

which also corresponds to the elementary horizontal length introduced in part $i.$ in Definition 2.13; i.e., $\varepsilon_m^m = (\varepsilon_m)^m = \dfrac{1}{N_b - 1}\dfrac{1}{N_b^m}$.

Observe that, clearly, ε_m itself – and not just ε_m^m – depends on m.

In addition, since $N_b > 1$, ε_m^m satisfies the following asymptotic behavior,

$$\varepsilon_m^m \to 0, \text{ as } m \to \infty,$$

which, naturally, results in the fact that the larger m, the smaller ε_m^m. It is for this reason that we call ε_m^m – or rather, the *sequence* $(\varepsilon_m^m)_{m=0}^\infty$ of positive numbers tending to zero as $m \to \infty$, with $\varepsilon_m^m = (\varepsilon_m)^m$, for each $m \in \mathbb{N}$ – an *infinitesimal*, called *the cohomology infinitesimal*. Note that this cohomology infinitesimal is the one naturally associated to the scaling relation of Proposition 2.2.

In the sequel, it is also useful to keep in mind that the sequence of positive numbers $(\varepsilon_m)_{m=0}^\infty$ itself satisfies

$$\varepsilon_m \sim \frac{1}{N_b}, \text{ as } m \to \infty;$$

i.e., $\varepsilon_m \to \dfrac{1}{N_b}$, as $m \to \infty$. In particular, $\varepsilon_m \not\to 0$, as $m \to \infty$, but, instead, ε_m tends to a strictly positive and finite limit.

We also introduce, given any $m \in \mathbb{N}$, the mth *intrinsic cohomology infinitesimal*, denoted by $\varepsilon^m > 0$, such that

$$\varepsilon^m = \frac{1}{N_b^m},$$

where

$$\varepsilon = \frac{1}{N_b}.$$

We call ε *the intrinsic scale*, or *intrinsic subdivision scale.*

Note that

$$\varepsilon_m^m = \frac{\varepsilon^m}{N_b - 1}$$

and that the mth intrinsic cohomology infinitesimal ε^m is asymptotic (when m tends to ∞, and up to a multiplicative constant) to the mth cohomology infinitesimal ε_m^m.

REMARK 3.2 (**Addressing numerical estimates**). From a practical point of view, an important question is the value of the ratio

$$\frac{\text{Cohomology infinitesimal}}{\text{Maximal height}} = \frac{\varepsilon_m^m}{h_m};$$

see relation (4).

Thanks to the estimates given in relation (7), we have that

$$\frac{\varepsilon_m^m}{h_m} = L_m^{1-D_{\mathcal{W}}}\,\mathcal{O}(1) = \varepsilon_m^{m(1-D_{\mathcal{W}})}\,\mathcal{O}(1),$$

with

$$0 < C_{inf} \leqslant \mathcal{O}(1) \leqslant C_{sup} < \infty.$$

Given $q \in \mathbb{N}^\star$, we then have

$$\frac{1}{10^q}\,C_{inf} \leqslant \frac{\varepsilon_m^m}{h_m} \leqslant \frac{1}{10^q}\,C_{sup}$$

when

$$\frac{C_{inf}}{10^q} \leqslant e^{(1-D_{\mathcal{W}})\ln L_m} \leqslant \frac{C_{sup}}{10^q},$$

or, equivalently, when

$$-\frac{1}{\ln N_b}\ln\left((N_b-1)\left(\frac{C_{sup}}{10^q}\right)^{\frac{1}{1-D_{\mathcal{W}}}}\right) \leqslant m \leqslant -\frac{1}{\ln N_b}\ln\left((N_b-1)\left(\frac{C_{inf}}{10^q}\right)^{\frac{1}{1-D_{\mathcal{W}}}}\right).$$

Numerical values for $N_b = 3$ and $\lambda = \frac{1}{2}$ yield:

i. For $q = 1$: $2 \leqslant m \leqslant 3$.
ii. For $q = 2$: $7 \leqslant m \leqslant 9$.
iii. For $q = 3$: $13 \leqslant m \leqslant 15$.

Hence, when m increases, the ratio decreases, and tends to 0. This numerical – but very practical and explicit argument – also accounts for our forthcoming neighborhoods, of width equal to the cohomology infinitesimal.

We refer to part iv. of Property 2.11 above, along with Figure 1, for the definition of the polygons $\mathcal{P}_{m,j}$ (resp., $\mathcal{Q}_{m,j}$) associated with the Weierstrass Curve in the next definition, as well as throughout the rest of this section. See also Definition 2.12 where the polygonal families are introduced.

DEFINITION 3.3 (**Power of a vertex of the prefractal graph $\Gamma_{\mathcal{W}_m}$, with respect to the polygonal families $\mathcal{P}_m$ and $\mathcal{Q}_m$**). Given a strictly positive integer m, a vertex X of the prefractal graph $\Gamma_{\mathcal{W}_m}$ will be said to be:

i. *of power one relative to the polygonal family $\mathcal{P}_m$* if X belongs to (to be understood in the sense that X is a vertex of) one and only one N_b-gon $\mathcal{P}_{m,j}$, $0 \leqslant j \leqslant N_b^m - 1$;
ii. *of power $\frac{1}{2}$ relative to the polygonal family $\mathcal{P}_m$* if X is a common vertex to two consecutive N_b-gons $\mathcal{P}_{m,j}$ and $\mathcal{P}_{m,j+1}$, $0 \leqslant j \leqslant N_b^m - 2$;
iii. *of power zero relative to the polygonal family $\mathcal{P}_m$* if X does not belong to (to be understood in the sense that X is not a vertex of) any N_b-gon $\mathcal{P}_{m,j}$, $0 \leqslant j \leqslant N_b^m - 1$.

Similarly, given $m \in \mathbb{N}$, a vertex X of the prefractal graph $\Gamma_{\mathcal{W}_m}$ is said to be:

i. *of power one relative to the polygonal family $\mathcal{Q}_m$* if X belongs to (as above, to be understood in the sense that X is a vertex of) one and only one N_b-gon $\mathcal{Q}_{m,j}$, $0 \leqslant j \leqslant N_b^m - 2$;
ii. *of power $\frac{1}{2}$ relative to the polygonal family $\mathcal{Q}_m$* if X is a common vertex to two consecutive N_b–gons $\mathcal{Q}_{m,j}$ and $\mathcal{Q}_{m,j+1}$, $0 \leqslant j \leqslant N_b^m - 3$;
iii. *of power zero relative to the polygonal family $\mathcal{Q}_m$* if X does not belong to (as previously, to be understood in the sense that X is not a vertex of) any N_b-gon $\mathcal{Q}_{m,j}$, $0 \leqslant j \leqslant N_b^m - 2$.

NOTATION 9. In the sequel, given a strictly positive integer m, the *power of a vertex X of the prefractal graph $\Gamma_{\mathcal{W}_m}$ relative to the polygonal families $\mathcal{P}_m$ and $\mathcal{Q}_m$* will be respectively denoted by

$$p(X, \mathcal{P}_m) \quad \text{and} \quad p(X, \mathcal{Q}_m).$$

NOTATION 10 (**Lebesgue measure (on $\mathbb{R}^2$)**). In the sequel, we denote by $\mu_{\mathcal{L}}$ the Lebesgue measure on $\mathbb{R}^2$.

NOTATION 11. For any $m \in \mathbb{N}$, and any vertex X of V_m, we set:

$$\mu^{\mathcal{L}}(X,\mathcal{P}_m,\mathcal{Q}_m) = \begin{cases} \dfrac{1}{N_b}\, p(X,\mathcal{P}_m) \displaystyle\sum_{0\leqslant j\leqslant N_b^m-1,X \text{ vertex of } \mathcal{P}_{m,j}} \mu_{\mathcal{L}}(\mathcal{P}_{m,j}), & \text{if } X \notin \mathcal{Q}_m, \\ \dfrac{1}{N_b}\, p(X,\mathcal{Q}_m) \displaystyle\sum_{1\leqslant j\leqslant N_b^m-2,X \text{ vertex of } \mathcal{Q}_{m,j}} \mu_{\mathcal{L}}(\mathcal{Q}_{m,j}), & \text{if } X \notin \mathcal{P}_m, \\ \dfrac{1}{2\,N_b}\Big\{ p(X,\mathcal{P}_m) \displaystyle\sum_{0\leqslant j\leqslant N_b^m-1,X \text{ vertex of } \mathcal{P}_{m,j}} \mu_{\mathcal{L}}(\mathcal{P}_{m,j}) + p(X,\mathcal{Q}_m)\times & \\ \displaystyle\sum_{1\leqslant j\leqslant N_b^m-2,X \text{ vertex of } \mathcal{Q}_{m,j}} \mu_{\mathcal{L}}(\mathcal{Q}_{m,j}) \Big\}, & \text{if } X \in \mathcal{P}_m \cap \mathcal{Q}_m. \end{cases}$$

NOTATION 12 (**Euclidean distance**). We hereafter denote by d_{Eucl} the Euclidean distance.

PROPERTY 3.4. *For any $m \in \mathbb{N}$, and any pair (X,Y) of adjacent vertices of V_m belonging to the same polygon $\mathcal{P}_{m,j}$, with $0 \leqslant j \leqslant N_b^m - 1$ (resp., $\mathcal{Q}_{m,j}$, with $0 \leqslant j \leqslant N_b^m - 2$), we have that*

$$d_{Eucl}(X,Y) = \sqrt{h_{j,j+1,m}^2 + L_m^2} > |h_{j,j+1,m}|,$$

which, due to the inequality given in Property 2.20, ensures that

$$\frac{1}{d_{Eucl}(X,Y)} < \frac{1}{|h_{j,j+1,m}|} \lesssim L_m^{D_{\mathcal{W}}-2} \lesssim N_b^{(2-D_{\mathcal{W}})\,m}.$$

At the same time, we also have that

$$d_{Eucl}(X,Y) \lesssim h_m \lesssim L_m^{2-D_{\mathcal{W}}} \lesssim N_b^{(D_{\mathcal{W}}-2)\,m}.$$

PROOF. This follows at once from Property 2.20. □

COROLLARY 3.5. *For any $m \in \mathbb{N}$, any integer j of $\{0,\ldots,N_b^m-1\}$, and any pair of points (X,Y) of $\mathcal{P}_{m,j}$ or of $\mathcal{Q}_{m,j}$, we have that*

$$\frac{1}{d_{Eucl}(X,Y)} \lesssim L_m^{D_{\mathcal{W}}-2} \lesssim N_b^{(2-D_{\mathcal{W}})\,m},$$

and

$$d_{Eucl}(X,Y) \lesssim h_m \lesssim L_m^{2-D_{\mathcal{W}}} \lesssim N_b^{(D_{\mathcal{W}}-2)\,m}.$$

PROPERTY 3.6. *For any $m \in \mathbb{N}$, and any vertex X of V_m:*

$$\mu^{\mathcal{L}}(X,\mathcal{P}_m,\mathcal{Q}_m) \lesssim h_m\,L_m \lesssim L_m^{3-D_{\mathcal{W}}} \lesssim N_b^{(D_{\mathcal{W}}-3)\,m},$$

and

$$\mu^{\mathcal{L}}(X,\mathcal{P}_m,\mathcal{Q}_m) \lesssim h_m\,L_m \lesssim L_m^{3-D_{\mathcal{W}}} \lesssim N_b^{(D_{\mathcal{W}}-3)\,m}.$$

PROOF. This also directly comes from Property 2.20. □

DEFINITION 3.7 (**Trace of a polygon on the Weierstrass Curve**). Given $m \in \mathbb{N}$, and $0 \leqslant j \leqslant N_b^m - 1$ (resp., $0 \leqslant j \leqslant N_b^m - 2$), and a polygon $\mathcal{P}_{m,j}$ (resp., $\mathcal{Q}_{m,j}$) of extreme vertices $\mathcal{V}_{initial}(\mathcal{P}_{m,j}) \in V_m$ and $\mathcal{V}_{end}(\mathcal{P}_{m,j}) \in V_m$ (resp., $\mathcal{V}_{initial}(\mathcal{Q}_{m,j}) \in V_m$ and $\mathcal{V}_{end}(\mathcal{Q}_{m,j}) \in V_m$; see Property 2.11), we define *the trace* of the polygon $\mathcal{P}_{m,j}$ (resp., $\mathcal{Q}_{m,j}$) on the Weierstrass Curve as the set $tr_{\gamma_{\mathcal{W}}}(\mathcal{P}_{m,j})$ (resp., $tr_{\gamma_{\mathcal{W}}}(\mathcal{Q}_{m,j})$) of points $\{\mathcal{V}_{initial}(\mathcal{P}_{m,j}), M_\star, \mathcal{V}_{end}(\mathcal{P}_{m,j})\}$ (resp., $\{\mathcal{V}_{initial}(\mathcal{Q}_{m,j}), M_\star, \mathcal{V}_{end}(\mathcal{Q}_{m,j})\}$), where we denote by $M_\star$ any point of the Weierstrass Curve strictly located between $\mathcal{V}_{initial}(\mathcal{P}_{m,j})$ and $\mathcal{V}_{end}(\mathcal{P}_{m,j})$ (resp., $\mathcal{V}_{initial}(\mathcal{Q}_{m,j})$ and $\mathcal{V}_{end}(\mathcal{Q}_{m,j})$).

DEFINITION 3.8 (**Sequence of domains delimited by the Weierstrass IFD**). We introduce *the sequence of domains delimited by the Weierstrass IFD* as the sequence $(\mathcal{D}(\Gamma_{\mathcal{W}_m}))_{m\in\mathbb{N}}$ of open, connected polygonal sets $(\mathcal{P}_m \cup \mathcal{Q}_m)_{m\in\mathbb{N}}$, where, for each $m \in \mathbb{N}$, $\mathcal{P}_m$ and $\mathcal{Q}_m$ respectively denote the polygonal sets introduced in Definition 2.12.

PROPERTY 3.9 (**Domain delimited by the Weierstrass IFD**). *We call* domain, delimited by the Weierstrass IFD, *the set, which is equal to the following limit,*

$$\mathcal{D}(\Gamma_{\mathcal{W}}) = \lim_{m\to\infty} \mathcal{D}(\Gamma_{\mathcal{W}_m}),$$

where the convergence can be interpreted in the set-theoretic sense, or in the sense of the Hausdorff metric on $\mathbb{R}^2$; see Remark 3.10 *below. In fact, we have that*

$$\mathcal{D}(\Gamma_{\mathcal{W}}) = \Gamma_{\mathcal{W}}.$$

the limit of the sequence

$$\left(\overline{\mathcal{D}}(\Gamma_{\mathcal{W}_m})\right)_{m\in\mathbb{N}}$$

PROOF. Note, first, that the sequence $(\mathcal{D}(\Gamma_{\mathcal{W}_m}))_{m\in\mathbb{N}}$ can be replaced by its closure $\left(\overline{\mathcal{D}}(\Gamma_{\mathcal{W}_m})\right)_{m\in\mathbb{N}}$, with closed polygons $\overline{\mathcal{P}}_{m,j}$ and $\overline{\mathcal{Q}}_{m,j}$, instead of open polygons $\mathcal{P}_{m,j}$ and $\mathcal{Q}_{m,j}$, used in the counterpart of Property 3.9 just above. We can then easily prove that the sequence $\left(\overline{\mathcal{D}}(\Gamma_{\mathcal{W}_m})\right)_{m\in\mathbb{N}}$ converges to the Weierstrass Curve $\Gamma_{\mathcal{W}}$. This simply comes from the fact that, given any (positive) infinitesimal $\epsilon = (\varepsilon_m^m)_{m\in\mathbb{N}}$ (in the sense of Definition 3.1), there exists an integer m_0 such that

$$\forall\, m \geqslant m_0 : \quad \overline{\mathcal{D}}(\Gamma_{\mathcal{W}_m}) \subset \mathcal{D}(\Gamma_{\mathcal{W}_m}, \varepsilon_m^m),$$

where $\mathcal{D}(\Gamma_{\mathcal{W}_m}, \varepsilon_m^m)$ denotes the (tubular) (m, ε_m^m)-neighborhood of the Weierstrass IFD introduced in [**DL25c**]; namely,

$$\mathcal{D}(\Gamma_{\mathcal{W}_m}, \varepsilon_m^m) = \left\{M \in \mathbb{R}^2,\ d(M, \Gamma_{\mathcal{W}_m}) \leqslant \varepsilon_m^m\right\}.$$

This also ensures that

$$\lim_{m\to\infty} \mu_{\mathcal{L}}\left(\overline{\mathcal{D}}(\Gamma_{\mathcal{W}_m})\right) = 0.$$

□

REMARK 3.10. In our proof, we have considered the limit of the sequence

$$\left(\overline{\mathcal{D}}(\Gamma_{\mathcal{W}_m})\right)_{m\in\mathbb{N}}$$

in the set-theoretic sense. In fact, we could have considered, instead, the Hausdorff limit of this sequence; i.e., by using the Hausdorff metric $d_{\mathcal{H}}$ on $\mathbb{R}^2$. This would not have changed our result, since

$$d_{\mathcal{H}}(\mathcal{D}(\Gamma_{\mathcal{W}_m}), \Gamma_{\mathcal{W}}) = d_{\mathcal{H}}\left(\overline{\mathcal{D}}(\Gamma_{\mathcal{W}_m}), \Gamma_{\mathcal{W}}\right).$$

As we explained in our proof just above, there exists an integer m_0 such that,

$$\forall\, m \geqslant m_0 : \quad \overline{\mathcal{D}}(\Gamma_{\mathcal{W}_m}) \subset \mathcal{D}(\Gamma_{\mathcal{W}_m}, \varepsilon_m^m).$$

This ensures, for all $m \geqslant m_0$, any j in $\{0, \ldots, N_b^m - 1\}$, any point P belonging to $\mathcal{P}_{m,j} \subset \overline{\mathcal{D}}(\Gamma_{\mathcal{W}_m})$ (resp., $Q \in \mathcal{Q}_{m,j} \subset \overline{\mathcal{D}}(\Gamma_{\mathcal{W}_m})$), and any point $M \in tr_{\gamma_{\mathcal{W}}}(\mathcal{P}_{m,j})$ (resp., $M \in tr_{\gamma_{\mathcal{W}}}(\mathcal{Q}_{m,j})$), that the Hausdorff distance $d_{\mathcal{H}}(P, M)$ (resp., $d_{\mathcal{H}}(Q, M)$) between P (resp., Q) and M is such that

$$d_{\mathcal{H}}(P, M) = \max\left\{ \underbrace{\sup_{M' \in tr_{\gamma_{\mathcal{W}}}(\mathcal{P}_{m,j})} d_{Eucl}(P, M')}_{\lesssim \varepsilon_m^m} , \underbrace{\sup_{P' \in \mathcal{P}_{m,j}} d_{Eucl}(P', M)}_{\lesssim \varepsilon_m^m} \right\} \lesssim \varepsilon_m^m$$

$\Bigg($resp.,

$$d_{\mathcal{H}}(Q, M) = \max\left\{ \underbrace{\sup_{M' \in tr_{\gamma_{\mathcal{W}}}(\mathcal{Q}_{m,j})} d_{Eucl}(Q, M')}_{\lesssim \varepsilon_m^m} , \underbrace{\sup_{Q' \in \mathcal{Q}_{m,j}} d_{Eucl}(Q', M)}_{\lesssim \varepsilon_m^m} \right\} \lesssim \varepsilon_m^m \Bigg),$$

as desired. It follows that

$$d_{\mathcal{H}}\left(\overline{\mathcal{D}}\left(\Gamma_{\mathcal{W}_m}\right), \Gamma_{\mathcal{W}}\right) \lesssim \varepsilon_m^m \underset{m \to \infty}{\to} 0.$$

Hence, $\mathcal{D}(\Gamma_{\mathcal{W}})$ is the Hausdorff limit of $(\mathcal{D}(\Gamma_{\mathcal{W}_m}))_{m \in \mathbb{N}}$, and so, by the uniqueness of such a limit, we deduce that $\mathcal{D}(\Gamma_{\mathcal{W}}) = \Gamma_{\mathcal{W}}$.

NOTATION 13 (**Minimal and maximal values of the Weierstrass function $\mathcal{W}$ on $[0,1]$**). We set

$$m_{\mathcal{W}} = \min_{t \in [0,1]} \mathcal{W}(t) = -\frac{1}{1-\lambda} \quad , \quad M_{\mathcal{W}} = \max_{t \in [0,1]} \mathcal{W}(t) = \frac{1}{1-\lambda}.$$

NOTATION 14. Henceforth, for a given $m \in \mathbb{N}$, the notation $\sum\limits_{X \in \mathcal{P}_m \bigcup \mathcal{Q}_m}$ means that the associated finite sum runs through all of the vertices of the polygons belonging to the sets $\mathcal{P}_m$ and $\mathcal{Q}_m$ introduced in Definition 2.12; see also Notation 7, following that definition.

PROPERTY 3.11. *Given a continuous function u on $[0,1] \times [m_{\mathcal{W}}, M_{\mathcal{W}}]$, we have that, for any $m \in \mathbb{N}$, and any vertex X of V_m:*

$$\left|\mu^{\mathcal{L}}(X, \mathcal{P}_m, \mathcal{Q}_m)\, u(X)\right| \leqslant \mu^{\mathcal{L}}(X, \mathcal{P}_m, \mathcal{Q}_m) \left(\max_{[0,1]\times[m_{\mathcal{W}}, M_{\mathcal{W}}]} |u| \right) \lesssim N_b^{-(3-D_{\mathcal{W}})m}.$$

Consequently, with the notation of Definition 3.1, we have that

$$\varepsilon_m^{m(D_{\mathcal{W}}-2)} \left|\mu^{\mathcal{L}}(X, \mathcal{P}_m, \mathcal{Q}_m)\, u(X)\right| \lesssim \varepsilon_m^{-m}.$$

Since the sequence $\left(\sum\limits_{X \in \mathcal{P}_m \bigcup \mathcal{Q}_m} \varepsilon_m^{-m} \right)_{m \in \mathbb{N}}$ *is a positive and increasing sequence (note that the number of vertices involved increases as m increases), this ensures the existence of the finite limit*

$$\lim_{m \to \infty} \varepsilon_m^{m(D_{\mathcal{W}}-2)} \sum_{X \in \mathcal{P}_m \bigcup \mathcal{Q}_m} \mu^{\mathcal{L}}(X, \mathcal{P}_m, \mathcal{Q}_m)\, u(X),$$

where we have used Notation 14.

PROOF. Thanks to Property 3.6, for any $m \in \mathbb{N}$, and any vertex X of V_m, we have that

$$\mu^{\mathcal{L}}(X, \mathcal{P}_m, \mathcal{Q}_m) \lesssim N_b^{(D_{\mathcal{W}}-3)\,m} \quad \text{and} \quad \mu^{\mathcal{L}}(X, \mathcal{P}_m, \mathcal{Q}_m) \lesssim N_b^{(D_{\mathcal{W}}-3)\,m}.$$

We then recall from Section 2 that, for any $m \in \mathbb{N}$, the total number of polygons $\mathcal{P}_m$ is N_b^m, while the total number of polygons $\mathcal{Q}_m$ is equal to $N_b^m - 1$; see Property 2.11. We then have that

$$\sum_{X \in \mathcal{P}_m \bigcup \mathcal{Q}_m} \mu^{\mathcal{L}}(X, \mathcal{P}_m, \mathcal{Q}_m) \lesssim N_b^m\, N_b^{(D_{\mathcal{W}}-3)\,m};$$

i.e.,

$$\sum_{X \in \mathcal{P}_m \bigcup \mathcal{Q}_m} \mu^{\mathcal{L}}(X, \mathcal{P}_m, \mathcal{Q}_m) \lesssim N_b^{(D_{\mathcal{W}}-2)\,m},$$

or, equivalently, due to the relation between the mth *cohomology infinitesimal* ε_m^m introduced in Definition 3.1, and the Weierstrass parameter N_b,

$$\sum_{X \in \mathcal{P}_m \bigcup \mathcal{Q}_m} \mu^{\mathcal{L}}(X, \mathcal{P}_m, \mathcal{Q}_m) \lesssim \varepsilon_m^{m\,(2-D_{\mathcal{W}})},$$

which, as desired, ensures the existence of the finite limit

$$\left(\max_{[0,1]\times[m_{\mathcal{W}}, M_{\mathcal{W}}]} |u|\right) \lim_{m\to\infty} \varepsilon_m^{m\,(D_{\mathcal{W}}-2)} \sum_{X \in \mathcal{P}_m \bigcup \mathcal{Q}_m} \mu^{\mathcal{L}}(X, \mathcal{P}_m, \mathcal{Q}_m).$$

□

PROPERTY 3.12 (**Polyhedral measure on the Weierstrass IFD**). *We introduce* the polyhedral measure on the Weierstrass IFD, *denoted by* μ, *such that for any continuous function* u *on the Weierstrass Curve, with the use of Notation* 11, *and Notation* 14,

$$\int_{\Gamma_{\mathcal{W}}} u\, d\mu = \lim_{m\to\infty} \varepsilon_m^{m\,(D_{\mathcal{W}}-2)} \sum_{X \in \mathcal{P}_m \bigcup \mathcal{Q}_m} \mu^{\mathcal{L}}(X, \mathcal{P}_m, \mathcal{Q}_m)\, u(X), \quad (\star) \tag{8}$$

which, thanks to Property 3.9, *can also be understood in the following way,*

$$\int_{\Gamma_{\mathcal{W}}} u\, d\mu = \int_{\mathcal{D}(\Gamma_{\mathcal{W}})} u\, d\mu.$$

REMARK 3.13. In a sense, our polyhedral measure can be seen as a measure which is an extension of the Riemann integral, where the step functions are replaced by upper and lower affine functions which approximate the Weierstrass Curve.

THEOREM 3.14. *The polyhedral measure* μ *is well defined, positive, as well as a bounded, nonzero, Borel measure on* $\mathcal{D}(\Gamma_{\mathcal{W}})$. *The associated total mass is given by*

$$\mu(\mathcal{D}(\Gamma_{\mathcal{W}})) = \mu(\Gamma_{\mathcal{W}}) = \lim_{m\to\infty} \varepsilon_m^{m\,(D_{\mathcal{W}}-2)} \sum_{X \in \mathcal{P}_m \bigcup \mathcal{Q}_m} \mu^{\mathcal{L}}(X, \mathcal{P}_m, \mathcal{Q}_m), \quad (\star\star) \tag{9}$$

and satisfies the following estimate:

$$\mu(\mathcal{D}(\Gamma_{\mathcal{W}})) \leqslant \frac{2}{N_b}(N_b - 1)^2\, C_{sup}. \quad (\star\star\star) \tag{10}$$

Furthermore, the support of μ *coincides with the entire curve:*

$$\operatorname{supp} \mu = \mathcal{D}(\Gamma_{\mathcal{W}}) = \Gamma_{\mathcal{W}}.$$

In addition, μ is the weak limit as $m \to \infty$ of the following discrete measures *(or Dirac Combs), given, for each $m \in \mathbb{N}$, by*

$$\mu_m = \varepsilon_m^{m\,(D_{\mathcal{W}}-2)} \sum_{X \in \mathcal{P}_m \bigcup \mathcal{Q}_m} \mu^{\mathcal{L}}\left(X, \mathcal{P}_m, \mathcal{Q}_m\right) \delta_X,$$

where ε_m^m denotes the mth cohomology infinitesimal introduced in Definition 3.1, *δ_X is the Dirac measure concentrated at X, and we have used Notation* 11, *for $\mu^{\mathcal{L}}\left(X, \mathcal{P}_m, \mathcal{Q}_m\right)$, along with Notation* 14.

PROOF. For the sake of simplicity, we restrict ourselves to the case when $N_b < 7$. When $N_b \geqslant 7$, there are reentrant interior angles (see Property 2.26), which result in the fact that for $m \in \mathbb{N}^\star$ and $0 \leqslant j \leqslant N_b^m - 1$, each polygon $\mathcal{P}_{m,j}$ is not necessarily convex. This significantly lengthens the corresponding proof which, however, consists of a mere, but cumbersome adaptation to this configuration of the present proof.

i. μ is a well-defined measure.

Indeed, according to Proposition 3.11, the map φ

$$u \mapsto \varphi(u) = \int_{\Gamma_{\mathcal{W}}} u \, d\mu$$

is a well-defined linear functional on the space $C\left(\Gamma_{\mathcal{W}}\right)$ of real-valued, continuous functions on $\Gamma_{\mathcal{W}}$. Hence, by a well-known argument, it is a continuous linear functional on $C\left(\Gamma_{\mathcal{W}}\right)$, equipped with the *sup* norm. Since $\Gamma_{\mathcal{W}}$ is compact, and in light of relation (8) in Property 3.12, μ is a bounded, Radon measure, with total mass $\varphi(1) = \mu\left(\mathcal{D}\left(\Gamma_{\mathcal{W}}\right)\right)$, also given by inequality (9), and where 1 denotes the constant function equal to 1 on $\Gamma_{\mathcal{W}}$.

Then, according to the Riesz representation theorem, the associated positive Borel measure (still denoted by μ) is a bounded and positive Borel measure with the same total mass $\mu\left(\mathcal{D}\left(\Gamma_{\mathcal{W}}\right)\right) = \mu\left(\Gamma_{\mathcal{W}}\right)$.

ii. The nonzero measure – Estimates for the total mass of μ.

For $0 \leqslant j \leqslant N_b^m - 1$, each polygon $\mathcal{P}_{m,j}$ is contained in a rectangle of height at most equal to $(N_b - 1)\, h_m$ (where h_m is the maximal height introduced in part *iv.* of Definition 2.13), and of width at most equal to $(N_b - 1)\, L_m$. This ensures that the Lebesgue measure of each polygon $\mathcal{P}_{m,j}$ is at most equal to $(N_b - 1)^2\, h_m\, L_m$. We now recall that, thanks to Property 2.20, for any $m \in \mathbb{N}$, we have the following estimate:

$$h_m \leqslant C_{sup}\, L_m^{2-D_{\mathcal{W}}},$$

where

$$C_{sup} = (N_b - 1)^{2-D_{\mathcal{W}}} \left(\max_{0 \leqslant j \leqslant N_b - 1} \left| \mathcal{W}\left(\frac{j+1}{N_b - 1}\right) - \mathcal{W}\left(\frac{j}{N_b - 1}\right) \right| + \frac{2\,\pi}{(N_b - 1)\,(\lambda\, N_b - 1)} \right).$$

Consequently, the Lebesgue measure $\mu_{\mathcal{L}}\left(\mathcal{P}_{m,j}\right)$ of each polygon $\mathcal{P}_{m,j}$ is such that

$$\mu_{\mathcal{L}}\left(\mathcal{P}_{m,j}\right) \leqslant (N_b - 1)^2\, C_{sup}\, L_m^{3-D_{\mathcal{W}}}. \tag{11}$$

In the same way, for $0 \leqslant j \leqslant N_b^m - 2$, the Lebesgue measure $\mu_{\mathcal{L}}\left(\mathcal{P}_{m,j}\right)$ of each polygon $\mathcal{Q}_{m,j}$ is such that

$$\mu_{\mathcal{L}}\left(\mathcal{Q}_{m,j}\right) \leqslant (N_b - 1)^2\, C_{sup}\, L_m^{3-D_{\mathcal{W}}}. \tag{12}$$

We then deduce that, for any vertex X of V_m,

$$\mu(X, \mathcal{P}_m, \mathcal{Q}_m) \leqslant \frac{1}{N_b}(N_b - 1)^2 C_{sup} L_m^{3-D_{\mathcal{W}}}.$$

Hence, since the total number of polygons involved is at most equal to $2N_b^m - 1 \leqslant 2N_b^m$, we can deduce that

$$\sum_{X \in \mathcal{P}_m \bigcup \mathcal{Q}_m} \mu^{\mathcal{L}}(X, \mathcal{P}_m, \mathcal{Q}_m) \leqslant \frac{2N_b^m}{N_b}(N_b - 1)^2 C_{sup} L_m^{3-D_{\mathcal{W}}},$$

or, equivalently,

$$\sum_{X \in \mathcal{P}_m \bigcup \mathcal{Q}_m} \mu^{\mathcal{L}}(X, \mathcal{P}_m, \mathcal{Q}_m) \leqslant 2\frac{\varepsilon^{-m}}{N_b}(N_b - 1)^2 C_{sup}\, \varepsilon_m^{m(3-D_{\mathcal{W}})}.$$

We then have that

$$\varepsilon_m^{m(D_{\mathcal{W}}-2)} \sum_{X \in \mathcal{P}_m \bigcup \mathcal{Q}_m} \mu^{\mathcal{L}}(X, \mathcal{P}_m, \mathcal{Q}_m) \leqslant \frac{2}{N_b}(N_b - 1)^2 C_{sup} < \infty,$$

from which we can deduce that the polyhedral measure is a bounded measure.

For $0 \leqslant j \leqslant N_b^m - 1$, each polygon $\mathcal{P}_{m,j}$ (which is convex) contains an inscribed circle, whose Lebesgue measure is greater than $\dfrac{h_m^{inf} L_m}{C_{N_b}}$, where h_m^{inf} is the minimal height introduced in part *iv.* of Definition 2.13, and where C_{N_b} is a strictly positive constant, which depends on the value of the integer N_b (depending on the number of sides of the polygon, i.e., depending on the value of this integer, we can express the radius of this circle in function of the side lengths of the polygon). We now recall that, thanks to Property 2.20, for any $m \in \mathbb{N}$, we have the following estimate:

$$C_{inf} L_m^{2-D_{\mathcal{W}}} \leqslant h_m^{inf},$$

where

$$C_{inf} = (N_b - 1)^{2-D_{\mathcal{W}}} \min_{0 \leqslant j \leqslant N_b - 1,\, \mathcal{W}\left(\frac{j+1}{N_b-1}\right) \neq \mathcal{W}\left(\frac{j}{N_b-1}\right)} \left| \mathcal{W}\left(\frac{j+1}{N_b - 1}\right) - \mathcal{W}\left(\frac{j}{N_b - 1}\right) \right|.$$

Consequently, the Lebesgue measure $\mu_{\mathcal{L}}(\mathcal{P}_{m,j})$ of each polygon $\mathcal{P}_{m,j}$ is such that

$$\mu_{\mathcal{L}}(\mathcal{P}_{m,j}) \geqslant \frac{h_m^{inf} L_m}{C_{N_b}} \geqslant \frac{C_{inf} L_m^{3-D_{\mathcal{W}}}}{C_{N_b}}.$$

In the same way, for $0 \leqslant j \leqslant N_b^m - 2$, the Lebesgue measure $\mu_{\mathcal{L}}(\mathcal{P}_{m,j})$ of each polygon $\mathcal{Q}_{m,j}$ is such that

$$\mu_{\mathcal{L}}(\mathcal{Q}_{m,j}) \geqslant \frac{h_m^{inf} L_m}{C_{N_b}} \geqslant \frac{C_{inf} L_m^{3-D_{\mathcal{W}}}}{C_{N_b}}.$$

We then deduce that, for any vertex X of V_m,

$$\mu(X, \mathcal{P}_m, \mathcal{Q}_m) \geqslant \frac{1}{N_b} \frac{C_{inf} L_m^{3-D_{\mathcal{W}}}}{C_{N_b}}.$$

Hence, since the total number of polygons involved is greater than $N_b^m - 1 \geqslant \dfrac{N_b^m}{2}$, we can deduce that

$$\sum_{X \in \mathcal{P}_m \bigcup \mathcal{Q}_m} \mu^{\mathcal{L}}(X, \mathcal{P}_m, \mathcal{Q}_m) \geqslant \frac{N_b^m}{2} \frac{C_{inf} L_m^{3-D_{\mathcal{W}}}}{N_b C_{N_b}},$$

or, equivalently,

$$\sum_{X \in \mathcal{P}_m \bigcup \mathcal{Q}_m} \mu^{\mathcal{L}}(X, \mathcal{P}_m, \mathcal{Q}_m) \geqslant \frac{\varepsilon_m^{-m}}{2(N_b - 1)} \frac{C_{inf} \varepsilon_m^{m(3-D_{\mathcal{W}})}}{N_b C_{N_b}}.$$

We then have that

$$\varepsilon_m^{m(D_{\mathcal{W}}-2)} \sum_{X \in \mathcal{P}_m \bigcup \mathcal{Q}_m} \mu^{\mathcal{L}}(X, \mathcal{P}_m, \mathcal{Q}_m) \geqslant \frac{1}{2(N_b - 1)} \frac{C_{inf}}{N_b C_{N_b}} > 0,$$

from which, upon passing to the limit when $m \to \infty$, we can deduce that the polyhedral measure is a nonzero measure, and that its total mass satisfies inequality (10).

iii. The support of μ coincides with the entire curve $\Gamma_{\mathcal{W}}$.

This simply comes from the proof given in *ii.* just above that the measure μ is nonzero. In the case of a positive, continuous function u defined on the Weierstrass Curve, we have that

$$\varepsilon_m^{m(D_{\mathcal{W}}-2)} \sum_{X \in \mathcal{P}_m \bigcup \mathcal{Q}_m} \mu^{\mathcal{L}}(X, \mathcal{P}_m, \mathcal{Q}_m)\, u(X) \geqslant \frac{1}{2(N_b - 1)} \frac{C_{inf}}{N_b C_{N_b}} \left(\min_{\Gamma_{\mathcal{W}}} |u|\right) > 0.$$

Hence, upon passing to the limit when $m \to \infty$, we deduce that

$$\varphi(u) = \int_{\Gamma_{\mathcal{W}}} u \, d\mu > 0,$$

and thus, $\varphi(u) \neq 0$, from which the claim follows easily.

Indeed, otherwise, if $\operatorname{supp} \mu \neq \Gamma_{\mathcal{W}}$, there exists $M \in \Gamma_{\mathcal{W}} \setminus \operatorname{supp} \mu$, and thus, by Urisohn's lemma (see, e.g., [**Fol99**], [**Rud87**] or [**Rud91**]), there exists $u \in C(\Gamma_{\mathcal{W}})$ and an open neighborhood $\mathcal{V}(M)$ of M in $\Gamma_{\mathcal{W}}$ disjoint from $\operatorname{supp} \mu$ and such that

$$u(M) = 1 \quad , \quad 0 \leqslant u \leqslant 1 \quad , \quad \text{and } u_{|\Gamma_{\mathcal{W}} \setminus \mathcal{V}(M)} = 0.$$

Hence, by the above argument, $\varphi(u) \neq 0$, which contradicts the fact that $M \notin \operatorname{supp} \mu$ (see, e.g., *loc. cit.*).

iv. μ is a singular measure.

First, note that

$$\mu^{\mathcal{L}}(\Gamma_{\mathcal{W}}) = 0,$$

because $D_{\mathcal{W}} < 2$, and, up to a multiplicative positive constant, $\mu^{\mathcal{L}}$ coincides with the 2-dimensional measure on $\mathbb{R}^2$. Now, since $\operatorname{supp} \mu \subset \Gamma_{\mathcal{W}}$, and $\mu^{\mathcal{L}}(\Gamma_{\mathcal{W}}) = 0$, it follows that μ is supported on a set of Lebesgue measure zero, which precisely implies that μ (viewed as a Borel measure on the rectangle $[0,1] \times [m_{\mathcal{W}}, M_{\mathcal{W}}]$ in the obvious way), is singular with respect to the restriction of $\mu^{\mathcal{L}}$ to this rectangle.

v. μ is the weak limit of the discrete measures μ_m.

Indeed, this follows at once from the latter part of Property 3.11, according to which, for every $u \in \mathcal{C}(\Gamma_{\mathcal{W}})$,

$$\int_{\Gamma_{\mathcal{W}}} u\, d\mu = \lim_{m \to \infty} \int_{\Gamma_{\mathcal{W}}} u\, d\mu_m,$$

as desired.

This completes the proof of Theorem 3.14. □

REMARK 3.15. This choice of measure is obtained in the same manner as the one we introduced while working on the Arrowhead Curve [**Dav20**]. In our present case, however we define the measure more precisely, as well as establish several new properties.

Considering a two-dimensional measure is both essential and natural in so far as we will consider geometric conditions and two-dimensional nets in Section 4.

4. Atomic decompositions – trace theorems, and consequences

We begin this section by discussing Markov's Inequality and a closely related smoothness condition. Our goal, in this first part of Section 4, is to show that the Weierstrass Curve satifsfies Markov's Inequality (see Property 4.5).

DEFINITION 4.1 (**Markov's Inequality on a subset $\mathcal{F}$ of $\mathbb{R}^n$ (see [JW84], Chapter II, Definition 2, on pages** 34-35)). A closed, nonempty subset $\mathcal{F}$ of $\mathbb{R}^n$ preserves Markov's Inequality if, for all positive integers N, any real polynomial P with d_P variables, of degree at mots equal to N, any point $X \in \mathcal{F}$, and any real number $r \in]0,1]$:

$$\max_{\mathcal{F} \cap \mathcal{B}(X,r)} |\nabla P| \leqslant \frac{c(P,n,N,\mathcal{F})}{r} \max_{\mathcal{F} \cap \mathcal{B}(X,r)} |P|,$$

where $c(P,n,N,\mathcal{F})$ denotes a positive constant that depend on n, P, N and $\mathcal{F}$.

DEFINITION 4.2 (**Smoothness geometric condition (see [JW84], Chapter II, Theorem 2, on page** 38)). A closed, nonempty subset $\mathcal{F}$ of $\mathbb{R}^n$ satisfies the *smoothness geometric condition* $\mathcal{G}_{smooth}$ if, for any strictly positive number ϵ, there exists a point $X \in \mathcal{F}$, a real number $r \in]0,1]$, and a normed vector $b \in \mathbb{R}^n$ such that the intersection

$$\mathcal{F} \cap \mathcal{B}(X,r)$$

is contained in a band of the type

$$\left\{x \in \mathbb{R}^d,\ |b.(x-X)| < \varepsilon\, r\right\}.$$

REMARK 4.3. A smooth curve satisfies the geometric condition $\mathcal{G}_{smooth}$. In a sense, the Markov inequality means that locally, the curve is sufficiently flat.

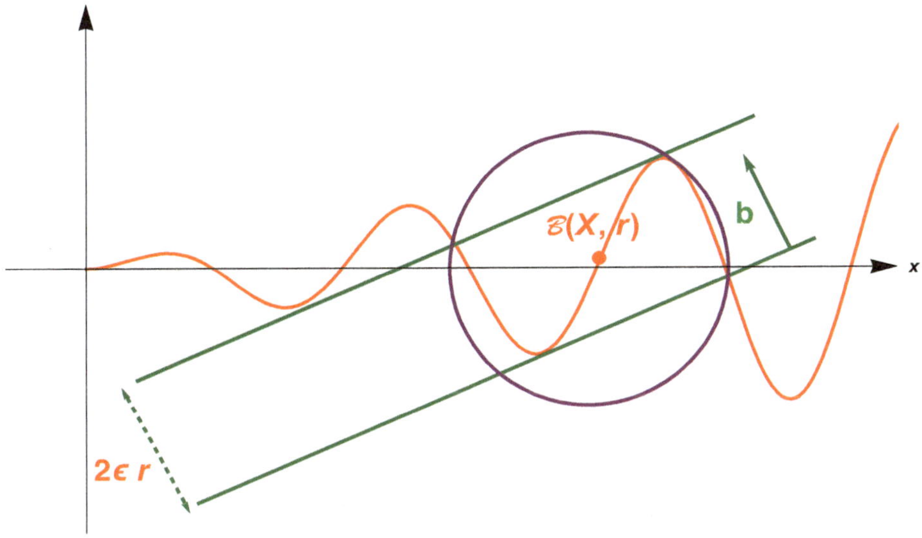

FIGURE 6. A smooth curve satisfying the geometric condition $\mathcal{G}_{smooth}$

THEOREM 4.4 (**Necessary and sufficient condition for a subset $\mathcal{F}$ of $\mathbb{R}^n$ to satisfy Markov's Inequality (see [JW84], Chapter II, Theorem 2, on pages 38)**). *A closed, nonempty subset $\mathcal{F}$ of $\mathbb{R}^n$ preserves Markov's Inequality if and only if, for every strictly positive number ϵ, the geometric condition $\mathcal{G}_{smooth}$ does not hold.*

PROOF. The following proof is inspired by the proof given in [**JW84**], Chapter II, on pages 38-39, but rewritten and containing much more detail and precision.

i. Let us assume that the geometric condition $\mathcal{G}_{smooth}$ holds. Then, for all $\epsilon > 0$, there exists $X = (x_1, \dots, x_n) \in \mathcal{F}$, along with $r \in]0, 1]$ and a unit vector $b = (b_1 \dots, b_n) \in \mathbb{R}^n$ such that

$$\mathcal{F} \cap \mathcal{B}(X, r) \subset \left\{x \in \mathbb{R}^n,\ |b \cdot (x - X)| < \epsilon\, r\right\},$$

which of course, in the case when $\epsilon < 1$ implies that

$$\mathcal{F} \cap \mathcal{B}(X, r) \subset \left\{x \in \mathbb{R}^n,\ |b \cdot (x - X)| \leqslant \epsilon < 1\right\}.$$

By considering the polynomial $P \in \mathbb{R}[X_1, \dots, X_n]$ such that

$$P(x) = b \cdot (x - X) = b_1\,(x_1 - X_1) + \dots + b_n\,(x_n - X_n),$$

we obtain that

$$0 < \max_{\mathcal{F} \cap \mathcal{B}(X,r)} |P| \leqslant \max_{x \in \mathbb{R}^n} |b \cdot (x - X)| < 1.$$

At the same time, we have that

$$\forall\, x = (x_1, \dots, x_n) \in \mathbb{R}^n \ : \quad \nabla P(x) = b,$$

which, since b is a unit vector (i.e., $|b| = 1$) ensures that

$$\max_{\mathcal{F} \cap \mathcal{B}(X,r)} |\nabla P| = |b| = 1 > \max_{\mathcal{F} \cap \mathcal{B}(X,r)} |P|.$$

Consequently, there cannot exist any positive constant $c(P,n,N,\mathcal{F})$ depending on n, P, N and $\mathcal{F}$ such that

$$\max_{\mathcal{F}\cap\mathcal{B}(X,r)} |\nabla P| \leqslant \frac{c(P,n,N,\mathcal{F})}{r} \max_{\mathcal{F}\cap\mathcal{B}(X,r)} |P|.$$

Indeed, note that, in such a case, we would have that

$$\max_{\mathcal{F}\cap\mathcal{B}(X,r)} |P| \geqslant \frac{r}{c(P,n,N,\mathcal{F})} \max_{\mathcal{F}\cap\mathcal{B}(X,r)} |\nabla P| \geqslant \frac{\max_{\mathcal{F}\cap\mathcal{B}(X,r)} |\nabla P|}{c(P,n,N,\mathcal{F})} > \frac{\max_{\mathcal{F}\cap\mathcal{B}(X,r)} |P|}{c(P,n,N,\mathcal{F})},$$

where $\max_{\mathcal{F}\cap\mathcal{B}(X,r)} |P| > 0$.

ii. Conversely, let us assume the existence of a strictly positive real number ϵ such that the geometric condition $\mathcal{G}_{smooth}$ does not hold.

By considering a first degree polynomial $P \in \mathbb{R}[X_1,\ldots,X_n]$ such that

$$\forall\, x=(x_1,\ldots,x_n)\in\mathbb{R}^n:\quad P(x) = b\cdot(x-X_0)+a = \sum_{j=1}^{n} b_j\,(x_j - X_{j,0}) + a,$$

where $b=(b_1\ldots,b_n)\in\mathbb{R}^n$ is a unit vector, $a\in\mathbb{R}$ and $X_0=\left(X_{0,1}\ldots,X_{0,n}\right)$ belongs to $\mathcal{F}$, we then obtain that

$$\forall\, x=(x_1,\ldots,x_n)\in\mathbb{R}^n:\quad \nabla P(x) = b$$

and, obviously,

$$\forall\, x=(x_1,\ldots,x_n)\in\mathbb{R}^n:\quad |\nabla P(x)| = |b| = 1.$$

Since the geometric condition $\mathcal{G}_{smooth}$ does not hold, there exists $\epsilon>0$ such that, given $r\in\,]0,1]$, we can then find a point

$$X_1 = \left(X_{1,1}\ldots,X_{1,n}\right)\in\mathcal{F}\cap\mathcal{B}(X_0,r),$$

satisfying

$$|b\cdot(X_1-X_0)| \geqslant \epsilon\, r,$$

i.e., equivalently,

$$|P(X_1)-P(X_0)| \geqslant \epsilon\, r.$$

Consequently,

$$|P(X_1)| \geqslant \epsilon\, r + |P(X_0)| = \epsilon\, r + |a| \Rightarrow \frac{1}{r}\,|P(X_1)| \geqslant \epsilon + \frac{|a|}{r}.$$

We then have to consider the following two configurations:

⇝ $\frac{r}{k} \leqslant |a| \leqslant r$, where $k>1$:

In this case, we have that

$$\frac{1}{r}\,|P(X_1)| \geqslant \epsilon + \frac{|a|}{r} \geqslant \epsilon + \frac{1}{k}.$$

We then obtain that

$$\frac{1}{\left(\varepsilon+\frac{1}{k}\right)r}\max_{\mathcal{F}\cap\mathcal{B}(X,r)} |P| \geqslant \frac{1}{\left(\varepsilon+\frac{1}{k}\right)r}\,|P(X_1)| \geqslant 1,$$

or, equivalently,

$$\frac{1}{\left(\varepsilon+\frac{1}{k}\right)r}\max_{\mathcal{F}\cap\mathcal{B}(X,r)} |P| \geqslant \underbrace{\max_{\mathcal{F}\cap\mathcal{B}(X,r)} |\nabla P|}_{1},$$

in which case the inequality holds.

⇝ $|a| > r$:

In this case, we have that

$$\frac{1}{r}\,|P(X_1)| \geqslant \varepsilon + 1.$$

We then have that

$$\frac{1}{(1+\varepsilon)\,r}\max_{\mathcal{F}\cap\mathcal{B}(X,r)}|P| \geqslant \frac{1}{(1+\varepsilon)\,r}\,|P(X_1)| \geqslant 1,$$

or, equivalently,

$$\frac{1}{(1+\varepsilon)\,r}\max_{\mathcal{F}\cap\mathcal{B}(X,r)}|P| \geqslant \underbrace{\max_{\mathcal{F}\cap\mathcal{B}(X,r)}|\nabla P|}_{1},$$

in which case the inequality also holds.

The remainder of the proof, i.e., when $P \in \mathbb{R}[X_1,\dots,X_n]$ is not a first degree polynomial $P \in \mathbb{R}[X_1,\dots,X_n]$, is obtained by induction. For the sake of concision, we only present the case when P is of degree $N \geqslant 2$ and is of the following form:

$$\forall\, x = (x_1,\dots,x_n) \in \mathbb{R}^n : \quad P(x) = (b\cdot(x-X_0))^N,$$

where $b = (b_1\dots,b_n) \in \mathbb{R}^n$ is a unit vector and $X_0 = (X_{0,1}\dots,X_{0,n})$ belongs to $\mathcal{F}$. In this case, we have that

$$\forall\, x = (x_1,\dots,x_n) \in \mathbb{R}^n : \quad \nabla P(x) = N\,(b\cdot(x-X_0))^{N-1}\,b$$

and, consequently,

$$\forall\, x = (x_1,\dots,x_n) \in \mathbb{R}^n : \quad |\nabla P(x)| = N\,|b\cdot(x-X_0)|^{N-1}.$$

Again, since the geometric condition $\mathcal{G}_{smooth}$ does not hold, there exists $\epsilon > 0$ such that, given $r \in]0,1]$, we can then find a point

$$X_1 = (X_{1,1}\dots,X_{1,n}) \in \mathcal{F}\cap\mathcal{B}(X_0,r)$$

satisfying

$$|b\cdot(X_1-X_0)| \geqslant \epsilon\, r.$$

We then have that, for all $x = (x_1,\dots,x_n) \in \mathcal{F}\cap\mathcal{B}(X_0,r)$,

$$\max_{\mathcal{F}\cap\mathcal{B}(X_0,r)}|P| \geqslant |b\cdot(X_1-X_0)|\;|b\cdot(x-X_0)|^{N-1},$$

which ensures that

$$\max_{\mathcal{F}\cap\mathcal{B}(X_0,r)}|P| \geqslant |b\cdot(X_1-X_0)|\max_{\mathcal{F}\cap\mathcal{B}(X_0,r)}|b\cdot(x-X_0)|^{N-1},$$

i.e., equivalently,

$$\max_{\mathcal{F}\cap\mathcal{B}(X_0,r)}|P| \geqslant \frac{\epsilon\, r}{N}\max_{\mathcal{F}\cap\mathcal{B}(X_0,r)}|\nabla P|,$$

in which case the inequality holds.

When, more generally, P is of degree $N \geqslant 2$, of the following form: $\forall\, x = (x_1,\dots,x_n) \in \mathbb{R}^n$:

$$P(x) = \sum_{k=0}^{N}\beta_k\,(b_k\cdot(x-X_0))^k = \sum_{k=0}^{N}\beta_k\,(b_k\cdot(x-X_0))^k,$$

where, for $0 \leqslant k \leqslant N$, β_k denotes a real number, while $b_k = \left(b_{k,1} \ldots, b_{k,n}\right) \in \mathbb{R}^n$ is a unit vector and $X_0 = \left(X_{0,1} \ldots, X_{0,n}\right)$ belongs to $\mathcal{F}$, we apply the result given in [**JW84**], on page 37 (Proposition 4), which states that if Markov's Inequality is valid for all polynomials P of degree 1, then $\mathcal{F}$ preserves Markov's Inequality. It is based on the fact that, since the geometric condition $\mathcal{G}_{smooth}$ does not hold, for $0 \leqslant k \leqslant N$, there exists $\epsilon > 0$ such that, given $r \in]0,1]$, we can then find a point

$$X_1 = \left(X_{1,k} \ldots, X_{1,n}\right) \in \mathcal{F} \cap \mathcal{B}(X_0, r)$$

satisfying

$$|b_k \cdot (X_1 - X_0)| \geqslant \epsilon\, r$$

and hence, obviously,

$$|\beta_k|\, |b_k \cdot (X_1 - X_0)|^k \geqslant |\beta_k|\, \epsilon^k\, r^k .$$

This ensures that

$$\max_{\mathcal{F}\cap\mathcal{B}(X_0,r)} |P| \geqslant \sum_{k=0}^{N} |\beta_k|\, |b_k \cdot (X_1 - X_0)|^k \geqslant \sum_{k=0}^{N} |\beta_k|\, \epsilon^k\, r^k .$$

We also have that

$$\max_{\mathcal{F}\cap\mathcal{B}(X_0,r)} |P| \lesssim \sum_{k=0}^{N} |\beta_k|\, r^k .$$

Since, at the same time,

$$\forall\, x = (x_1, \ldots, x_n) \in \mathbb{R}^n : \quad \nabla P(x) = \sum_{k=0}^{N} k\, \left(b_k \cdot (x - X_{0,k})\right)^{k-1} \beta_k$$

and, consequently,

$$\forall\, x = (x_1, \ldots, x_n) \in \mathbb{R}^n : \quad |\nabla P(x)| \leqslant \sum_{k=0}^{N} k\, |\beta_k|\, |b_k \cdot (x - X_{0,k})|^{k-1} ,$$

we have, in the same manner, that

$$\max_{\mathcal{F}\cap\mathcal{B}(X_0,r)} |\nabla P| \lesssim \sum_{k=0}^{N} k\, |\beta_k|\, r^{k-1} \lesssim \frac{1}{r} \max_{\mathcal{F}\cap\mathcal{B}(X_0,r)} |P| ,$$

which is the desired result. □

PROPERTY 4.5 (**The Weierstrass Curve satisfies Markov's Inequality**). *The Weierstrass Curve does not satisfy the geometric condition $\mathcal{G}_{smooth}$ and, therefore, satisfies Markov's Inequality.*

REMARK 4.6. A general result that could directly be applied to the Weierstrass Curve is given in the book by A. Jonsson and H. Wallin ([**JW84**], page 39):

THEOREM 4.7. *If $\mathcal{F} \subset \mathbb{R}^n$ is a d-dimensional set such that $d > n - 1$, then $\mathcal{F}$ preserves Markov's Inequality.*

Yet, the proof requires the use of d-measures. We propose a specific geometric proof that applies to fractal curves.

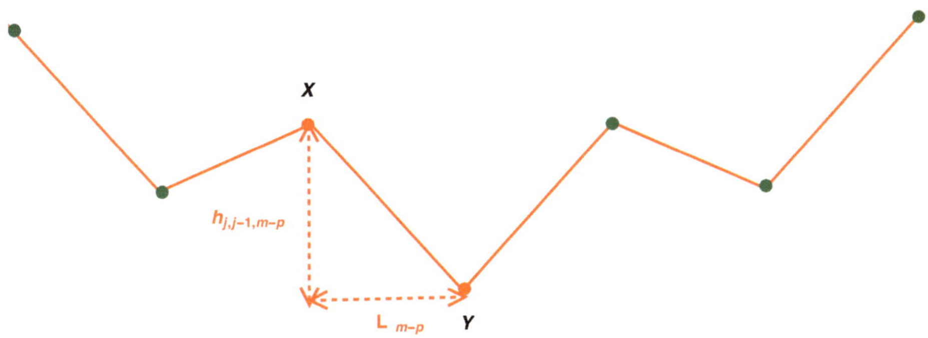

FIGURE 7. The adjacent vertices $X \in V_{\mathcal{W}_{m+p}}$ and $Y \in V_{\mathcal{W}_{m+p}}$.

PROOF. Building upon our results on the polygonal neighborhoods of the Weierstrass Curve (see [**DL25a**]), we can restrict ourseves to the mth prefractal approximations $\Gamma_{\mathcal{W}_m}$, for $m \in \mathbb{N}^\star$ sufficienly large. Indeed, thanks to Property 2.9, the set $V^\star = \bigcup_{n\in\mathbb{N}} V_n$ is dense in the Weierstrass Curve $\Gamma_{\mathcal{W}_m}$.

Let us consider a point $X = X_m$ belonging to the set of vertices V_m of the mth prefractal approximation $\Gamma_{\mathcal{W}_m}$, for $m \in \mathbb{N}^\star$ sufficiently large. We presently aim at determining a strictly positive number ϵ such that the intersection $\mathcal{B}(X_m, L_m) \cap \Gamma_{\mathcal{W}_m}$ (where L_m denotes the elementary horizontal length, introduced in part i. of Property 2.13) cannot be contained in a band of the type

$$D = \left\{x \in \mathbb{R}^2,\ |b\cdot(x-X)| < \epsilon\, L_m\right\} = \left\{x \in \mathbb{R}^2,\ |b\cdot(x-X_m)| < \epsilon\, L_m\right\},$$

where b denotes a unit vector of $\mathbb{R}^2$; see Figure 8.

Intuitively, one understands that the geometric condition cannot hold, because of the specific properties of the maps T_i, $0 \leqslant i \leqslant N_b - 1$, which extend distances vertically (see [**Dav19**]).

Thus, the simplest way to show that the condition cannot be satisfied could be achieved by finding a nonnegative integer p and an adjacent vertex $Y \in V_{\mathcal{W}_{m+p}}$ of X such that Y does not belong to the band D; see Figure 7.

The Euclidean distance $d_{Eucl}(X,Y)$ is given by

$$d_{Eucl}(X,Y) = \sqrt{h^2_{j,j+1,m+p} + L^2_{m+p}} < \sqrt{h^2_{m+p} + L^2_{m+p}} < \sqrt{2}\, h_{m+p},$$

where h_{m+p} is the $(m+p)^{th}$ maximal height introduced in relation (4).

Thanks to inequality (5), there exists a nonnegative integer p such that

$$\sqrt{2}\, h_{m+p} < L_m,$$

since

$$\frac{h_{m+p}}{L_m} \leqslant \frac{C_{sup}\, L_{m+p}^{2-D_{\mathcal{W}}}}{L_m},$$

where the constant C_{sup} is the constant involved in Property 2.20, and where

$$\frac{L_{m+p}^{2-D_{\mathcal{W}}}}{L_m} = \frac{L_m^{1-D_{\mathcal{W}}}}{N_b^{(3-D_{\mathcal{W}})\,p}} = \frac{1}{(N_b-1)^{1-D_{\mathcal{W}}}\, N_b^{(3-D_{\mathcal{W}})\,p+(1-D_{\mathcal{W}})\,m}}.$$

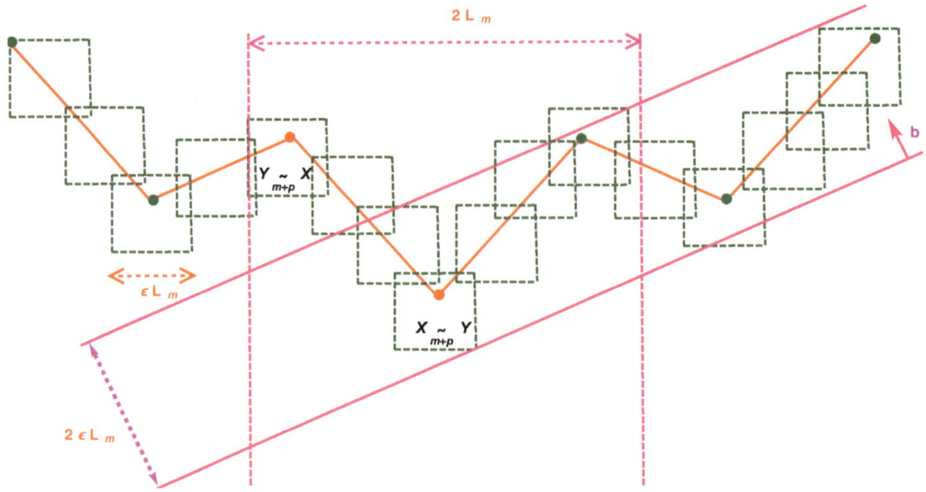

FIGURE 8. The intersection $\mathcal{B}(X_m, L_m) \cap \Gamma_{\mathcal{W}_m}$

In the case when $N_b - 1 > 1$, the positive integer p has thus to be choosen such that:

$$(3 - D_{\mathcal{W}})\, p > (D_{\mathcal{W}} - 1)\, m.$$

As is done in the previous work of the first author in [**Dav22**], we recall that a convenient cover of $\Gamma_{\mathcal{W}_m}$ between X and an adjacent vertex $Y \in \Gamma_{\mathcal{W}_{m+p}}$ of X requires at least:

$$\frac{h_{m+p} \times L_{m+p}}{\epsilon^2\, L_m^2} \leqslant \frac{C_{inf}\, L_{m+p}^{3-D_{\mathcal{W}}}}{\epsilon^2\, L_m^2} \quad \text{squares of side length } \epsilon\, L_m,$$

where

$$\frac{L_{m+p}^{3-D_{\mathcal{W}}}}{L_m^2} = \frac{L_m^{3-D_{\mathcal{W}}}}{L_m^2\, N_b^{(3-D_{\mathcal{W}})\, p}} = \frac{L_m^{1-D_{\mathcal{W}}}}{N_b^{(3-D_{\mathcal{W}})\, p}}.$$

We thus deal with the restriction of the band D to a vertical band of width $2\, L_m$, which can be thus covered by at most

$$C \times \frac{2\, \epsilon\, L_m \times 2\, L_m}{(\epsilon\, L_m)^2} = \frac{4\, C}{\epsilon}, \quad \text{with } C > 0,$$

squares of side length equal to $\epsilon\, L_m$.

Therefore, provided that

$$\frac{4\, C}{\epsilon} < \frac{C_1\, L_{m+p}^{3-D_{\mathcal{W}}}}{\epsilon^2\, L_m^2},$$

i.e.,

$$\epsilon < \frac{C_{inf}}{4\, C}\, \frac{L_{m+p}^{3-D_{\mathcal{W}}}}{L_m^2}$$

or, equivalently

$$\epsilon < \frac{C_{inf}}{4\, C}\, \frac{L_m^{1-D_{\mathcal{W}}}}{N_b^{(3-D_{\mathcal{W}})\, p}},$$

the vertex Y does not belong to the band D, which is the expected result. □

REMARK 4.8. We can check that our upper bound

$$\frac{C_{inf}}{4C}\,\frac{L_m^{1-D_{\mathcal{W}}}}{N_b^{(3-D_{\mathcal{W}})p}}$$

is explicitly a very small one, in fact as small as one chooses, whereas the one given in the book [**JW84**], on page 40, is of the following form:

$$\text{Constant}^{\frac{1}{d-(n-1)}} \quad , \quad 1 < d < n,$$

where the real positive constant is not explicitly known, in particular, it is not specified wether the constant is greater or less than 1.

NOTATION 15 (**Set of polynomials in two real variables**). In the sequel, $\mathbb{R}[X,Y]$ denotes the set of all real polynomials in two real variables.

Given $k \in \mathbb{N}$, we will denote by $\mathcal{P}ol_k \subset \mathbb{R}[X,Y]$ the set of all real polynomials in two real variables of degree at most equal to k.

DEFINITION 4.9 (**Two-dimensional π_r-net ([Wal91], on page 119)**). Given a strictly positive real number r, we will call *two-dimensional π_r-net* a tessellation of $\mathbb{R}^2$ into half-open, non-overlapping squares of side lengths r, obtained by intersecting $\mathbb{R}^2$ with lines orthogonal to the coordinate axes.

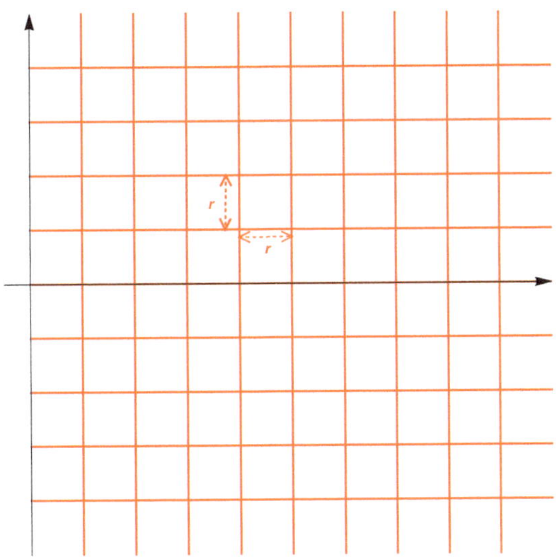

FIGURE 9. A two-dimensional π_r-net

DEFINITION 4.10 (**Two-dimensional polygonal $\pi_{\mathcal{W},m}$-net, $m \in \mathbb{N}$**). Given a strictly positive integer m, we call *two-dimensional polygonal $\pi_{\mathcal{W},m}$-net* a tessellation of $\mathbb{R}^2$ into half-open N_b-gons of side lengths at most equal to $\sqrt{2}\,h_m$ which contains the set of polygons

$$\left\{\bigcup_{j=0}^{N_b^m-1}\mathcal{P}_{m,j}\right\}\bigcup\left\{\bigcup_{k=1}^{N_b^m-2}\mathcal{Q}_{m,k}\right\}.$$

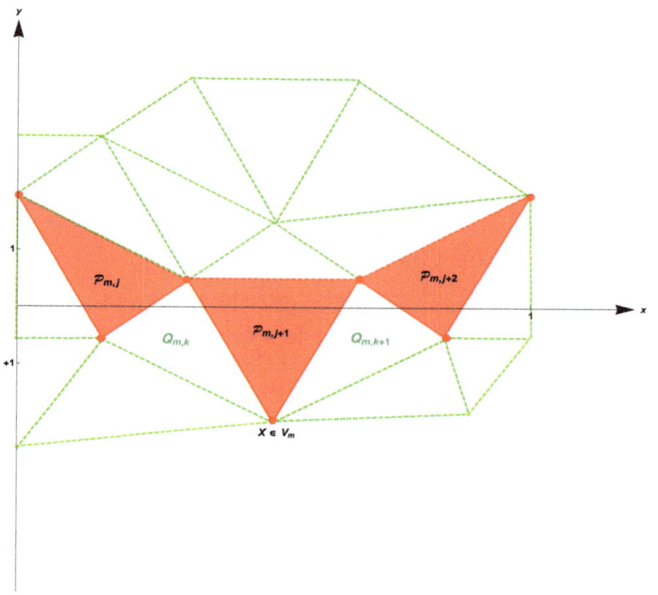

FIGURE 10. A two-dimensional polygonal $\pi_{\mathcal{W},m}$-net. Note that the polygons are not necessarily isometric.

PROPERTY 4.11. *Given a strictly positive integer m, the following properties hold:*

i. *For any integer $j \in \{0, \ldots, N_b^m - 1\}$, and any pair of vertices $(X,Y) \in (V_m \cap \mathcal{P}_{m,j})^2$:*

$$d_{Eucl}(X,Y) \lesssim N_b \, h_m \lesssim N_b^{-m\,(2-D_{\mathcal{W}})}.$$

ii. *For any integer $j \in \{1, \ldots, N_b^m - 2\}$, and any pair of vertices $(X,Y) \in (V_m \cap \mathcal{Q}_{m,j})^2$:*

$$d_{Eucl}(X,Y) \lesssim N_b \, h_m \lesssim N_b^{-m\,(2-D_{\mathcal{W}})}.$$

PROOF. This simply follows from the fact that the polygons have N_b sides, and that two adjacent vertices are distant from at most an Euclidean distance equal to h_m. □

NOTATION 16 (**Set of piecewise polynomial functions on a polygonal net**). Given a pair of nonnegative integers (k,m), and a polygonal $\pi_{\mathcal{W},m}$-net, we denote by $\mathcal{P}ol_k\left(\pi_{\mathcal{W},m}\right)$ the set of non-smooth splines of degree k on $\pi_{\mathcal{W},m}$, i.e., piecewise polynomial functions on $\pi_{\mathcal{W},m}$:

$$\mathcal{P}ol_k\left(\pi_{\mathcal{W},m}\right) = \Big\{ \textit{spline} \text{ such that for any polygon } \mathcal{P} \in \pi_{\mathcal{W},m},$$
$$\text{there exists } P \in \mathcal{P}ol_k : \textit{spline}_{|\mathcal{P}} = P_{|\mathcal{P}} \Big\}.$$

DEFINITION 4.12 (**Atoms (generalization of [Kab12])**). Given a strictly positive real number $s < 1$, a real number $p > 1$, two nonnegative integers m

and $j \in \{0,\dots,N_b^m-1\}$, a function $f_{m,j}$ defined on the prefractal graph $\Gamma_{\mathcal{W}_m}$ will be called a $(\mathcal{P}_{m,j},s,p)$*–atom* if the following three conditions are satisfied:

i. $\operatorname{Supp} f_{m,j} \subset \mathcal{P}_{m,j}$;

ii. $\forall\, X \in V_m \cap \mathcal{P}_{m,j}$:

$$\left|f_{m,j}(X)\right| \lesssim \mu_{\mathcal{L}}\left(\mathcal{P}_{m,j}\right)^{\frac{s}{D_{\mathcal{W}}}-\frac{1}{p}}$$

or, equivalently,

$$\left|f_{m,j}(X)\right| \lesssim \left(N_b^{(D_{\mathcal{W}}-3)\,m}\right)^{\frac{s}{D_{\mathcal{W}}}-\frac{1}{p}} ;$$

iii. $\forall\,(X,Y) \in \left(V_m \cap \mathcal{P}_{m,j}\right)^2$:

$$\left|f_{m,j}(X)-f_{m,j}(Y)\right| \lesssim d_{Eucl}(X,Y)\,\mu_{\mathcal{L}}\left(\mathcal{P}_{m,j}\right)^{\frac{s-1}{D_{\mathcal{W}}}-\frac{1}{p}} ,$$

or, equivalently,

$$\left|f_{m,j}(X)-f_{m,j}(Y)\right| \lesssim d_{Eucl}(X,Y)\,\left(N_b^{(D_{\mathcal{W}}-3)\,m}\right)^{\frac{s-1}{D_{\mathcal{W}}}-\frac{1}{p}} .$$

Similarly, given a strictly positive real number $s<1$, a real number $p>1$, two nonnegative integers m and $j \in \{1,\dots,N_b^m-2\}$, a function $g_{m,j}$ on the prefractal graph $\Gamma_{\mathcal{W}_m}$ will be called a $(\mathcal{Q}_{m,j},s,p)$*–atom* if the following three conditions are satisfied:

i. $\operatorname{Supp} g_{m,j} \subset \mathcal{Q}_{m,j}$;

ii. $\forall\, X \in V_m \cap \mathcal{Q}_{m,j}$:

$$\left|g_{m,j}(X)\right| \lesssim \mu_{\mathcal{L}}\left(\mathcal{Q}_{m,j}\right)^{\frac{s}{D_{\mathcal{W}}}-\frac{1}{p}} ,$$

or, equivalently,

$$\left|g_{m,j}(X)\right| \lesssim \left(N_b^{(D_{\mathcal{W}}-3)\,m}\right)^{\frac{s}{D_{\mathcal{W}}}-\frac{1}{p}} ;$$

iii. $\forall\,(X,Y) \in \left(V_m \cap \mathcal{Q}_{m,j}\right)^2$:

$$\left|g_{m,j}(X)-g_{m,j}(Y)\right| \lesssim d_{Eucl}(X,Y)\,\mu_{\mathcal{L}}\left(\mathcal{Q}_{m,j}\right)^{\frac{s-1}{D_{\mathcal{W}}}-\frac{1}{p}} ,$$

or, equivalently,

$$\left|g_{m,j}(X)-g_{m,j}(Y)\right| \lesssim d_{Eucl}(X,Y)\,\left(N_b^{(D_{\mathcal{W}}-3)\,m}\right)^{\frac{s-1}{D_{\mathcal{W}}}-\frac{1}{p}} .$$

REMARK 4.13 (**Atoms as Hölder functions**). Insofar as,

$$\forall\,(X,Y) \in \left(V_m \cap \mathcal{P}_{m,j}\right)^2 : \quad d_{Eucl}(X,Y) \lesssim N_b^{-m\,(2-D_{\mathcal{W}})} ,$$

the above condition *ii.* for a $(\mathcal{P}_{m,j},s,p)$-atom can be also written as $\forall\,(X,Y) \in \left(V_m \cap \mathcal{P}_{m,j}\right)^2$:

$$\left|f_{m,j}(X)-f_{m,j}(Y)\right| \lesssim N_b^{-m\,(2-D_{\mathcal{W}})+(D_{\mathcal{W}}-3)\,m\left(\frac{s-1}{D_{\mathcal{W}}}-\frac{1}{p}\right)} ,$$

which corresponds to a Hölder exponent of

$$1-\frac{(3-D_{\mathcal{W}})}{2-D_{\mathcal{W}}}\left(\frac{s-1}{D_{\mathcal{W}}}-\frac{1}{p}\right).$$

Note that, due to their definition, $(\mathcal{P}_{m,j},s,p)$-atoms are necessarily continuous.

An entirely similar property holds if the polygons $\mathcal{P}_{m,j}$ are replaced by their counterpart $\mathcal{Q}_{m,j}$, and the $(\mathcal{P}_{m,j}, s, p)$-atoms are replaced by the $(\mathcal{Q}_{m,j}, s, p)$-atoms.

REMARK 4.14 (**Atoms associated with the Weierstrass function**). In the case of the Weierstrass function $\mathcal{W}$, for any $m \in \mathbb{N}$, and any pair (X, Y) of adjacent vertices in V_m, we have that

$$|\mathcal{W}(X) - \mathcal{W}(Y)| \leqslant d_{Eucl}(X, Y)^{2-D_{\mathcal{W}}}.$$

By the triangle inequality, we immediately deduce that,
$\forall\, (X, Y) \in \left(V_m \cap \mathcal{P}_{m,j}\right)^2$ $\left(\text{resp., } \left(V_m \cap \mathcal{Q}_{m,j}\right)^2\right)$:

$$|\mathcal{W}(X) - \mathcal{W}(Y)| \leqslant d_{Eucl}(X, Y)^{2-D_{\mathcal{W}}}.$$

At the same time, thanks to the estimates given in Property 2.20, and in part *ii.* of the proof of Theorem 3.14 (see relation (11) and relation (12)), for any polygon $\mathcal{P}_{m,j}$ (resp., $\mathcal{Q}_{m,j}$), we have that

$$N_b^{(D_{\mathcal{W}}-3)\,m} \lesssim \mu_{\mathcal{L}}\left(\mathcal{P}_{m,j}\right) \lesssim N_b^{(D_{\mathcal{W}}-3)\,m}$$

$$\left(\text{resp., } N_b^{(D_{\mathcal{W}}-3)\,m} \lesssim \mu_{\mathcal{L}}\left(\mathcal{Q}_{m,j}\right) \lesssim N_b^{(D_{\mathcal{W}}-3)\,m}\right),$$

or, equivalently,

$$d_{Eucl}(X, Y)^{\frac{3-D_{\mathcal{W}}}{2-D_{\mathcal{W}}}} \lesssim \mu_{\mathcal{L}}\left(\mathcal{P}_{m,j}\right) \lesssim d_{Eucl}(X, Y)^{\frac{3-D_{\mathcal{W}}}{2-D_{\mathcal{W}}}}$$

$$\left(\text{resp., } d_{Eucl}(X, Y)^{\frac{3-D_{\mathcal{W}}}{2-D_{\mathcal{W}}}} \lesssim \mu_{\mathcal{L}}\left(\mathcal{Q}_{m,j}\right) \lesssim d_{Eucl}(X, Y)^{\frac{3-D_{\mathcal{W}}}{2-D_{\mathcal{W}}}}\right).$$

We then deduce that
$\forall\, (X, Y) \in \left(V_m \cap \mathcal{P}_{m,j}\right)^2$ $\left(\text{resp., } \left(V_m \cap \mathcal{Q}_{m,j}\right)^2\right)$:

$$|\mathcal{W}(X) - \mathcal{W}(Y)| \lesssim d_{Eucl}(X, Y)\, \mu_{\mathcal{L}}\left(\mathcal{P}_{m,j}\right)^{(1-D_{\mathcal{W}})\,\frac{2-D_{\mathcal{W}}}{3-D_{\mathcal{W}}}}$$

$$\left(\text{resp., } |\mathcal{W}(X) - \mathcal{W}(Y)| \lesssim d_{Eucl}(X, Y)\, \mu_{\mathcal{L}}\left(\mathcal{Q}_{m,j}\right)^{(1-D_{\mathcal{W}})\,\frac{2-D_{\mathcal{W}}}{3-D_{\mathcal{W}}}}\right).$$

It then follows that

$$\frac{s}{D_{\mathcal{W}}} - \frac{1}{p} = (1 - D_{\mathcal{W}})\,\frac{2 - D_{\mathcal{W}}}{3 - D_{\mathcal{W}}}, \quad \text{i.e.,} \quad s = \frac{D_{\mathcal{W}}}{p} + D_{\mathcal{W}}\,\frac{(1 - D_{\mathcal{W}})\,(2 - D_{\mathcal{W}})}{3 - D_{\mathcal{W}}},$$

and that the restriction of the Weierstrass function to each polygon $\mathcal{P}_{m,j}$, (resp., $\mathcal{Q}_{m,j}$) is a $(\mathcal{P}_{m,j}, s, p)$-*atom* (resp., a $(\mathcal{Q}_{m,j}, s, p)$-*atom*).

DEFINITION 4.15 (**Atomic decomposition of a function defined on the Weierstrass Curve**). Given a continuous function f on the Weierstrass Curve, we will say that f admits *an atomic decomposition* in the following form:

$$\begin{aligned} f = \lim_{m\to\infty} \Bigg\{ & \sum_{j=0}^{N_b^m-1} \sum_{X \text{ vertex of } \mathcal{P}_{m,j}} p(X,\mathcal{P}_m)\, \lambda_{f,m,j,X}\, f_{m,j,X} \\ & + \sum_{j=1}^{N_b^m-2} \sum_{X \text{ vertex of } \mathcal{Q}_{m,j}} p(X,\mathcal{Q}_m)\, \lambda_{g,m,j,X}\, g_{m,j} \Bigg\} \\ = \lim_{m\to\infty} \Bigg\{ & \sum_{j=0}^{N_b^m-1} \sum_{X \text{ vertex of } \mathcal{P}_{m,j},\, X\notin \mathcal{Q}_m} \lambda_{f,m,j,X}\, f_{m,j,X} \\ & + \sum_{j=1}^{N_b^m-2} \sum_{X \text{ vertex of } \mathcal{P}_{m,j},\, X\in \mathcal{P}_m\cap\mathcal{Q}_m} \{\lambda_{f,m,j,X}\, f_{m,j,X} + \lambda_{g,m,j,X}\, g_{m,j}\} \\ & + \sum_{j=1}^{N_b^m-2} \sum_{X \text{ vertex of } \mathcal{Q}_{m,j},\, X\notin \mathcal{P}_m} \lambda_{g,m,j,X}\, g_{m,j} \Bigg\}, \end{aligned}$$

where, for any $m \in \mathbb{N}$, the functions $f_{m,j}$, $0 \leqslant j \leqslant N_b^m - 1$ and $g_{m,j}$, $1 \leqslant j \leqslant N_b^m - 2$ are respectively $(\mathcal{P}_{m,j}, s, p)$ and $(\mathcal{Q}_{m,j}, s, p)$-atoms, $s < 1$, $p > 1$, while the coefficients $\lambda_{f,m,j}$, $0 \leqslant j \leqslant N_b^m - 1$ and $\lambda_{g,m,j}$, $1 \leqslant j \leqslant N_b^m - 2$, denote real numbers.

For the sake of simplicity, we will write the above decomposition in the following briefer form, where the two summation signs $\sum_{j=0}^{N_b^m-1}$ and $\sum_{X \text{ vertex of } \mathcal{P}_{m,j}}$ have been condensed in the equivalent summation sign $\sum_{X\in\mathcal{P}_m\bigcup\mathcal{Q}_m}$ as follows:

$$f = \lim_{m\to\infty} \sum_{X\in\mathcal{P}_m\bigcup\mathcal{Q}_m} \tilde{\lambda}_{f,m,X}\, \tilde{f}_{m,X} = \lim_{m\to\infty} \sum_{X\in\mathcal{P}_m\bigcup\mathcal{Q}_m} \tilde{\lambda}_{f,m}\, \tilde{f}_m,$$

where

$$\tilde{\lambda}_{f,m,X}\, \tilde{f}_{m,X} = \begin{cases} \lambda_{f,m,j,X}\, f_{m,j,X}, \text{if } X \notin \mathcal{Q}_m,\ X \text{ is a vertex of } \mathcal{P}_{m,j},\ 0 \leqslant j \leqslant N_b^m - 1, \\ \lambda_{g,m,j,X}\, g_{m,j,X}, \text{if } X \notin \mathcal{P}_m,\ X \text{ is a vertex of } \mathcal{Q}_{m,j},\ 1 \leqslant j \leqslant N_b^m - 2, \\ \lambda_{f,m,j,X}\, f_{m,j,X} + \lambda_{g,m,j,X}\, g_{m,j,X}\, f_{m,j,X},\ \text{if } X \in \mathcal{P}_m \cap \mathcal{Q}_m, \\ \qquad X \text{ is a vertex of } \mathcal{P}_{m,j} \text{ and } \mathcal{Q}_{m,j},\ 1 \leqslant j \leqslant N_b^m - 2. \end{cases}$$

For any $m \in \mathbb{N}$, we say that $\tilde{\lambda}_{f,m}$ is the mth*-atomic coefficient.*

From now on, the functions $\tilde{f}_{m,X}$ and $\tilde{f}_m$ will be called (m, s, p')*-atoms*. In the sequel, we will use the most suitable notation among these two possibilities.

PROPERTY 4.16 (**Atomic decomposition of spline functions in** $\mathcal{P}ol_k\left(\pi_{N_b^{(D_{\mathcal{W}}-3)n}}\right)$, $n \in \mathbb{N}$). *Given a pair (n, k) of nonnegative integers, a spline function (denoted by* spline*) in $\mathcal{P}ol_k\left(\pi_{N_b^{(D_{\mathcal{W}}-3)n}}\right)$ admits* an atomic decomposition *of the form*

$$spline = \lim_{m\to\infty} \sum_{X\in\mathcal{P}_m\bigcup\mathcal{Q}_m} \tilde{\lambda}_{s,m,X}\, \widetilde{spline}_{m,X}\ .$$

PROOF. This directly comes from the definition of functions of $\mathcal{P}ol_k\left(\pi_{N_b^n}\right)$ as piecewise polynomial functions. □

PROPERTY 4.17. *Given the polyhedral measure μ on the Weierstrass Curve $\Gamma_{\mathcal{W}}$, and a continuous function f on $\Gamma_{\mathcal{W}}$, of atomic decomposition*

$$f = \lim_{m\to\infty} \sum_{X\in\mathcal{P}_m\bigcup\mathcal{Q}_m} \tilde{\lambda}_{f,m,X}\,\tilde{f}_{m,X},$$

we have that

$$\int_{\mathcal{D}(\Gamma_{\mathcal{W}})} f\,d\mu = \lim_{m\to\infty} \varepsilon_m^{m(D_{\mathcal{W}}-2)} \sum_{X\in\mathcal{P}_m\bigcup\mathcal{Q}_m} \tilde{\lambda}_{f,m,X}\,\tilde{f}_{m,X}\,\mu(X,\mathcal{P}_m,\mathcal{Q}_m).$$

Note that since the Weierstrass Curve $\Gamma_{\mathcal{W}}$ satisfies Markov's Inequality (see Property 4.5 *above), this ensures* a good quality *of our atomic decomposition, in the sense that the corresponding and forthcoming characterization of Besov spaces on $\Gamma_{\mathcal{W}}$ will be obtained in a constructive way, along the lines of the work by Alf Jonsson in* [**Jon09**].

PROOF. This simply follows from the definition of the polyhedral measure on the Weierstrass Curve provided (and justified) by Property 3.12. □

REMARK 4.18. Such a decomposition makes sense since the set of vertices $(V_m)_{m\in\mathbb{N}}$ is dense in $\Gamma_{\mathcal{W}}$. Thus, because we deal with continuous functions, given any point X of the Weierstrass Curve, there exists a sequence $(X_m)_{m\in\mathbb{N}}$ such that

$$f(X) = \lim_{m\to\infty} f(X_m),$$

where, for any $m\in\mathbb{N}$, X_m belongs to the prefractal graph $\Gamma_{\mathcal{W}_m}$.

We can naturally write $f(X_m)$ as

$$f(X_m) = \sum_{Y_m\in V_m} f(Y_m)\,\delta_{X_mY_m}(X_m), \tag{13}$$

where δ is the classical Kronecker symbol; i.e.,

$$\forall\, Y_m \in V_m\ :\quad \delta_{X_mY_m}(Y_m) = \begin{cases} ccc1, & \text{if } Y_m = X_m,\\ 0, & \text{else.}\end{cases} \tag{14}$$

This, of course, yields

$$f(X) = \lim_{m\to\infty} \sum_{Y_m\in V_m} f(Y_m)\,\delta_{X_mY_m}(Y_m).$$

Now, we can go a little further and, as in [**Str06**], introduce spline functions $\psi_{X_m}^m$ such that

$$\forall\, Y\in\Gamma_{\mathcal{W}}\ :\quad \psi_{X_m}^m(Y) = \begin{cases} ccc\delta_{X_mY_m}, & \forall\, Y\in V_m\\ 0, & \forall\, Y\notin V_m,\end{cases}$$

and write

$$f(X) = \lim_{m\to\infty} \sum_{Y_m\in V_m} f(Y_m)\,\psi_{X_m}^m(Y_m),$$

which is nothing but an application in this context of the Weierstrass approximation theorem. In particular, spline functions are a natural choice for atoms.

CONVENTION. In the sequel, all functions f considered on the Weierstrass Curve are implicitly supposed to be Lebesgue measurable.

DEFINITION 4.19 (**L^p-norm of a function on the Weierstrass Curve**). Given a function f on the Weierstrass Curve, we define its L^p-norm as follows:

$$\begin{aligned} \|f\|_{L^p(\Gamma_{\mathcal{W}})} &= \left(\int_{\mathcal{D}(\Gamma_{\mathcal{W}})} |f|^p \, d\mu \right)^{\frac{1}{p}} \\ &= \left(\lim_{m\to\infty} \varepsilon_m^{m(D_{\mathcal{W}}-2)} \sum_{X \in \mathcal{P}_m \bigcup \mathcal{Q}_m} \mu^{\mathcal{L}}(X, \mathcal{P}_m, \mathcal{Q}_m) \, |f^p(X)| \right)^{\frac{1}{p}} . \end{aligned}$$

PROPERTY 4.20 (**L^p-norm of a function on the Weierstrass Curve defined by means of an atomic decomposition**). *Given a positive integer p, and a continuous function f on $\Gamma_{\mathcal{W}}$, whose absolute value $|f|$ is defined by means of an atomic decomposition as*

$$|f| = \lim_{m\to\infty} \sum_{X \in \mathcal{P}_m \bigcup \mathcal{Q}_m} \tilde{\lambda}_{|f|,m,X} \, \widetilde{|f|}_{m,X},$$

the L^p-norm of f with respect to the measure μ is given by

$$\begin{aligned} \|f\|_{L^p(\Gamma_{\mathcal{W}})} &= \left(\int_{\mathcal{D}(\Gamma_{\mathcal{W}})} |f|^p \, d\mu \right)^{\frac{1}{p}} \\ &= \left(\lim_{m\to\infty} \varepsilon_m^{m(D_{\mathcal{W}}-3)} \sum_{X \in \mathcal{P}_m \bigcup \mathcal{Q}_m} \mu^{\mathcal{L}}(X, \mathcal{P}_m, \mathcal{Q}_m) \, \tilde{\lambda}^p_{|f|,m,j,X} \, \widetilde{|f|}^p_{m,j,X} \right)^{\frac{1}{p}} . \end{aligned}$$

In particular, we have that

$$\begin{aligned} \|f\|_{L^1(\Gamma_{\mathcal{W}})} &= \int_{\mathcal{D}(\Gamma_{\mathcal{W}})} |f| \, d\mu \\ &= \lim_{m\to\infty} \varepsilon_m^{m(D_{\mathcal{W}}-2)} \sum_{X \in \mathcal{P}_m \bigcup \mathcal{Q}_m} \mu^{\mathcal{L}}(X, \mathcal{P}_m, \mathcal{Q}_m) \, \tilde{\lambda}_{|f|,m,j,X} \, \widetilde{|f|}_{m,j,X}. \end{aligned}$$

REMARK 4.21. In the above definition, two limits are a priori considered at the same time; the limit associated to the integral, with respect to the polyhedral measure μ, namely,

$$\lim_{m\to\infty} \varepsilon_m^{m(D_{\mathcal{W}}-2)} \sum_{X \in \mathcal{P}_m \bigcup \mathcal{Q}_m} \mu^{\mathcal{L}}(X, \mathcal{P}_m, \mathcal{Q}_m) \, |f(X)|^p,$$

and the limit associated to the atomic decomposition

$$\lim_{m\to\infty} \sum_{X \in \mathcal{P}_m \bigcup \mathcal{Q}_m} \tilde{\lambda}_{|f|,m,j,X} \, \widetilde{|f|}^p_{m,j,X}.$$

In fact, these two limits coincide.

DEFINITION 4.22 (**Besov space on the Weierstrass Curve (extension to the Weierstrass Curve of the result given by Theorem 6, on page 135, in [JW84]**)). Given $k \in \mathbb{N}$, a real number α such that

$$k < \alpha \leqslant k+1,$$

and two real numbers p and q greater or equal than 1, the *Besov space* $B_\alpha^{p,q}(\Gamma_{\mathcal{W}})$ is defined as the set of functions $f \in L^p(\mu)$ such that there exists a sequence $(c_m)_{m\in\mathbb{N}}$ (in ℓ^q) of nonnegative real numbers such that for every $\pi_{N_b^{(D_{\mathcal{W}}-3)m}}$-net, one can find a spline function $spline\left(\pi_{N_b^{(D_{\mathcal{W}}-3)m}}\right) \in \mathcal{P}ol_{[\alpha]}\left(\pi_{N_b^{(D_{\mathcal{W}}-3)m}}\right)$ satisfying, for all $m \in \mathbb{N}$,

$$\left\|f - spline\left(\pi_{N_b^{(D_{\mathcal{W}}-3)m}}\right)\right\|_{L^p(\mu)} \leqslant N_b^{(D_{\mathcal{W}}-3)m\,\alpha}\, c_m,, \quad (\mathcal{C}ond_{Besov\,spline})$$

where, if we write the respective atomic decompositions of f and $spline\left(\pi_{N_b^{(D_{\mathcal{W}}-3)m}}\right)$ as

$$f = \lim_{m\to\infty} \sum_{X\in\mathcal{P}_m \bigcup \mathcal{Q}_m} \tilde{\lambda}_{f,m,X}\, \tilde{f}_{m,X},$$

and

$$spline\left(\pi_{\pi_{N_b^{(D_{\mathcal{W}}-3)m}}}\right)$$
$$= \lim_{m\to\infty} \sum_{X\in\mathcal{P}_m \bigcup \mathcal{Q}_m} \tilde{\lambda}_{spline\left(\pi_{N_b^{(D_{\mathcal{W}}-3)m}}\right),m,X}\, \widetilde{spline}\left(\pi_{N_b^{(D_{\mathcal{W}}-3)m}}\right)_{m,X},$$

we then have that

$$\left\|f - spline\left(\pi_{\pi_{N_b^{(D_{\mathcal{W}}-3)m}}}\right)\right\|_{L^p(\mu)}^p = \lim_{m\to\infty} \varepsilon_m^{m\,(D_{\mathcal{W}}-2)} \times$$

$$\sum_{X\in\mathcal{P}_m \bigcup \mathcal{Q}_m} \mu^{\mathcal{L}}(X, \mathcal{P}_m, \mathcal{Q}_m)\, \tilde{\lambda}^p_{\left|f-spline\left(\pi_{N_b^{(D_{\mathcal{W}}-3)m}}\right)\right|,m,X} \left|\widetilde{f - spline\left(\pi_{N_b^{(D_{\mathcal{W}}-3)m}}\right)}\right|_{m,X}^p;$$

here, for the sake of simplicity, we have introduced the atomic decomposition of $\left|f - spline\left(\pi_{N_b^{(D_{\mathcal{W}}-3)m}}\right)\right|$ as

$$\left|f - spline\left(\pi_{\pi_{N_b^{(D_{\mathcal{W}}-3)m}}}\right)\right|$$
$$= \lim_{m\to\infty} \sum_{X\in\mathcal{P}_m \bigcup \mathcal{Q}_m} \tilde{\lambda}_{\left|f-spline\left(\pi_{N_b^{(D_{\mathcal{W}}-3)m}}\right)\right|,m,X} \left|\widetilde{f - spline\left(\pi_{N_b^{(D_{\mathcal{W}}-3)m}}\right)}\right|_{m,X}.$$

REMARK 4.23. The atomic decomposition used in [**Kab12**] is obtained by introducing small neighborhoods of the curve under study (union of balls). Our polygonal domain introduced in Definition 4.10 appears to be a more natural choice. Indeed, unlike the aforementioned balls, the polygons involved do not overlap with each other, which works better for the required nets.

DEFINITION 4.24 (**Besov norm**). Given $k \in \mathbb{N}$, a real number α such that

$$k < \alpha \leqslant k+1,$$

and two real numbers p and q greater or equal than 1, one can define, as in [**Wal91**], the $B_\alpha^{p,q}(\Gamma_{\mathcal{W}})$*-norm* of a function f defined on the Weierstrass Curve as

$$\|f\|_{B_\alpha^{p,q}(\Gamma_{\mathcal{W}})} = \|f\|_{L^p(\Gamma_{\mathcal{W}})} + \inf\left\{\sum_{n\in\mathbb{N}} c_n^q\right\}^{\frac{1}{q}},$$

where the infimum is taken over all the sequences $(c_m)_{m\in\mathbb{N}} \in \ell^q$ of nonnegative real numbers involved in condition $(\mathcal{C}ond_{Besov\,spline})$ in Definition 4.22 just above.

REMARK 4.25 (**Alternative definition of a Besov space on the Weierstrass Curve**). One of the interesting properties of Besov spaces is that they can be defined in two different ways: first, as given previously in Definition 4.22, by means of a polynomial approximation, which provides information on the degree of regularity of the functions involved; also, as the set of functions of a finite specific norm as in Definition 4.24. This latter definition is all the more interesting, because it enables one to establish connections with discrete and Laplace-type operators, by means of the fractional difference quotients involved. Our present case is, of course, distinct from the classical one on $\mathbb{R}^N$, with $N \in \mathbb{N}$. Yet, the case of $\mathbb{R}^N$, as it can be found in [**JW84**], Chapter 1, Section 1. 5, on page 7, enables us to understand the underlying connection; namely, for $\alpha > 0$, $1 \leqslant p \leqslant \infty$ and $1 \leqslant q < \infty$,

$$B_\alpha^{p,q}\left(\mathbb{R}^N\right)$$

$$= \left\{ f \in L^p(\mathbb{R}^N),\ \sum_{|j| \leqslant [\alpha]} \left\| D^j f \right\|_{L^p(\Gamma_{\mathcal{W}})} + \sum_{|j|=[\alpha]} \left\{ \int_{\mathbb{R}^N} \frac{\left\| \Delta_h^{first\, diff.} D^j f \right\|_{L^p(\Gamma_{\mathcal{W}})}^q}{|h|^{N+(\alpha-[\alpha])\, q}}\, dh \right\}^{\frac{1}{q}} < \infty \right\},$$

where $[\alpha]$ denotes the integer part of the positive number α, while $\Delta_h^{first\, diff.}$ is the usual first difference, defined here by

$$\forall\, t \in \Gamma_{\mathcal{W}},\ \forall\, h \in \mathbb{R}^n\,, \quad \Delta_h^{first\, diff.} f(t) = f(t+h) - f(t).$$

The difficulty, in our case, is to obtain an equivalent formulation. For our forthcoming purpose of determining the order of a Laplace-type operator on the (fractal) Weierstrass Curve $\Gamma_{\mathcal{W}}$, in Corollary 4.36 below, we will restrict ourselves to the case where $\alpha < 2$. We thus have to deal with expressions of the following form,

$$\left\{ \int_{\mathcal{D}(\Gamma_{\mathcal{W}})} \frac{|f(T) + f(Y) - 2\, f(X)|^q}{|h|^{N+(\alpha-[\alpha])\, q}}\, dh \right\}^{\frac{1}{q}}.$$

Thanks to the Minkowski inequality, we of course have that, for all points (X,Y,T) in $(\mathcal{D}(\Gamma_{\mathcal{W}}))^3$,

$$\left\{ \int_{\mathcal{D}(\Gamma_{\mathcal{W}})} \frac{|f(T) + f(Y) - 2\, f(X)|^q}{|h|^{N+(\alpha-[\alpha])\, q}}\, dh \right\}^{\frac{1}{q}}$$

$$\leqslant \left\{ \int_{\mathcal{D}(\Gamma_{\mathcal{W}})} \frac{|f(T) - f(X)|^q}{|h|^{N+(\alpha-[\alpha])\, q}}\, dh \right\}^{\frac{1}{q}} + \left\{ \int_{\mathcal{D}(\Gamma_{\mathcal{W}})} \frac{|f(Y) - f(X)|^q}{|h|^{N+(\alpha-[\alpha])\, q}}\, dh \right\}^{\frac{1}{q}}.$$

In our present context, the term

$$\left\{ \int_{\mathcal{D}(\Gamma_{\mathcal{W}})} \frac{|f(T) + f(Y) - 2\, f(X)|^q}{|h|^{N+(\alpha-[\alpha])\, q}}\, dh \right\}^{\frac{1}{q}}, \tag{15}$$

where N has to be replaced by $D_{\mathcal{W}}$, is obtained thanks to the atomic decompositions by means of the sum of the following positive series,

$$\begin{aligned}\varepsilon_m^{3\,m\,(D_{\mathcal{W}}-2)} \sum_{(X,Y,T)\in(\mathcal{P}_m \bigcup \mathcal{Q}_m)^3,\, Y \underset{m}{\sim} X,\, T \underset{m}{\sim} X} & \Bigg(\mu^{\mathcal{L}}(X,\mathcal{P}_m,\mathcal{Q}_m)\, \mu^{\mathcal{L}}(Y,\mathcal{P}_m,\mathcal{Q}_m) \\ & \mu^{\mathcal{L}}(T,\mathcal{P}_m,\mathcal{Q}_m)\, \tilde{\lambda}_{\star,m} \frac{\left|\tilde{f}_m(T) + \tilde{f}_m f(Y) - 2\,\tilde{f}_m(Y)\right|^q}{d_{Eucl}^{D_{\mathcal{W}}+(\alpha-k)\,q}(X,Y)} \Bigg),\end{aligned} \tag{16}$$

since, for vertices $Y \underset{m}{\sim} X$ or $T \underset{m}{\sim} X$, we have, thanks to Property 2.20, that

$$d_{Eucl}(X,Y) \lesssim d(X,T))$$

and where, for the sake of simplicity, we have denoted by $\tilde{\lambda}_{\star,m}$ the atomic coefficients involved.

As for the terms

$$\int_{\mathcal{D}(\Gamma_{\mathcal{W}})} \frac{|f(Y)-f(X)|^q}{|h|^{N+(\alpha-[\alpha])\,q}}\,dh,$$

or, equivalently,

$$\int_{\mathcal{D}(\Gamma_{\mathcal{W}})} \frac{|f(T)-f(X)|^q}{|h|^{N+(\alpha-[\alpha])\,q}}\,dh,$$

they are obtained by means of the sum of the following positive series,

$$\begin{aligned}\varepsilon_m^{2\,m\,(D_{\mathcal{W}}-2)} \sum_{(T,Y)\in(\mathcal{P}_m \bigcup \mathcal{Q}_m)^2,\, Y \underset{m}{\sim} T} & \Bigg(\mu^{\mathcal{L}}(Y,\mathcal{P}_m,\mathcal{Q}_m) \\ & \mu^{\mathcal{L}}(T,\mathcal{P}_m,\mathcal{Q}_m)\, \tilde{\lambda}_{\star,m} \frac{\left|\tilde{f}_m(T) - \tilde{f}_m(Y)\right|^q}{d_{Eucl}^{D_{\mathcal{W}}+(\alpha-k)\,q}(T,Y)} \Bigg).\end{aligned} \tag{17}$$

This ensures that, if the (positive) series in expression (17) just above converges, the (positive) series in expression (16) also converges, which ensures the existence of the integral given in expression (15) .

Going further, we may note that by applying the same reasoning, we could extend the validity to higher values of α.

REMARK 4.26. This enables us to obtain the first side of the comparison relation required in order to establish the equivalence of norms. It is the side that plays the most significant part in the definition of Besov spaces on the Weierstrass Curve. Thus, characterizing Besov spaces on $\Gamma_{\mathcal{W}}$ by means of the norm introduced in Definition 4.24, is directly associated to the definition of a sequence of (suitably renormalized) discrete graph Laplacians $(\Delta_m)_{m\in\mathbb{N}}$ on the sequence of prefractal approximations $(\Gamma_{\mathcal{W}_m})_{m\in\mathbb{N}}$. In a sense, it is also connected to the existence of the limit

$$\lim_{m\to\infty} \Delta_m$$

by means of an equivalent pointwise formula expressed in terms of integrals, somehow the counterpart, in a way, of the one which is well known in the case of the fractal Laplacian on the Sierpiński Gasket [**Kig01**], [**Str06**].

PROPERTY 4.27 (**Characterization of Besov spaces: sufficient condition**). *Given a real number α such that*

$$1 < \alpha \leqslant 2,$$

two real numbers p and q greater or equal than 1, and a continuous function f given by means of an atomic decomposition of the form

$$f = \lim_{m\to\infty} \sum_{X\in\mathcal{P}_m \bigcup \mathcal{Q}_m} \tilde{\lambda}_{f,m,X}\, \tilde{f}_{m,X}$$

belongs to the Besov space $B^{p,q}_{\alpha}(\Gamma_{\mathcal{W}})$ *if the following two conditions are satisfied,*

$$(3 - D_{\mathcal{W}}) \left\{ q \left(\frac{1}{p} - \frac{s-1}{D_{\mathcal{W}}} \right) \right\} + (2 - D_{\mathcal{W}})\,(D_{\mathcal{W}} + (\alpha - 1)\, q) < 2, \quad (Cond_{Besov})$$

and

$$\frac{D_{\mathcal{W}}}{3 - D_{\mathcal{W}}} + \frac{D_{\mathcal{W}}}{p} \leqslant s, \quad (Cond_{L^p})$$

where the real number $s \in]0,1[$ *has been introduced in Definition* 4.12.

REMARK 4.28. As was explained in the introduction, since the Weierstrass Curve preserves Markov's Inequality, this ensures the *quality* of the approximation, when it comes to characterize Besov spaces by means of atomic decompositions (see [**Jon09**]), since Besov spaces are then obtained in a constructive manner.

PROOF. (of Property 4.27) Since $1 < \alpha \leqslant 2$, we have that

$$\|f\|^q_{B^{p,q}_{\alpha}(\Gamma_{\mathcal{W}})}$$

$$= \quad \|f\|^p_{L^p(\Gamma_{\mathcal{W}})} + \left(\iiint_{(X,Y,T)\in(\mathcal{D}(\Gamma_{\mathcal{W}}))^3} \frac{|f(T)+f(Y)-2\,f(X)|^q}{d_{Eucl}^{D_{\mathcal{W}}+\alpha\, q}(T,Y)}\, d\mu^3 \right)^{\frac{1}{q}}.$$

By applying the results at the end of Remark 4.25, we can restrict ourselves to expressions of the following form:

$$\left(\iint_{(T,Y)\in(\mathcal{D}(\Gamma_{\mathcal{W}}))^2} \frac{|f(T)-f(Y)|^q}{d_{Eucl}^{D_{\mathcal{W}}+\alpha\, q}(T,Y)}\, d\mu^2 \right)^{\frac{1}{q}},$$

where, via the atomic decomposition of f,

$$\iint_{(T,Y)\in\Gamma^2_{\mathcal{W}}} \frac{|f(T)-f(Y)|^q}{d_{Eucl}^{D_{\mathcal{W}}+\alpha\, q}(T,Y)}\, d\mu^2$$

$$= \lim_{m\to\infty} \varepsilon_m^{2\,m\,(D_{\mathcal{W}}-2)} \sum_{(T,Y)\in(\mathcal{P}_m \bigcup \mathcal{Q}_m)^2} \mu^{\mathcal{L}}(T,\mathcal{P}_m,\mathcal{Q}_m)\, \mu^{\mathcal{L}}(Y,\mathcal{P}_m,\mathcal{Q}_m)\, \frac{|f(T)-f(Y)|^q}{d_{Eucl}^{D_{\mathcal{W}}+\alpha\, q}(T,Y)}$$

$$= \lim_{m\to\infty} \varepsilon_m^{2\,m\,(D_{\mathcal{W}}-2)} \sum_{(T,Y)\in(\mathcal{P}_m \bigcup \mathcal{Q}_m)^2} \mu^{\mathcal{L}}(T,\mathcal{P}_m,\mathcal{Q}_m)\, \mu^{\mathcal{L}}(Y,\mathcal{P}_m,\mathcal{Q}_m)\, \tilde{\lambda}_{f,m}\, \frac{\left|\tilde{f}_m(T)-\tilde{f}_m(Y)\right|^q}{d_{Eucl}^{D_{\mathcal{W}}+\alpha\, q}(T,Y)}$$

and where, for the sake of simplicity, we have denoted by $\tilde{\lambda}_{\star,m}$ the atomic coefficients involved.

Note that since the function f is continuous, the atomic coefficients $\lambda_{\star,m}$ are necessarily bounded (since $\Gamma_{\mathcal{W}}$ is compact, and hence, f is bounded).

Since we deal with the atomic decomposition of f, due to part *iii.* of Definition 4.12, we have that

$$\mu^{\mathcal{L}}(T,\mathcal{P}_m,\mathcal{Q}_m)\,\mu^{\mathcal{L}}(Y,\mathcal{P}_m,\mathcal{Q}_m)\,\tilde{\lambda}_{f,m}\,\frac{\left|\tilde{f}_n(T)-\tilde{f}_m(Y)\right|^q}{d_{Eucl}^{D_{\mathcal{W}}+\alpha q}(T,Y)}$$
$$\lesssim \mu^{\mathcal{L}}(T,\mathcal{P}_m,\mathcal{Q}_m)\,\mu^{\mathcal{L}}(Y,\mathcal{P}_m,\mathcal{Q}_m)\,\tilde{\lambda}_{f,m}\,\frac{\left(N_b^{-m\,q\,(3-D_{\mathcal{W}})}\right)^{\frac{s-1}{d}-\frac{1}{p}}}{d_{Eucl}^{D_{\mathcal{W}}+\alpha q-q}(T,Y)}.$$

We also have that

$$\frac{1}{d_{Eucl}(T,Y)}<\frac{1}{|h_{jm}|}\lesssim L_m^{D_{\mathcal{W}}-2}\lesssim N_b^{(2-D_{\mathcal{W}})\,m},$$

and

$$d_{Eucl}(T,Y)\lesssim h_m\lesssim L_m^{2-D_{\mathcal{W}}}\lesssim N_b^{(D_{\mathcal{W}}-2)\,m}.$$

Moreover,

$$\mu^{\mathcal{L}}(T,\mathcal{P}_m,\mathcal{Q}_m)\lesssim h_m\,L_m\lesssim L_m^{3-D_{\mathcal{W}}}\lesssim N_b^{(D_{\mathcal{W}}-3)\,m}$$

and, in the same way,

$$\mu^{\mathcal{L}}(Y,\mathcal{P}_m,\mathcal{Q}_m)\lesssim N_b^{(D_{\mathcal{W}}-3)\,m}.$$

We therefore deduce that

$$\mu^{\mathcal{L}}(T,\mathcal{P}_m,\mathcal{Q}_m)\,\mu^{\mathcal{L}}(Y,\mathcal{P}_m,\mathcal{Q}_m)\,\tilde{\lambda}_{f,m}\,\frac{\left(N_b^{-m\,q\,(3-D_{\mathcal{W}})}\right)^{\frac{s-1}{d}-\frac{1}{p}}}{d_{Eucl}^{D_{\mathcal{W}}+\alpha q-q}(T,Y)}$$
$$\lesssim N_b^{2\,m\,(D_{\mathcal{W}}-3)}\,\tilde{\lambda}_{f,m}\,\left(N_b^{-m\,q\,(3-D_{\mathcal{W}})}\right)^{\frac{s-1}{D_{\mathcal{W}}}-\frac{1}{p}}\,N_b^{(2-D_{\mathcal{W}})\,m\,(D_{\mathcal{W}}+\alpha q-q)},$$

and

$$\varepsilon_m^{2\,m\,(D_{\mathcal{W}}-2)}\,\mu^{\mathcal{L}}(T,\mathcal{P}_m,\mathcal{Q}_m)\,\mu^{\mathcal{L}}(Y,\mathcal{P}_m,\mathcal{Q}_m)\,\tilde{\lambda}_{f,m}\,\frac{\left(N_b^{-m\,q\,(3-D_{\mathcal{W}})}\right)^{\frac{s-1}{d}-\frac{1}{p}}}{d_{Eucl}^{D_{\mathcal{W}}+\alpha q-q}(T,Y)}$$
$$\lesssim \varepsilon_m^{2\,m\,(D_{\mathcal{W}}-2)}\,N_b^{2\,m\,(D_{\mathcal{W}}-3)}\,\tilde{\lambda}_{f,m}\,\left(N_b^{-m\,q\,(3-D_{\mathcal{W}})}\right)^{\frac{s-1}{D_{\mathcal{W}}}-\frac{1}{p}}\,N_b^{(2-D_{\mathcal{W}})\,m\,(D_{\mathcal{W}}+\alpha q-q)},$$

along with

$$\begin{aligned}
&\int_{T\in\Gamma_{\mathcal{W}}}\int_{Y\in\Gamma_{\mathcal{W}}} \frac{\left(N_b^{-m\,q\,(3-D_{\mathcal{W}})}\right)^{\frac{s-1}{d}-\frac{1}{p}}}{d_{Eucl}^{D_{\mathcal{W}}+\alpha\,q-q}(T,Y)}\,d\mu^2 \\
&\quad = \lim_{m\to\infty} \varepsilon_m^{2\,m\,(D_{\mathcal{W}}-2)} \sum_{(T,Y)\in(\mathcal{P}_m\bigcup\mathcal{Q}_m)^2} \mu^{\mathcal{L}}(T,\mathcal{P}_m,\mathcal{Q}_m)\,\mu^{\mathcal{L}}(Y,\mathcal{P}_m,\mathcal{Q}_m) \\
&\qquad\qquad \times \frac{\left(N_b^{-m\,q\,(3-D_{\mathcal{W}})}\right)^{\frac{s-1}{D_{\mathcal{W}}}-\frac{1}{p}}}{d_{Eucl}^{D_{\mathcal{W}}+\alpha\,q-a}(T,Y)} \\
&\lesssim \lim_{m\to\infty} \varepsilon_m^{2\,m\,(D_{\mathcal{W}}-2)} \sum_{(T,Y)\in(\mathcal{P}_m\bigcup\mathcal{Q}_m)^2} \mu^{\mathcal{L}}(T,\mathcal{P}_m,\mathcal{Q}_m)\,\mu^{\mathcal{L}}(Y,\mathcal{P}_m,\mathcal{Q}_m) \\
&\qquad\qquad \times \left(N_b^{-m\,q\,(3-D_{\mathcal{W}})}\right)^{\frac{s-1}{D_{\mathcal{W}}}-\frac{1}{p}} N_b^{(2-D_{\mathcal{W}})\,m\,(D_{\mathcal{W}}+\alpha\,q-q)} \\
&\lesssim \lim_{m\to\infty} \varepsilon_m^{2\,m\,(D_{\mathcal{W}}-2)} \sum_{(T,Y)\in(\mathcal{P}_m\bigcup\mathcal{Q}_m)^2} N_b^{2\,m\,(D_{\mathcal{W}}-3)} \\
&\qquad\qquad \times \left(N_b^{-m\,q\,(3-D_{\mathcal{W}})}\right)^{\frac{s-1}{D_{\mathcal{W}}}-\frac{1}{p}} N_b^{(2-D_{\mathcal{W}})\,m\,(D_{\mathcal{W}}+\alpha\,q-q)}.
\end{aligned}$$

The function f will thus belong to the Besov space $B_\alpha^{p,q}(\Gamma_{\mathcal{W}})$ provided that

$$\sum_{m\in\mathbb{N}} \varepsilon_m^{2\,m\,(D_{\mathcal{W}}-2)}\, N_b^{2\,m\,(D_{\mathcal{W}}-3)} \left(N_b^{-m\,q\,(3-D_{\mathcal{W}})}\right)^{\frac{s-1}{D_{\mathcal{W}}}-\frac{1}{p}} N_b^{(2-D_{\mathcal{W}})\,m\,(D_{\mathcal{W}}+\alpha\,q-q)} < \infty,$$

or, equivalently,

$$\sum_{m\in\mathbb{N}} N_b^{-2\,m\,(D_{\mathcal{W}}-2)}\, N_b^{2\,m\,(D_{\mathcal{W}}-3)} \left(N_b^{-m\,q\,(3-D_{\mathcal{W}})}\right)^{\frac{s-1}{D_{\mathcal{W}}}-\frac{1}{p}} N_b^{(2-D_{\mathcal{W}})\,m\,(D_{\mathcal{W}}+\alpha\,q-q)} < \infty,$$

converges, which requires that

$$-2\,(D_{\mathcal{W}}-2)+2\,(D_{\mathcal{W}}-3)-q\,(3-D_{\mathcal{W}})\left(\frac{s-1}{D_{\mathcal{W}}}-\frac{1}{p}\right)+(2-D_{\mathcal{W}})\,(D_{\mathcal{W}}+\alpha\,q-q) < 0,$$

i.e.,

$$(3-D_{\mathcal{W}})\left\{q\left(\frac{1}{p}-\frac{s-1}{D_{\mathcal{W}}}\right)\right\} + (2-D_{\mathcal{W}})\,(D_{\mathcal{W}}+(\alpha-1)\,q) < 2. \quad (\mathcal{C}ond_{Besov})$$

At the same time, thanks to Property 4.20, and in light of Definition 4.12, we have that

$$\begin{aligned}
&\|f\|^p_{L^p(\Gamma_{\mathcal{W}})} \\
= &\lim_{m\to\infty} \varepsilon_m^{m\,(D_{\mathcal{W}}-3)} \sum_{X\in\mathcal{P}_m\bigcup\mathcal{Q}_m} \mu^{\mathcal{L}}(X,\mathcal{P}_m,\mathcal{Q}_m)\,\tilde{\lambda}^p_{|f|,m,j,X}\,\widetilde{|f|}^p_{m,j,X} \\
\lesssim &\lim_{m\to\infty} \varepsilon^{D_{\mathcal{W}}-3} \sum_{X\in\mathcal{P}_m\bigcup\mathcal{Q}_m} \mu^{\mathcal{L}}(X,\mathcal{P}_m,\mathcal{Q}_m)\,\tilde{\lambda}^p_{|f|,m,j,X}\,\underbrace{\mu_{\mathcal{L}}(\mathcal{P}_{m,j})}_{\lesssim\,\mu^{\mathcal{L}}(X,\mathcal{P}_m,\mathcal{Q}_m)}{}^{\frac{s}{D_{\mathcal{W}}}-\frac{1}{p}} \\
\lesssim &\lim_{m\to\infty} \varepsilon_m^{m\,(D_{\mathcal{W}}-3)} \sum_{X\in\mathcal{P}_m\bigcup\mathcal{Q}_m} \underbrace{\mu^{\mathcal{L}}(X,\mathcal{P}_m,\mathcal{Q}_m)}_{\lesssim\,N_b^{(D_{\mathcal{W}}-3)\,m}}{}^{1+\frac{s}{D_{\mathcal{W}}}-\frac{1}{p}} \\
\lesssim &\lim_{m\to\infty} N_b^{m\,(3-D_{\mathcal{W}})} \sum_{X\in\mathcal{P}_m\bigcup\mathcal{Q}_m} N_b^{(D_{\mathcal{W}}-3)\,m\left(1+\frac{s}{D_{\mathcal{W}}}-\frac{1}{p}\right)} \\
\lesssim &\lim_{m\to\infty} N_b^m\,N_b^{m\,(3-D_{\mathcal{W}})}\,N_b^{(D_{\mathcal{W}}-3)\,m\left(1+\frac{s}{D_{\mathcal{W}}}-\frac{1}{p}\right)}.
\end{aligned}$$

The convergence of this latter expression requires that

$$1+3-D_{\mathcal{W}}+(D_{\mathcal{W}}-3)\left(1+\frac{s}{D_{\mathcal{W}}}-\frac{1}{p}\right)=1+(D_{\mathcal{W}}-3)\left(\frac{s}{D_{\mathcal{W}}}-\frac{1}{p}\right)\leqslant 0;$$

i.e.,

$$1\leqslant(3-D_{\mathcal{W}})\left(\frac{s}{D_{\mathcal{W}}}-\frac{1}{p}\right),$$

or, equivalently,

$$\frac{D_{\mathcal{W}}}{3-D_{\mathcal{W}}}+\frac{D_{\mathcal{W}}}{p}\leqslant s. \quad (Cond_{L^p})$$

We can check that if the above conditions $(Cond_{Besov})$ and $(Cond_{L^p})$ are satisfied,

$$\|f\|_{L^p(\Gamma_{\mathcal{W}})}+\left\{\iiint_{(X,T,Y)\in(\mathcal{D}(\Gamma_{\mathcal{W}}))^3}\frac{|f(T)+f(Y)-2\,f(X)|^q}{d_{Eucl}^{D_{\mathcal{W}}+\alpha\,q}(T,Y)}\,d\mu^3\right\}^{\frac{1}{q}}<\infty,$$

as desired. □

COROLLARY 4.29 (**The specific case of $B^{p,p}_\beta(\Gamma_{\mathcal{W}})$**). *Given $k\in\mathbb{N}$, a real number p greater or equal than 1, we set*

$$\beta=k-\frac{2-D_{\mathcal{W}}}{p}.$$

A function f, given by means of an atomic decomposition of the form

$$f=\lim_{m\to\infty}\sum_{X\in\mathcal{P}_m\bigcup\mathcal{Q}_m}\tilde{\lambda}_{f,m,X}\,\tilde{f}_{m,X}$$

belongs to the Besov space $B^{p,p}_\beta(\Gamma_{\mathcal{W}})$ if

$$(3-D_{\mathcal{W}})\left\{p\left(\frac{1}{p}-\frac{s-1}{D_{\mathcal{W}}}\right)\right\}+(2-D_{\mathcal{W}})\,(D_{\mathcal{W}}+(\beta-1)\,p)<2,$$

and

$$\frac{D_{\mathcal{W}}}{3 - D_{\mathcal{W}}} + \frac{D_{\mathcal{W}}}{p} \leqslant s.$$

DEFINITION 4.30 (**Trace of an $L^1_{loc}(\mathbb{R}^2)$ function on the Weierstrass Curve**). Along the lines of [**JW84**], on page 15, or [**Wal91**], we will say that an $L^1_{loc}(\mathbb{R}^2)$ function f is *strictly defined* at a vertex X of the Weierstrass Curve if the following limit exists and is given by

$$\bar{f}(X) = \lim_{m \to \infty} \frac{1}{\mu^{\mathcal{L}}(X, \mathcal{P}_m, \mathcal{Q}_m)} \sum_{Y \sim X} \mu^{\mathcal{L}}(Y, \mathcal{P}_m, \mathcal{Q}_m)\, f(Y) < \infty.$$

This enables us to define *the trace* $f_{|\Gamma_{\mathcal{W}}}$ of the function f on the Weierstrass Curve, via

$$\forall\, X \in \Gamma_{\mathcal{W}} : f_{|\Gamma_{\mathcal{W}}}(X) = \bar{f}(X).$$

REMARK 4.31. The trace $\bar{f}$ of an $L^1_{loc}(\mathbb{R}^2)$ function thus naturally admits an atomic decomposition of the form

$$\bar{f} = \lim_{m \to \infty} \sum_{X \in \mathcal{P}_m \bigcup \mathcal{Q}_m} \tilde{\lambda}_{\bar{f},m,X}\, \tilde{\bar{f}}_{m,X}.$$

Also,

$$C_{inf}\, N_b^{(D_{\mathcal{W}}-2)\,m} \leqslant f(Y) \leqslant C_{sup}\, N_b^{(D_{\mathcal{W}}-2)\,m}.$$

$$C_{inf}\, N_b^{(D_{\mathcal{W}}-3)\,m}\, N_b^{(D_{\mathcal{W}}-2)\,m} \leqslant N_b^{(D_{\mathcal{W}}-3)\,m} f(Y) \leqslant C_{sup}\, N_b^{(D_{\mathcal{W}}-3)\,m}\, N_b^{(D_{\mathcal{W}}-2)\,m}.$$

The following simple property was already used implicitly when introducing the polyhedral measure, earlier in Section 3.

PROPERTY 4.32 (**The compact set of $\mathbb{R}^2$ which contains the Weierstrass Curve**). *The Weierstrass Curve $\Gamma_{\mathcal{W}}$ is contained in the following compact set of $\mathbb{R}^2$,*

$$\Omega_{\mathcal{W}} = [0,1] \times [m_{\mathcal{W}}, M_{\mathcal{W}}],$$

where $m_{\mathcal{W}}$ and $M_{\mathcal{W}}$ respectively denote the minimal and maximal values of the Weierstrass function $\mathcal{W}$ on $[0,1]$, introduced in Notation 13.

NOTATION 17 (**Interior of the compact set $\Omega_{\mathcal{W}}$**). We hereafter denote by $\mathring{\Omega}_{\mathcal{W}}$ the interior of the compact set $\Omega_{\mathcal{W}}$.

DEFINITION 4.33 (**Sobolev space on the open set $\mathring{\Omega}_{\mathcal{W}}$**). Given $k \in \mathbb{N}$, and $p \geqslant 1$, we recall that the Sobolev space on the open set $\mathring{\Omega}_{\mathcal{W}} \subset \mathbb{R}^2$ (which boundary is a rectangle), denoted by $W^p_k\left(\mathring{\Omega}_{\mathcal{W}}\right)$, is given by

$$W^p_k\left(\mathring{\Omega}_{\mathcal{W}}\right) = \left\{ f \in L^p\left(\mathring{\Omega}_{\mathcal{W}}\right)\ ,\ \forall\, \alpha \text{ such that } |\alpha| \leqslant k,\ D^\alpha f \in L^p\left(\mathring{\Omega}_{\mathcal{W}}\right) \right\},$$

where $L^p\left(\mathring{\Omega}_{\mathcal{W}}\right)$ denotes the Lebesgue space of order p on $\mathring{\Omega}_{\mathcal{W}}$, while, for the multi-index α, $D^\alpha f$ is the classical partial derivative of order α, interpreted in the weak sense.

The following result is the counterpart, in our context, of the corresponding one obtained in [**JW84**] (Theorem 1 of Chapter VI, on page 141), or, in a more powerful way, when $p > 1$, in the case of the von Koch Snowflake (which preserves Markov's Inequality), as shown in [**Wal91**], Proposition 4, on page 120.

THEOREM 4.34 (**The trace of Sobolev spaces viewed as Besov spaces**). *Given a positive integer k, and a real number $p \geqslant 1$, we set*

$$\beta_{k,p} = k - \frac{2 - D_{\mathcal{W}}}{p}.$$

We then have that

$$W_k^p \left(\mathring{\Omega}_{\mathcal{W}} \right)_{|\Gamma_{\mathcal{W}}} = B_{\beta_{k,p}}^{p,p} \left(\Gamma_{\mathcal{W}} \right).$$

REMARK 4.35. Note that, as was previoulsy explained in the introduction, the fact that the Weierstrass Curve preserves Markov's Inequality plays an important part in this theorem. Indeed, it ensures that the correspondance between the Sobolev space and its trace on the fractal boundary, is obtained in a constructive way.

COROLLARY 4.36 (**The specific case $k = 2$, and its consequences – order of the Laplace-type operator**). *Given a real number $p \geqslant 1$, we then have that*

$$W_2^p \left(\mathring{\Omega}_{\mathcal{W}} \right)_{|\Gamma_{\mathcal{W}}} = B_{\beta_{2,p}}^{p,p} \left(\Gamma_{\mathcal{W}} \right),$$

where

$$\beta_{2,p} = 2 - \frac{2 - D_{\mathcal{W}}}{p} = 2 + \frac{1}{p} \frac{\ln \lambda}{\ln N_b} < 2.$$

In particular, in the case when $p = 2$, which corresponds to

$$\beta_{2,2} = 2 - \frac{2 - D_{\mathcal{W}}}{2},$$

and provided that

$$(3 - D_{\mathcal{W}}) \left\{ 2 \left(\frac{1}{2} - \frac{s-1}{D_{\mathcal{W}}} \right) \right\} + (2 - D_{\mathcal{W}}) \left(D_{\mathcal{W}} - 2 \left(1 + \frac{2 - D_{\mathcal{W}}}{2} \right) \right) < 2,$$

i.e.,

$$1 - D_{\mathcal{W}} - 2\,(3 - D_{\mathcal{W}})\, \frac{s-1}{D_{\mathcal{W}}} + (2 - D_{\mathcal{W}})\,(2\,D_{\mathcal{W}} - 4) < 0,$$

or, equivalently,

$$s > 1 + D_{\mathcal{W}} \left(\frac{1 - D_{\mathcal{W}} + (2 - D_{\mathcal{W}})\,(2\,D_{\mathcal{W}} - 4)}{2\,(3 - D_{\mathcal{W}})} \right),$$

we then have that

$$W_2^2 \left(\Omega_{\mathcal{W}} \right)_{|\Gamma_{\mathcal{W}}} = B_{\beta_{2,2}}^{2,2} \left(\Gamma_{\mathcal{W}} \right),$$

where

$$\beta_{2,2} = 2 - \frac{1}{2} \frac{\ln \lambda}{\ln N_b} < 2.$$

Consequently, by analogy with the classical theories, the Laplacian on the Weierstrass Curve (see Remark 4.26 above) arises as a differential operator of order $\beta_{2,2} \in]1, 2[$.

REMARK 4.37. Indeed, a Laplace-type operator arising in this way is non-local. Our aim is to sketch a connection between such a Laplace-type operator and the fractal Laplacian in the sense of Kigami, Strichartz and others, which is defined for certain classes of fractals by means of discrete approximations. The point is that those operators each correspond to a specific order of differentiation, which is not the same. As it can be found in [**Str03**], when the fractal Laplacian in the sense of Kigami, Strichartz and others, is involved, the associated function spaces are distinct from the function spaces obtained via traces, as is the case in the work of A. Jonsson and H. Wallin, which can be found, for instance, in [**JW84**].

REMARK 4.38 (**Connection with the optimal exponent of Hölder continuity**). We note that

$$\beta_{2,2} = 2 - \frac{\alpha_{\mathcal{W}}}{2},$$

where the Codimension $\alpha_{\mathcal{W}} = 2 - D_{\mathcal{W}} = -\dfrac{\ln \lambda}{\ln N_b} \in]0,1[$ is the best (i.e., optimal) Hölder exponent for the Weierstrass function, as was initially obtained by G. H. Hardy in [**Har16**]), and then, by a completely different method – geometrically – in [**DL25c**] (this latter result is recalled in Theorem 2.21, and in Corollary 2.22).

PROPERTY 4.39 (**Connection with fractional derivatives**). *In Definition* 4.24, *the Besov norm of a function f defined on the Weierstrass Curve involves the integral*

$$\int\int_{(T,Y)\in\Gamma_{\mathcal{W}}^2} \frac{|f(T)-f(Y)|^q}{d_{Eucl}^{D_{\mathcal{W}}+\alpha\,q}(T,Y)}\, d\mu^2 . \tag{18}$$

For $p=2$, $q=1$ and $\alpha = 3-2\,D_{\mathcal{W}}$, i.e., if $f \in B^{2,1}_{3-2\,D_{\mathcal{W}}}(\Gamma_{\mathcal{W}})$, the integral in expression (18) *just above can be connected to the associated fractional derivative of order*

$$\gamma = 2 - D_{\mathcal{W}} \in]0,1[,$$

defined, subject to its existence, for any vertex $Y \in V_{m_0} \subset \Gamma_{\mathcal{W}}$, with $m_0 \in \mathbb{N}$, by the following expression:

$$D^{2-D_{\mathcal{W}}} f(Y) = \frac{2-D_{\mathcal{W}}}{\Gamma(D_{\mathcal{W}}-1)} \lim_{m\to\infty} \int_{T\in\mathcal{D}(\Gamma_{\mathcal{W}}),\, T \underset{m}{\sim} Y} \frac{|f(T)-f(Y)|}{d_{Eucl}^{3-D_{\mathcal{W}}}(T,Y)}\, d\mu,$$

i.e., equivalently,

$$D^{2-D_{\mathcal{W}}} f(Y) = \frac{2-D_{\mathcal{W}}}{\Gamma(D_{\mathcal{W}}-1)} \lim_{m\to\infty} \varepsilon_m^{m\,(D_{\mathcal{W}}-2)} \sum_{T \underset{m}{\sim} Y} \mu^{\mathcal{L}}(T,\mathcal{P}_m,\mathcal{Q}_m)\, \lambda_{\star,m}\, \frac{\left|\tilde f_m(T) - \tilde f_m f(Y)\right|}{d_{Eucl}^{3-D_{\mathcal{W}}}(T,Y)},$$

where, for any $m \in \mathbb{N}$, $\lambda_{\star,m}$ is the mth scalar coefficient involved, and where Γ denotes the usual Gamma function.

REMARK 4.40 (**Bounds for fractional derivatives**). Note that due in the forthcoming case of functions f satisfying the first condition given in Definition 5.1 below, there exist strictly positive constants $\tilde C_{inf}$ and $\tilde C_{sup}$ such that, for all vertices $(T,Y) \in V_m \times V_m \subset \Gamma_{\mathcal{W}} \times \Gamma_{\mathcal{W}}$,

$$\tilde C_{inf}\, d_{Eucl}^{2-D_{\mathcal{W}}}(T,Y) \leqslant \left|\tilde f_m(T) - \tilde f_m(Y)\right| \leqslant \tilde C_{sup}\, d_{Eucl}^{2-D_{\mathcal{W}}}(T,Y).$$

Hence, we have that, for all vertices $(T,Y) \in V_m \times V_m \subset \Gamma_{\mathcal{W}} \times \Gamma_{\mathcal{W}}$,

$$\frac{\tilde C_{inf}}{d_{Eucl}(T,Y)} \leqslant \frac{\left|\tilde f_m(T) - \tilde f_m f(Y)\right|}{d_{Eucl}^{3-D_{\mathcal{W}}}(T,Y)} \leqslant \frac{\tilde C_{sup}}{d_{Eucl}(T,Y)},$$

or, equivalently, expressed in terms of the mth cohomology infinitesimal ε_m^m (see Definition 3.1),

$$\frac{\tilde{C}_{inf}}{\varepsilon_m^m} \leqslant \frac{\left|\tilde{f}_m(T) - \tilde{f}_m f(Y)\right|}{d_{Eucl}^{3-D_{\mathcal{W}}}(T,Y)} \leqslant \frac{\tilde{C}_{sup}}{\varepsilon_m^m},$$

which, in conjunction with the estimates given in Property 3.6, namely,

$$\mu^{\mathcal{L}}(X, \mathcal{P}_m, \mathcal{Q}_m) \lesssim \varepsilon_m^{m(3-D_{\mathcal{W}})},$$

yields

$$\varepsilon_m^{m(D_{\mathcal{W}}-2)} \varepsilon_m^{m(3-D_{\mathcal{W}})} \frac{\tilde{C}_{inf}}{\varepsilon_m^m} \lesssim \varepsilon_m^{m(D_{\mathcal{W}}-2)} \mu^{\mathcal{L}}(T, \mathcal{P}_m, \mathcal{Q}_m) \frac{\left|\tilde{f}_m(T) - \tilde{f}_m f(Y)\right|}{d_{Eucl}^{3-D_{\mathcal{W}}}(T,Y)}$$

and

$$\varepsilon_m^{m(D_{\mathcal{W}}-2)} \varepsilon_m^{m(3-D_{\mathcal{W}})} \frac{\tilde{C}_{inf}}{\varepsilon^m} \lesssim \varepsilon_m^{m(D_{\mathcal{W}}-2)} \varepsilon_m^{m(3-D_{\mathcal{W}})} \frac{\tilde{C}_{sup}}{\varepsilon_m^m};$$

i.e.,

$$\tilde{C}_{inf} \lesssim \varepsilon_m^{m(D_{\mathcal{W}}-2)} \mu^{\mathcal{L}}(T, \mathcal{P}_m, \mathcal{Q}_m) \frac{\left|\tilde{f}_m(T) - \tilde{f}_m f(Y)\right|}{d_{Eucl}^{3-D_{\mathcal{W}}}(T,Y)} \lesssim \tilde{C}_{sup}.$$

Since the vertex $Y \in V_m$ admits at most two adjacent vertices in V_m, we then deduce that

$$\tilde{C}_{inf} \left|\lambda_{\star,m}\right| \lesssim \varepsilon_m^{m(D_{\mathcal{W}}-2)} \mu^{\mathcal{L}}(T, \mathcal{P}_m, \mathcal{Q}_m) \left|\lambda_{\star,m}\right| \frac{\left|\tilde{f}_m(T) - \tilde{f}_m f(Y)\right|}{d_{Eucl}^{3-D_{\mathcal{W}}}(T,Y)} \lesssim \tilde{C}_{sup} \left|\lambda_{\star,m}\right|.$$

Note that in the case of a continuous function f, the atomic coefficients $\lambda_{\star,m}$ are necessarily bounded (since $\Gamma_{\mathcal{W}}$ is compact, and hence, f is bounded).

5. Towards an extension of Morse theory

Classical Morse theory (see, e.g., [**Bot82**], [**Bot88**] and [**Mil63**]) enables us to explore the shape (i.e., the topology) of a smooth manifold by means of the study of the critical points of suitable smooth functions defined on the manifold. Such functions are required to be nondegenerate – in the sense that their Hessian determinant is nonzero at critical points – and are then called Morse functions.

In the classical Morse theory, i.e., for smooth manifolds, the height function plays a major role. More precisely, along with its critical points and its level sets, it encodes the information that enables us to reconstruct the manifold.

For fractal curves or IFDs such as, for instance, the Weierstrass Curve, this does not make sense anymore. Since fractals are involved, a change of shape occurs at each vertex of each prefractal approximation.

In particular, we will show that the Weierstrass function $\mathcal{W}$ is a fractal Morse function. Note (as in [**DL25c**], [**DL24c**], [**DL24a**]) that $\mathcal{W}$ can be viewed as a function on $\Gamma_{\mathcal{W}}$, namely, the identity function; see Remark 5.19 below.

By using some of the results of [**DL25c**], [**DL24c**], [**DL24a**], we hereafter begin to lay the foundations of a fractal Morse theory that should eventually enable us to explore the shape of *fractal manifolds* (viewed as higher-dimensional IFDs). The important example of the Weierstrass Curve $\Gamma_{\mathcal{W}}$ – or, rather, of the associated Weierstrass IFD – sheds a useful light, especially when it comes to studying not only the (appropriately defined) critical points, but also other suitable (inflection) points of *fractal Morse functions*, and the remainaing points, which are themselves subject to a change of shape, in connection with the associated angle between the corresponding adjacent edges; i.e., a typical change of curvature.

In this light, the simple knowledge of the equivalent of the usual Morse indexes, along with a suitable analogue of the height function, does not appear as being sufficient when it comes to reconstruct the fractal IFD. The missing data can be obtained by means of the sequence of cohomological integers associated, at each step $m \in \mathbb{N}$ of the prefractal approximation, with the set of vertices V_m; see Definition 5.8.

Also, in our context, the sequence of sets of critical point introduced in Definition 5.12 is an increasing sequence. For this reason, instead of the height function, we choose to consider *the increasing reordering* of *the absolute heights*, as introduced in Definition 5.24 below. We believe that this increasing reordering still bears the fractality of the Weierstrass Curve.

As was previously encountered in Corollary 2.25, at any given level $m \in \mathbb{N}$ of the prefractal graph approximation of the Curve, the extreme and bottom vertices of the polygons $\mathcal{P}_{m,k}$, $0 \leqslant k \leqslant N_b^m - 1$, are respectively local maxima and minima of the Weierstrass function $\mathcal{W}$. Those points are isolated ones, a result that directly comes from the construction of the Weierstrass Curve. Then, there remains to identify points that could play the role of degenerate ones, and to define *fractal suited Morse functions*. Also, as was alluded to in the introduction, the respective notions of *minimal real Complex Dimension* and *cohomological vertex integer* naturally arise from our results obtained in [**DL24c**] on the fractal cohomology of the Weierstrass Curve. For the sake of a better understanding, we next briefly recall those results.

DEFINITION 5.1 (**Set of functions of the same nature as the Weierstrass function $\mathcal{W}$** [**DL24c**]).

i. We say that a continuous, complex-valued function f, defined on $\Gamma_{\mathcal{W}} \supset V^\star$, is *of the same nature as the Weierstrass function* $\mathcal{W}$, if it satisfies local Hölder and reverse-Hölder properties analogous to those satisfied by the Weierstrass function $\mathcal{W}$; i.e., for any $m \in \mathbb{N}$ and for any pair of adjacent vertices (M, M') with respective affixes $(z, z') \in \mathbb{C}^2$, of the prefractal graph $\Gamma_{\mathcal{W}_m}$, we have that

$$\tilde{C}_{inf} \, |z' - z|^{2-D_{\mathcal{W}}} \leqslant \left| f(z') - f(z) \right| \leqslant \tilde{C}_{sup} \, |z' - z|^{2-D_{\mathcal{W}}},$$

where $\tilde{C}_{inf}$ and $\tilde{C}_{sup}$ denote suitable positive and finite constants possibly depending on f. This can be written, equivalently, as

$$\left| z - z' \right|^{2-D_{\mathcal{W}}} \lesssim \left| f(z) - f(z') \right| \lesssim \left| z - z' \right|^{2-D_{\mathcal{W}}}. \quad (\blacklozenge) \tag{19}$$

(Compare with Theorem 2.21, and Corollary 2.22 above.)

Hereafter, we will denote by $\mathcal{H}\ddot{o}ld\,(\Gamma_{\mathcal{W}})$ the set consisting of the continuous, complex-valued functions f, defined on $\Gamma_{\mathcal{W}} \supset V^\star$ and satisfying relation (19).

ii. Moreover, we will denote by $\mathcal{H}\ddot{o}ld_{geom}\,(\Gamma_{\mathcal{W}}) \subset \mathcal{H}\ddot{o}ld\,(\Gamma_{\mathcal{W}})$ the subset of $\mathcal{H}\ddot{o}ld\,(\Gamma_{\mathcal{W}})$ consisting of the functions f of $\mathcal{H}\ddot{o}ld\,(\Gamma_{\mathcal{W}})$ which satisfy the following *additional geometric condition* (20), again, for any pair of adjacent vertices (M, M') of the prefractal graph V_m with respective affixes $(z, z') \in \mathbb{C}^2$, and for $m \in \mathbb{N}$ arbitrary; namely,

$$\left| \arg\left(f(z)\right) - \arg\left(f(z')\right) \right| \lesssim |z - z'|^{D_{\mathcal{W}}-1}. \quad (\spadesuit) \tag{20}$$

REMARK 5.2. Note that, according to the results of [**DL25c**] and [**DL24c**], the Weierstrass function $\mathcal{W}$ belongs to $\mathcal{H}\ddot{o}ld_{geom}(\Gamma_{\mathcal{W}})$ – and hence, also, to $\mathcal{H}\ddot{o}ld(\Gamma_{\mathcal{W}})$.

DEFINITION 5.3 (**(m,p)-Fermion** [**DL24c**]). By analogy with particle physics, given a pair of integers (m,p), with $m \in \mathbb{N}$ and $p \in \mathbb{N}^\star$, we will call *(m,p)-fermion* on V_m, with values in $\mathbb{C}$, any antisymmetric map f from V_m^{p+1} to $\mathbb{C}$. Note that these maps are not assumed to be multilinear (which would be meaningless here, anyway, since V_m is not a vector space).

For any $m \in \mathbb{N}$, an *$(m,0)$-fermion* on V_m (or a *0-fermion*, in short) is simply a map f from V_m to $\mathbb{C}$. We adopt the convention according to which a 0-fermion on V_m is a *0-antisymmetric map* on V_m.

In the sequel, for any $(m,p) \in \mathbb{N}^2$, we will denote by $\mathcal{F}^p(V_m,\mathbb{C})$ the complex vector space of (m,p)-fermions on V_m, which makes it an abelian group with respect to the addition, with an external law from $\mathbb{C}\times\mathcal{F}^p(V_m,\mathbb{C})$ to $\mathcal{F}^p(V_m,\mathbb{C})$.

DEFINITION 5.4 (**$(m-1,m)$-differentials** [**DL24c**]). Here and thereafter, for $(m,p) \in \mathbb{N}\times\mathbb{N}^\star$, V_m^p denotes the p-fold product of V_m by itself.

Given a strictly positive integer m, we define the $(m-1,m)$-*differential* $\delta_{m-1,m}$ from $\mathcal{F}^0(V_m,\mathbb{C})$ to $\mathcal{F}^{N_b+1}(V_m,\mathbb{C})$, for any f in $\mathcal{F}^0(\Gamma_{\mathcal{W}},\mathbb{C})$ and any

$$\left(M_{i,m-1},M_{i+1,m-1},M_{j+1,m},\ldots,M_{j+N_b-2,m}\right) \in V_m^{N_b+1}$$

such that

$$M_{i,m-1} = M_{j,m} \quad\text{and}\quad M_{i+1,m-1} = M_{j+N_b,m},$$

by

$$\begin{aligned}&\delta_{m-1,m}(f)\left(M_{i,m-1},M_{i+1,m-1},M_{j+1,m},\ldots,M_{j+N_b-1,m}\right)\\ &= c_{m-1,m}\left(\sum_{q=0}^{N_b}(-1)^q f\left(M_{j+q,m}\right)\right),\end{aligned}$$

where $c_{m-1,m}$ denotes a suitable positive constant. The value of this constant is unimportant for the remainder of this discussion.

THEOREM 5.5 (**Fractal cohomology of the Weierstrass Curve** [**DL24c**]). *Within the set $\mathcal{H}\ddot{o}ld_{geom}(\Gamma_{\mathcal{W}})$ of continuous, complex-valued functions f, defined on the Weierstrass Curve $\Gamma_{\mathcal{W}} \supset V^\star = \bigcup_{n\in\mathbb{N}} V_n$ (see part ii. of Definition* 5.1 *above), let us consider the Complex (which can be called* the Total Fractal Complex *of $\Gamma_{\mathcal{W}}$),*

$$H^\star = H^\bullet\left(\mathcal{F}^\bullet(\Gamma_{\mathcal{W}},\mathbb{C}),\delta^\bullet\right) = \bigoplus_{m=0}^{\infty} H^m,$$

where, for any positive integer m, with the additional convention $\delta_{-2,-1} = 0$ and $\delta_{-1,0} = 0$ (which ensures that $H^0 = \{0\}$), H^m is the cohomology group

$$H^m = \ker\delta_{m-1,m}/Im\,\delta_{m-2,m-1},$$

which consist in maps the expression of which is obtained as the difference of an antisymmetric map with respect to the set of vertices

$$\left(M_{i,0},M_{i+1,0},M_{j+1,m},\ldots,M_{j+N_b^m-1,m}\right)$$

and of an antisymmetric map with respect to the set of vertices

$$\left(M_{i,0},M_{i+1,0},M_{j+1,m-1},\ldots,M_{j+N_b^{m-1}-1,m-1}\right).$$

Then, $H^\star$ is the set consisting of functions f on $\Gamma_{\mathcal{W}}$, viewed as 0-fermions (in the sense of Definition 5.3), and, for any integer $m \geqslant 1$, of the restrictions to V_m of (m, N_b^m+1)-fermions, i.e., the restrictions to (the Cartesian product space) $V_m^{N_b^m+1}$ of antisymmetric maps on $\Gamma_{\mathcal{W}}$, with $N_b^m + 1$ variables (corresponding to the vertices of V_m), involving the restrictions to V_m of the continuous, complex-valued functions f on $\Gamma_{\mathcal{W}}$ – as, naturally, the aforementioned 0-fermions – satisfying the following convergent (and even, absolutely convergent) Taylor-like expansions (with $V^\star = \bigcup_{n\in\mathbb{N}} V_n$),

(21)

$$\forall\, M_{\star,\star} \in V^\star :$$
$$f(M_{\star,\star}) = \sum_{k=0}^{\infty} c_k(f, M_{\star,\star})\, \varepsilon_k^{k\,(2-D_{\mathcal{W}})}\, \varepsilon^{i\,\ell_k\,\mathbf{p}} = \sum_{k=0}^{\infty} c_k(f, M_{\star,\star})\, \varepsilon_k^{k\,(2-D_{\mathcal{W}})+i\,\ell_k\,\mathbf{p}},$$

where, for each integer $k \geqslant 0$, $c_k(\star,\star) = c_k(f,\star) \in \mathbb{C}$, the number $\varepsilon_k^k > 0$ is the kth cohomology infinitesimal introduced in Definition 3.1, and where ℓ_k denotes an integer (in $\mathbb{Z}$) such that

$$0 \leqslant \left\{\ell_k\, \frac{\ln \varepsilon_k^k}{\ln N_b}\right\} \lesssim \frac{\varepsilon_k^{k\,(D_{\mathcal{W}}-1)}}{2\,\pi},$$

with $\{.\}$ being the fractional part.

Note that since the functions f involved are uniformly continuous on the Weierstrass Curve $\Gamma_{\mathcal{W}} \supset V^\star$, and since the set $V^\star$ is dense in $\Gamma_{\mathcal{W}}$, they are uniquely determined by their restriction to $V^\star$, as given by relation (21). We caution the reader, however, that at this stage of our investigations, we do not know wether $f(M)$ is given by an expansion analogous to the one in relation (21), for every $M \in \Gamma_{\mathcal{W}}$, rather than just for all $M \in V^\star$.

For each $M_\star = M_{\star,m} \in V^\star$, as is shown in [**DL24c**], *the coefficients $c_k(\star,\star)$ (for any $k \in \mathbb{N}$) are the residues at the possible* cohomological Complex Dimensions $-(k\,(2-D_{\mathcal{W}}) + i\,k\,\ell_k\,\mathbf{p}))$ *of a suitable global scaling zeta function evaluated at $M_\star$.*

The group $H^\star = \bigoplus_{m=0}^{\infty} H^m$ is called the total fractal cohomology group *of the Weierstrass Curve $\Gamma_{\mathcal{W}}$ (or else, of the Weierstrass function $\mathcal{W}$).*

DEFINITION 5.6 (**Minimal (real) complex dimension of a prefractal approximation**). Given $m \in \mathbb{N}$, we define the *minimal real Complex Dimension* of the prefractal approximation $\Gamma_{\mathcal{W}_m}$ as

$$\omega_m = D_{\mathcal{W}} - m\,(2 - D_{\mathcal{W}}).$$

REMARK 5.7. Clearly, the successive prefractal approximations play the role of level sets, in our present context.

DEFINITION 5.8 (**Cohomological vertex integer**). Given $m \in \mathbb{N}$, and a vertex $M_{j,m} = M_{(N_b-1)\,k'+k'',m} \in V_m$, of abscissa $\left((N_b-1)\,k' + k''\right)\varepsilon_m^m$, where $0 \leqslant k' \leqslant N_b^m - 1$ and $0 \leqslant k'' \leqslant N_b - 1$, we introduce the *cohomological vertex integer* $\ell_{j,m} \in \mathbb{N}$ associated to the vertex $M_{j,m}$ (which is also the $(k'')^{th}$ vertex of the polygon $\mathcal{P}_{m,k'}$; see part *iv.* of Property 2.11), as

$$\ell_{j,m} = \ell_{k',k'',m} = (N_b - 1)\, k' + k''. \tag{22}$$

Depending on the context; that is,

i. when the cohomological vertex integer enables one to locate the vertex $M_{j,m}$

ii. when it is used in a more general framework, i.e., in order to describe the generators of cohomology groups (see [**DL24b**]);

we will use the best suited notation between $\ell_{j,m}$, in case *i.*, or $\ell_{k',k'',m}$, in case *ii.*

Proposition 5.9 (**Cross-scales paths, and associated sequence of vertex integers**). *Given $m \in \mathbb{N}$, $0 \leq j \leq \# V_m - 1$ and a vertex $M_{j,m} = M_{(N_b-1)\,k'+k'',m}$ in V_m, with $0 \leq k' \leq N_b^m - 1$ and $0 \leq k'' \leq N_b - 1$, we introduce* the cross-scales path $\mathcal{P}ath\,(P_{k''}, M_{j,m})$, *where $P_{k'}$ is the $(k')^{th}$ fixed point of the map $T_{k'}$ (see Proposition 2.6, along with Notation 6), as the ordered set*

$$\left(M_{j_{k,m},k}\right)_{0 \leq k \leq m}$$

such that:

i. For $0 \leq k \leq m$, each vertex $M_{j_{k,m},k}$ strictly belongs to V_k, i.e., it is in the kth prefractal approximation $\Gamma_{\mathcal{W}_k}$, and not in $\Gamma_{\mathcal{W}_{k+1}}$.

ii. For $1 \leq k \leq m$, each vertex $M_{j_{k,m},k} = M_{(N_b-1)\,k'_{k,m}+k'',k}$, with $0 \leq k'_{k,m} \leq N_b^k - 1$, is the image of the point $M_{j_{k-1,m},k-1}$ under the map T_i (see again Proposition 2.6), where $i \in \{0, \ldots, N_b - 1\}$ is the smallest admissible value. We thus also have that

$$M_{j_{k-1,m},k-1}$$
$$= \left(\frac{(N_b - 1)\left(k'_{k,m} - i\,(N_b - 1)\,N_b^{k-1}\right) + k''}{(N_b - 1)\,N_b^{k-1}}, \mathcal{W}\left(\frac{(N_b - 1)\left(k'_{k,m} - i\,(N_b - 1)\,N_b^{k-1}\right) + k''}{(N_b - 1)\,N_b^{k-1}} \right) \right).$$

This latter point is also ***the*** $(k'')^{th}$ ***vertex of the polygon*** $k'_{k,m} - i\,(N_b - 1)\,N_b^{k-1}$ *(see part iv. of Property 2.11).*

The sequence of vertex integers associated with the cross-scales path $\mathcal{P}ath\,(P_{k''}, M_{j,m})$ (or, in short, and equivalently, also refered to as the sequence of vertex integers associated with $M_{j,m}$) is the sequence $\left(\ell_{j_{k,m},k}\right)_{0 \leq k \leq m}$, where, for $0 \leq k \leq m$, $\ell_{j_{k,m},k}$ is the cohomological vertex integer *associated with the* vertex $M_{j_{k,m},k}$ *(see Definition 5.8).*

We now recall the following result, obtained in [**DL24c**].

Theorem 5.10 (**Complex dimensions series expansion of the complexified Weierstrass function $\mathcal{W}_{comp}$ and of the Weierstrass function $\mathcal{W}$**). *For any sufficiently large positive integer m (i.e., for all $m \geq m_0$, with $m_0 \in \mathbb{N}^\star$ optimal) and for any j in $\{0, \ldots, \# V_m - 1\}$, we have the following fractal power series* exact *expansion, indexed by the Complex Codimensions $k\,(D_{\mathcal{W}} - 2) + i\,k\,\ell_{j_{k,m},k}\,\mathbf{p}$,*

with $0 \leqslant k \leqslant m$,
(23)

$$\begin{aligned}\mathcal{W}_{comp}\left(j\,\varepsilon_m^m\right) &= \mathcal{W}_{comp}\left(\frac{j\,\varepsilon^m}{N_b-1}\right)\\ &= \varepsilon^{m\,(2-D_{\mathcal{W}})}\,\mathcal{W}_{comp}\left(\frac{j}{N_b-1}\right)+\sum_{k=0}^{m-1} c_{k,j,m}\,\varepsilon^{k\,(2-D_{\mathcal{W}})}\,\varepsilon^{i\,\ell_{k,j,m}\,\mathbf{p}}\\ &= \sum_{k=0}^{m} c_{k,j,m}\,\varepsilon^{k\,(2-D_{\mathcal{W}})}\,\varepsilon^{i\,\ell_{j_{k,m},k}\,\mathbf{p}},\end{aligned}$$

where, for $0 \leqslant k \leqslant m$, ε^k *is the* k^{th} *intrinsic cohomology infinitesimal, introduced in Definition* 3.1, *with* $\mathbf{p} = \dfrac{2\,\pi}{\ln N_b}$ *denoting the oscillatory period of the Weierstrass Curve, as introduced in* [**DL25c**] *and where:*

i. $\ell_{j_{k,m},k} \in \mathbb{N}$ *is the cohomological vertex integer associated with the vertex* $M_{j_{k,m},k}$ *(see Definition* 5.8, *along with Proposition* 5.9*).*

ii. $c_{m,j,m} = \mathcal{W}_{comp}\left(\dfrac{j}{N_b-1}\right)$ *and, for* $0 \leqslant k \leqslant m-1$, $c_{k,j,m} \in \mathbb{C}$ *is given by*

$$c_{k,j,m} = \exp\left(\frac{2\,i\,\pi}{N_b-1}\,j\,\varepsilon^{m-k}\right). \tag{24}$$

For $0 \leqslant k \leqslant m$, *the coefficient* $c_{k,j,m}$ *will also be referred to as* the kth Weierstrass coefficient associated with the vertex $M_{j_{k,m},k} \in V_k$.

For any $m \in \mathbb{N}$, *the complex numbers* $\{c_{0,j,m+1},\ldots,c_{m+1,j,m+1}\}$ *satisfy the following recurrence relations:*

$$c_{m+1,j,m+1} = \mathcal{W}\left(\frac{j}{N_b-1}\right) = c_{m,j,m} \tag{25}$$

and

$$\forall\, k \in \{1,\ldots,m\}:\quad c_{k,j,m+1} = c_{k-1,j,m}. \tag{26}$$

In addition, since relation (23) *is valid for any* $m \in \mathbb{N}^\star$ *(and since, clearly, relation* (24) *implies that the coefficients* $c_{k,j,m}$ *are nonzero for* $0 \leqslant k \leqslant m$*), we deduce that the associated Complex Dimensions (i.e., in fact, the Complex Dimensions associated with the complexified Weierstrass function) are*

$$D_{\mathcal{W}} - k\,(2-D_{\mathcal{W}}) + i\,\ell_{j_{k,m},k}\,\mathbf{p}, \tag{27}$$

where $0 \leqslant j \leqslant \#V_m - 1$, $0 \leqslant k \leqslant m$ *and* $\ell_{j_{k,m},k} \in \mathbb{N}$ *is the cohomological vertex integer associated with the vertex* $M_{j_{k,m},k}$ *(see Definition* 5.8, *along with Proposition* 5.9*).*

Note that the Complex Dimension in (27) *is real if and only if it is equal to* $D_{\mathcal{W}} - k\,(2-D_{\mathcal{W}})$ *(i.e., if and only if* $\ell_{j_{k,m},k} = 0$*).*

This immediately ensures, for the Weierstrass function (i.e., the real part of the Complexified Weierstrass function $\mathcal{W}_{comp}$*), that, for any strictly positive integer* m

and for any j in $\{0, \ldots, \#V_m - 1\}$,
(28)

$$\begin{aligned} \mathcal{W}\left(j\,\varepsilon_m^m\right) &= \varepsilon^{m\,(2-D_{\mathcal{W}})}\,\mathcal{W}_{comp}\left(\frac{j}{N_b-1}\right) + \sum_{k=0}^{m-1} \varepsilon^{k\,(2-D_{\mathcal{W}})}\,\mathcal{R}e\left(c_{k,j,m}\,\varepsilon_k^{\,i\,\ell_{j_{k,m},k}\,\mathbf{p}}\right) \\ &= \frac{1}{2}\sum_{k=0}^{m} \varepsilon^{k\,(2-D_{\mathcal{W}})}\left(c_{k,j,m}\,\varepsilon^{i\,\ell_{j_{k,m},k}\,\mathbf{p}} + \overline{c_{k,j,m}}\,\varepsilon^{-\,i\,\ell_{j_{k,m},k}\,\mathbf{p}}\right). \end{aligned}$$

More generally, for any strictly positive integer m and for any integer j,

$$\mathcal{W}_{comp}\left(j\,\varepsilon^m\right) = \sum_{k=0}^{\infty} \varepsilon^{k\,(2-D_{\mathcal{W}})}\,c_{k,j,m}\,\varepsilon^{k\,(2-D_{\mathcal{W}})}\,\varepsilon^{i\,\ell_{j_{k,m},k}\,\mathbf{p}}, \tag{29}$$

where, for all $k \in \mathbb{N}$,

$$c_{k,j,m} = \varepsilon^{2\,i\,\pi\,N_b^k\,j\,\varepsilon^m}. \tag{30}$$

In addition, the Complex Dimensions associated with the (actual) Weierstrass function $\mathcal{W}$, are all exact and simple. Furthermore, they are given as in (27) above, except for the fact that $i\,\ell_{j_{k,m},k}\,\mathbf{p}$ should be replaced by $\pm i\,\ell_{j_{k,m},k}\,\mathbf{p}$. Accordingly, as should be the case, the nonreal Complex Dimensions of the Weierstrass function come in complex conjugate pairs.

In particular, for any $k \in \mathbb{N}$, the only real Complex Dimension of $\mathcal{W}$ with abscissa $d_k = D_{\mathcal{W}} - k\,(2 - D_{\mathcal{W}})$ is d_k itself, while the other Complex Dimensions of $\mathcal{W}$ with abscissa d_k consist of countably many nonreal complex conjugate pairs.

REMARK 5.11. Note that each integer $\ell_{k,j,m}$ (choosen as in Definition 5.8 just above) carries the information associated to the vertex point $M_{((N_b-1)\,k+j),m} \in V_m$ (and, hence, to a change of shape) when switching from $M_{((N_b-1)\,k+j),m} \in V_m$ to its consecutive neighbor $M_{((N_b-1)\,k+j+1),m} \in V_m$.

DEFINITION 5.12 (**Sequence of sets of critical points of the Weierstrass Curve**). We define *the sequence of sets of critical points* of the Weierstrass function $\mathcal{W}$ – or, equivalently, of the Weierstrass IFD – as the sequence $\left(Crit\left(\Gamma_{\mathcal{W}_m}\right)\right)_{m\in\mathbb{N}}$ such that, for any $m \in \mathbb{N}$, the set $Crit\left(\Gamma_{\mathcal{W}_m}\right)$ is obtained as the union of the set of local extrema of $\mathcal{W}$ given in Corollary 2.25, and of the set of vertices with a reentrant angle of the prefractal approximation $\Gamma_{\mathcal{W}_m}$, as given in Property 2.26.

DEFINITION 5.13 (**Topological Laplacian of level $m \in \mathbb{N}^\star$**). For any $m \in \mathbb{N}^\star$, and any real-valued function f, defined on the set V_m of the vertices of the prefractal graph $\Gamma_{\mathcal{W}_m}$, we introduce *the topological Laplacian of level m*, $\Delta_m^\tau(f)$, as applied to f, as follows:

$$\forall\, X \in V_m \setminus \partial V_m : \quad \Delta_m^\tau f(X) = \sum_{Y \in V_m,\, Y \underset{m}{\sim} X} \left(f(Y) - f(X)\right).$$

As a consequence, in the case of the Weierstrass function $\mathcal{W}$, we also have that

$$\forall\, X \in V_m \setminus \partial V_m : \quad \Delta_m^\tau \mathcal{W}(X) = \sum_{Y \in V_m,\, Y \underset{m}{\sim} X} \left(\mathcal{W}(Y) - \mathcal{W}(X)\right).$$

Here, ∂V_m consists of the two (identified) endpoints of V_m.

Note that we are excluding the case when $m = 0$ here, because $V_0 = \partial V_0$.

PROPOSITION 5.14 (**Topological Laplacian of the Weierstrass function at vertices of a prefractal graph approximation**). *For any $m \in \mathbb{N}^\star$, any integer k in $\{0,\ldots,N_b^m-1\}$, and any j in $\{1,\ldots,N_b-2\}$, we have that*

$$\begin{aligned}&\Delta^\tau_m \mathcal{W}\left(\frac{k\,(N_b-1)+j}{(N_b-1)\,N_b^m}\right)\\&= \mathcal{W}\left(\frac{k\,(N_b-1)+j+1}{(N_b-1)\,N_b^m}\right)+\mathcal{W}\left(\frac{k\,(N_b-1)+j-1}{(N_b-1)\,N_b^m}\right)-2\,\mathcal{W}\left(\frac{k\,(N_b-1)+j}{(N_b-1)\,N_b^m}\right).\end{aligned}$$

PROPERTY 5.15 (**Sign of the topological Laplacian of level $m \in \mathbb{N}^\star$**).

i. For any positive integer m, and any k in $\{0,\ldots,N_b^m-1\}$, we have that, for the initial vertex of a polygon $\mathcal{P}_{m,k}$,

$$\mathcal{W}\left(\frac{k\,(N_b-1)-1}{(N_b-1)\,N_b^m}\right)<\mathcal{W}\left(\frac{k\,(N_b-1)}{(N_b-1)\,N_b^m}\right),$$

along with

$$\mathcal{W}\left(\frac{k\,(N_b-1)+1}{(N_b-1)\,N_b^m}\right)<\mathcal{W}\left(\frac{k\,(N_b-1)}{(N_b-1)\,N_b^m}\right),$$

which enables us to deduce that

$$\Delta^\tau_m \mathcal{W}\left(\frac{k\,(N_b-1)}{(N_b-1)\,N_b^m}\right)<0.$$

ii. When $N_b<7$, for any positive integer m, any k in $\{0,\ldots,N_b^m-1\}$, and any j in $\{1,\ldots,N_b-2\}$, one has:

⇝ *For the left-side vertices in Definition 2.16, distinct from the initial one, of a polygon $\mathcal{P}_{m,k}$, we have that*

$$\mathcal{W}\left(\frac{k\,(N_b-1)+j+1}{(N_b-1)\,N_b^m}\right)<\mathcal{W}\left(\frac{k\,(N_b-1)+j}{(N_b-1)\,N_b^m}\right),$$

along with

$$\begin{aligned}&\left|\mathcal{W}\left(\frac{k\,(N_b-1)+j-1}{(N_b-1)\,N_b^m}\right)-\mathcal{W}\left(\frac{k\,(N_b-1)+j}{(N_b-1)\,N_b^m}\right)\right|\\&>\left|\mathcal{W}\left(\frac{k\,(N_b-1)+j+1}{(N_b-1)\,N_b^m}\right)-\mathcal{W}\left(\frac{k\,(N_b-1)+j}{(N_b-1)\,N_b^m}\right)\right|,\end{aligned}$$

which enables us to obtain that

$$\Delta^\tau_m \mathcal{W}\left(\frac{k\,(N_b-1)+j}{(N_b-1)\,N_b^m}\right)>0.$$

⇝ *For the right-side vertices in Definition 2.16, distinct from the last one, of a polygon $\mathcal{P}_{m,k}$, we have that*

$$\mathcal{W}\left(\frac{k\,(N_b-1)+j+1}{(N_b-1)\,N_b^m}\right)>\mathcal{W}\left(\frac{k\,(N_b-1)+j}{(N_b-1)\,N_b^m}\right),$$

along with

$$\begin{aligned}&\left|\mathcal{W}\left(\frac{k\,(N_b-1)+j-1}{(N_b-1)\,N_b^m}\right)-\mathcal{W}\left(\frac{k\,(N_b-1)+j}{(N_b-1)\,N_b^m}\right)\right|\\&<\left|\mathcal{W}\left(\frac{k\,(N_b-1)+j+1}{(N_b-1)\,N_b^m}\right)-\mathcal{W}\left(\frac{k\,(N_b-1)+j}{(N_b-1)\,N_b^m}\right)\right|,\end{aligned}$$

which enables us to obtain that

$$\Delta_m^{\tau}\mathcal{W}\left(\frac{k\,(N_b-1)+j}{(N_b-1)\,N_b^m}\right)>0.$$

iii. When $N_b \geqslant 7$, for any $m \in \mathbb{N}^\star$, any integer k in $\{0,\ldots,N_b^m-1\}$, and any j in $\{1,\ldots,N_b-2\}$ such that

$$0<j\leqslant\frac{N_b-3}{4}\quad or\quad \frac{3\,N_b-1}{4}\leqslant j<N_b-1,$$

we respectively have that:

⇝ *For the left-side vertices in Definition* 2.16, *distinct from the initial one, of a polygon $\mathcal{P}_{m,k}$,*

$$\mathcal{W}\left(\frac{k\,(N_b-1)+j+1}{(N_b-1)\,N_b^m}\right)<\mathcal{W}\left(\frac{k\,(N_b-1)+j}{(N_b-1)\,N_b^m}\right),$$

along with

$$\begin{aligned}&\left|\mathcal{W}\left(\frac{k\,(N_b-1)+j-1}{(N_b-1)\,N_b^m}\right)-\mathcal{W}\left(\frac{k\,(N_b-1)+j}{(N_b-1)\,N_b^m}\right)\right|\\&<\left|\mathcal{W}\left(\frac{k\,(N_b-1)+j+1}{(N_b-1)\,N_b^m}\right)-\mathcal{W}\left(\frac{k\,(N_b-1)+j}{(N_b-1)\,N_b^m}\right)\right|,\end{aligned}$$

which yields

$$\Delta_m^{\tau}\mathcal{W}\left(\frac{k\,(N_b-1)+j}{(N_b-1)\,N_b^m}\right)<0.$$

⇝ *For the right-side vertices in Definition* 2.16, *distinct from the last one, of a polygon $\mathcal{P}_{m,k}$,*

$$\mathcal{W}\left(\frac{k\,(N_b-1)+j+1}{(N_b-1)\,N_b^m}\right)>\mathcal{W}\left(\frac{k\,(N_b-1)+j}{(N_b-1)\,N_b^m}\right),$$

with

$$\begin{aligned}&\left|\mathcal{W}\left(\frac{k\,(N_b-1)+j-1}{(N_b-1)\,N_b^m}\right)-\mathcal{W}\left(\frac{k\,(N_b-1)+j}{(N_b-1)\,N_b^m}\right)\right|\\&>\left|\mathcal{W}\left(\frac{k\,(N_b-1)+j+1}{(N_b-1)\,N_b^m}\right)-\mathcal{W}\left(\frac{k\,(N_b-1)+j}{(N_b-1)\,N_b^m}\right)\right|,\end{aligned}$$

which enables us to obtain that

$$\Delta_m^{\tau}\mathcal{W}\left(\frac{k\,(N_b-1)+j}{(N_b-1)\,N_b^m}\right)>0.$$

PROOF. This is an immediate consequence of the proof of Property 2.26 above, given in [**DL25c**]. □

DEFINITION 5.16 (m^{th}**-level discrete Hessian**). Given $m \in \mathbb{N}^\star$, any k in $\{0,\ldots,N_b^m-1\}$, and any j in $\{1,\ldots,N_b-2\}$, we define the *m^{th}-level discrete Hessian $\mathcal{H}_m$* as follows:

$$\mathcal{H}_m\left(\frac{k\,(N_b-1)+j}{(N_b-1)\,N_b^m}\right)=\Delta_m^{\tau}\mathcal{W}\left(\frac{k\,(N_b-1)+j}{(N_b-1)\,N_b^m}\right).$$

The vertex

$$\left(\frac{k\,(N_b-1)+j}{(N_b-1)\,N_b^m},\mathcal{W}\left(\frac{k\,(N_b-1)+j}{(N_b-1)\,N_b^m}\right)\right)$$

is said to be *nondegenerate*, with respect to $\mathcal{H}_m$, if

$$\mathcal{H}_m\left(\frac{k\,(N_b-1)+j}{(N_b-1)\,N_b^m}\right) \neq 0.$$

PROPERTY 5.17 (**Absence of degenerate points for the sequence of discrete Hessians**). *Given $m \in \mathbb{N}^\star$, any integer k in $\{0,\ldots,N_b^m-1\}$, and any j in $\{1,\ldots,N_b-2\}$, the mth-level discrete Hessian $\mathcal{H}_m$ introduced in Definition 5.16 just above, never vanishes at the critical points of f.*

PROOF. This is a direct consequence of Proposition 5.14. □

DEFINITION 5.18 (**Fractal Morse function**). A function f defined on the Weierstrass Curve $\Gamma_{\mathcal{W}}$ will be said to be *a fractal Morse function* it its critical points are nondegenerate; i.e., if, for any $m \in \mathbb{N}^\star$, its discrete Hessian $\mathcal{H}_m$ (see Definition 5.16), never vanishes at the critical points of f.

REMARK 5.19 (**The Weierstrass function viewed as the identity function on $\Gamma_{\mathcal{W}}$**). As in [**DL24c**], we set, for any real number t in $[0,1]$,

$$\gamma_{\mathcal{W}}(t) = (t, \mathcal{W}(t)).$$

We then obtain the identity function on the Weierstrass Curve $\Gamma_{\mathcal{W}}$ as $\mathbb{1}_{\Gamma_{\mathcal{W}}}$. In this manner, the Weierstrass function can be viewed as the identity map on $\Gamma_{\mathcal{W}}$.

PROPERTY 5.20 (**The Weierstrass function viewed as a fractal Morse function**). *Since the discrete Hessian introduced in Definition 5.16, never vanishes, the Weierstrass function $\mathcal{W}$ is a fractal Morse function, in the sense of Definition 5.18 just above.*

PROOF. This is a direct consequence of Property 5.17, according to which the Weierstrass function $\mathcal{W}$ does not have any degenerate point. □

DEFINITION 5.21 (**m^{th}-level fractal Morse index**). Given $m \in \mathbb{N}^\star$, and any j in $\{1,\ldots,\#V_m-1\}$, the *m^{th}-level fractal Morse index* $\iota_{j,m}$ of the vertex $M_{j,m}$ is defined as follows:

$$\iota_{j,m} = \begin{cases} ccc1, & \text{if } \mathcal{H}_m\left(\dfrac{j}{(N_b-1)\,N_b^m}\right) < 0\,, \\ 0\,, & \text{if } \mathcal{H}_m\left(\dfrac{j}{(N_b-1)\,N_b^m}\right) > 0. \end{cases}$$

REMARK 5.22. An overview of the values of the indexes for the different types of vertices involved is given in Table 1.

Vertex	**Junction point** (between consecutive polygons)	**Bottom point**	**Plain interior point (obtuse angle)**
Index	1	0	0

Vertex	**Left-side acute corner**	**Right-side acute corner**
Index	1	0

TABLE 1. An overview of the values of the indexes for the different types of vertices involved.

REMARK 5.23. One should note that, as in the classical Morse theory, the index of a nondegenerate critical point is equal to the dimension of the largest subspace of what plays the role of a *tangent space* between two points, i.e., in our context, the edge that connects them (a line segment), where the Hessian is negative definite. It thus takes the value 1 at local maxima, and zero at local minima. It also takes the value zero at the right-side vertices with reentrant interior angles provided in Property 2.26. This specific configuration corresponds, in a sense, to a sign change in the *curvature*.

DEFINITION 5.24 (**Absolute height sequence of the Weierstrass Curve**). We define *the absolute heights sequence* of the Weierstrass function $\mathcal{W}$ – or, equivalently, of the Weierstrass IFD – as the sequence of positive numbers

$$\left(Abs_{\mathcal{H}}\left(\Gamma_{\mathcal{W}_m}\right)\right)_{m\in\mathbb{N}},$$

where, for any $m \in \mathbb{N}$,

$$Abs_{\mathcal{H}}\left(\Gamma_{\mathcal{W}_m}\right) = \left\{\mathcal{W}\left(\frac{j}{N_b^m}\right) + \frac{1}{1-\lambda},\ 0 \leqslant j \leqslant \#V_m - 1\right\}.$$

REMARK 5.25. The fact that, for any $m \in \mathbb{N}$ and any j in $\{0,\ldots,\#V_m - 1\}$, the value $Abs_{\mathcal{H}}\left(\Gamma_{\mathcal{W}_m}\right)$ is positive simply comes from the fact that the minimum value of the Weierstrass function on $[0,1]$ is equal to $m_{\mathcal{W}} = -\dfrac{1}{1-\lambda}$; see Notation 13 in Section 3.

PROPERTY 5.26 (**Fractal Morse height increasing reordered sequence**). *Given $m \in \mathbb{N}$, the set $Abs_{\mathcal{H}}\left(\Gamma_{\mathcal{W}_m}\right)$ introduced in Definition 5.24, admits an associated increasing reordered set.*

PROOF. This simply comes from the fact that, for any $m \in \mathbb{N}$, the set $Abs_{\mathcal{H}}\left(\Gamma_{\mathcal{W}_m}\right)$ is a finite set of positive numbers, therefore allowing us to reorder the points of $Abs_{\mathcal{H}}\left(\Gamma_{\mathcal{W}_m}\right)$ in increasing order. □

DEFINITION 5.27 (**Fractal Morse height increasing reordered sequence**). We define *the fractal Morse height reordered sequence* of the Weierstrass function $\mathcal{W}$ – or, equivalently, of the Weierstrass IFD – as the sequence of positive increasing reordered numbers $\left(Abs_{\mathcal{H}}^{reord}\left(\Gamma_{\mathcal{W}_m}\right)\right)_{m\in\mathbb{N}}$ where, for any $m \in \mathbb{N}$,

$$Abs_{\mathcal{H}}^{reord}\left(\Gamma_{\mathcal{W}_m}\right)$$

is the increasing reordered set associated to the finite set $Abs_{\mathcal{H}}\left(\Gamma_{\mathcal{W}_m}\right)$.

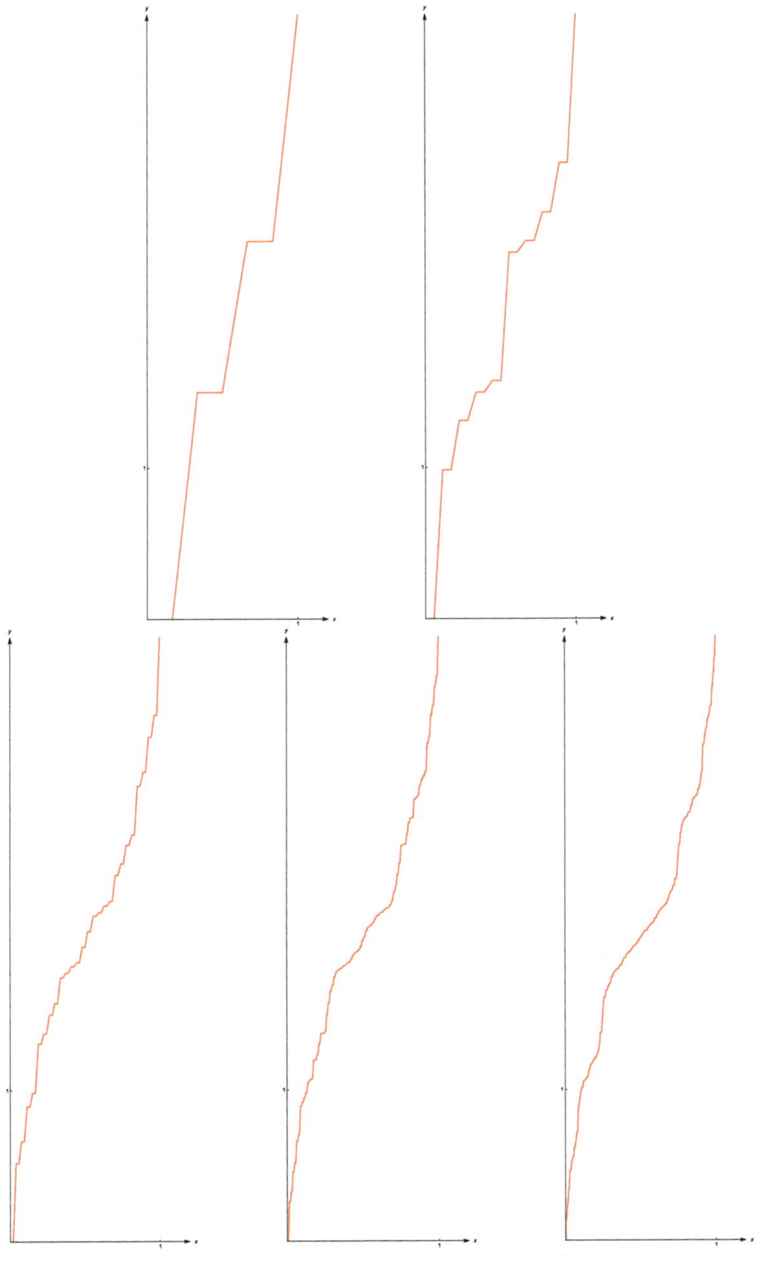

FIGURE 11. Plot of the fractal Morse heights $Abs_{\mathcal{H}}^{reord}(\Gamma_{\mathcal{W}_1})$, $Abs_{\mathcal{H}}^{reord}(\Gamma_{\mathcal{W}_2})$, $Abs_{\mathcal{H}}^{reord}(\Gamma_{\mathcal{W}_3})$, $Abs_{\mathcal{H}}^{reord}(\Gamma_{\mathcal{W}_4})$ and $Abs_{\mathcal{H}}^{reord}(\Gamma_{\mathcal{W}_5})$, presented from top to bottom and from left to right.

6. Further perspectives: The Weierstrass Curve as the projection of a vertical comb

In this section, we place ourselves in the Euclidean plane of dimension 3, equipped with a direct orthonormal frame. The usual Cartesian coordinates are denoted by (x,y). The usual axes will be respectively referred to as $(x'x)$, $(y'y)$ and $(z'z)$.

Thanks to Property 2.2, for any strictly positive integer m and any j in $\{0,\dots,\#V_m\}$, we have that

$$\mathcal{W}\left(\frac{j}{(N_b-1)\,N_b^m}\right)=\lambda^m\,\mathcal{W}\left(\frac{j}{N_b-1}\right)+\sum_{k=0}^{m-1}\lambda^k\,\cos\left(\frac{2\,\pi\,N_b^k\,j}{(N_b-1)\,N_b^m}\right),\tag{31}$$

or, equivalently, expressed in terms of the kth cohomology infinitesimal ε_k^k (see Definition 3.1),

$$\frac{\mathcal{W}\left(\frac{j}{(N_b-1)\,N_b^m}\right)}{(N_b-1)^{2-D_{\mathcal{W}}}}=\varepsilon_m^{m\,(2-D_{\mathcal{W}})}\,\mathcal{W}\left(\frac{j}{N_b-1}\right)+\sum_{k=0}^{m-1}\varepsilon_k^{k\,(2-D_{\mathcal{W}})}\,\cos\left(\frac{2\,\pi\,N_b^k\,j}{(N_b-1)\,N_b^m}\right).\tag{32}$$

If we consider the three-dimensional *vertical comb*, respectively comprised of the set of points

$$\left(\frac{j}{(N_b-1)\,N_b^m},\frac{k}{m},\varepsilon_k^{k\,(2-D_{\mathcal{W}})}\,\cos\left(\frac{2\,\pi\,N_b^k\,j}{(N_b-1)\,N_b^m}\right)\right),$$

for $0\leqslant j\leqslant \#V_m-1$ and $0\leqslant k\leqslant m-1$, along with the periodic set of points

$$\left(\frac{j}{(N_b-1)\,N_b^m},0,\varepsilon_m^{m\,(2-D_{\mathcal{W}})}\,\mathcal{W}\left(\frac{j}{N_b-1}\right)\right),\ \text{for } 0\leqslant j\leqslant \#V_m-1,$$

we note that the expression in relation (32) just above corresponds to the superposition of the teeth of the comb; see Figures 12 and Figure 13.

We can thus envision the following three-dimensional *vertical comb*, respectively comprised of the set of *horizontal (rear) rows*,

$$\left(\frac{j}{(N_b-1)\,N_b^m},\frac{k}{m},\varepsilon_k^{k\,(2-D_{\mathcal{W}})}\right),\ \text{for } 0\leqslant k\leqslant m-1,$$

along with *the front row*

$$\left(\frac{j}{(N_b-1)\,N_b^m},0,\varepsilon_m^{m\,(2-D_{\mathcal{W}})}\,\mathcal{W}\left(\frac{j}{(N_b-1)}\right)\right),\ \text{for } 0\leqslant k\leqslant m-1,$$

and a moving observer who moves on this latter set of points (the front teeth of the comb): when it comes to a specific value of the integer k in $0,\dots,m-1$, the observer looks at the associated comb under an angle of value

$$\vartheta_{j,k,m}=\frac{2\,\pi\,N_b^k\,j}{(N_b-1)\,N_b^m}=\frac{2\,\pi\,j\,\varepsilon^{(m-k)\,(2-D_{\mathcal{W}})}}{(N_b-1)},$$

so that the expression in relation (32) just above corresponds to the successive superposition of the projections of the teeth of the combs; see Figure 14, and Figure 15. Namely, each horizontal row of the comb corresponds to a prefractal level set; see Figure 16.

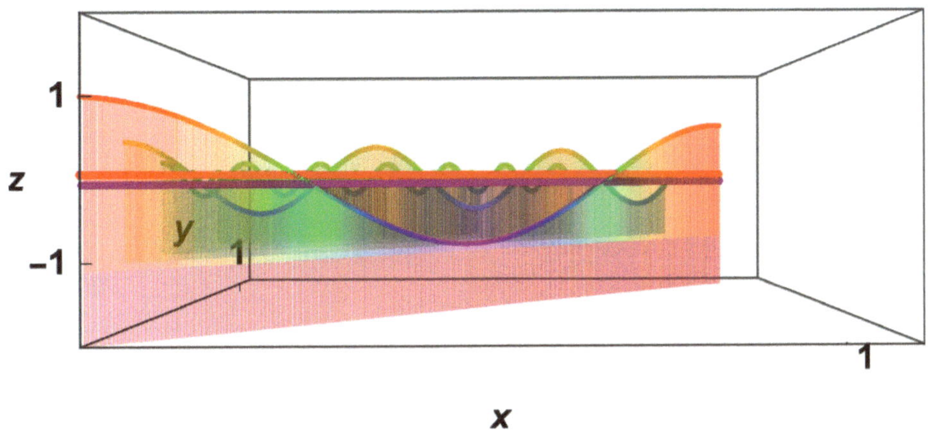

FIGURE 12. The vertical comb, for $m = 5$ – front view

We note that

$$\vartheta_{j,k+1,m} = \frac{\vartheta_{j,k,m}}{\varepsilon},$$

which means that the projection angle increases as one gets closer to the front row.

Note also that in connection with the results of Section 5, the angle $\dfrac{2\pi N_b^k j}{(N_b-1)N_b^m}$ can be represented by a cohomological vertex integer $\ell_{j,k}$, as given in Definition 5.8; see Figure 16.

In later work on the subject, it would be interesting to investigate the following open problem.

PROBLEM 1. *To what extent does the knowledge of the fractal Morse height reordered sequence given in Definition* 5.27, *along with the fractal Morse indexes introduced in Definition* 5.21, *the maximal real Complex Dimension introduced in Definition* 5.6, *and the cohomological vertex integers given in Definition* 5.8, *enable us to reconstruct the fractal (i.e., in our present setting, the Weierstrass Curve)?*

7. Concluding comments

All of the results obtained in this paper for the Weierstrass Curve (and for the associated Weierstrass IFD) are expected to extend naturally to a large class of fractal curves and their higher-dimensional analogues. Our methods and results should apply, in particular, to the Koch Curve (and the associated Koch IFD), for example, whose exact Complex Dimensions have been precisely determined by the authors in [**DL25b**] – from the new perspective introduced in [**DL24a**], [**DL24b**], and especially, in [**DL23a**], [**DL25a**] – thereby extending the earlier results in [**LP06**].

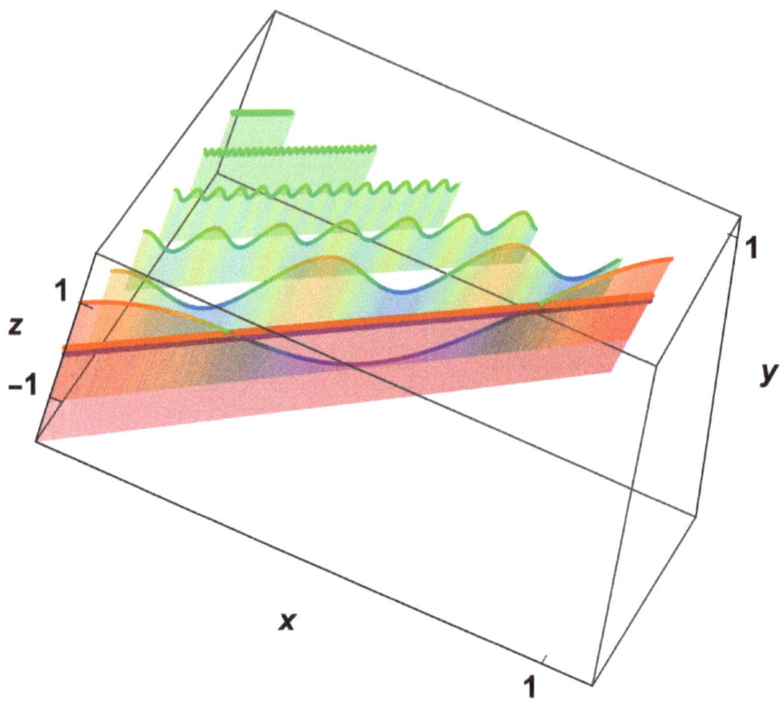

FIGURE 13. The vertical comb, for $m = 5$ – side view

We expect to consider these extensions to a more general setting of the notion of polyhedral measure, Sobolev and Besov spaces, along with the associated trace theorem, and of fractal Morse theory, in later work.

Acknowledgments

The authors would like to thank the anonymous referees for their judicious remarks about the first version of this paper, which contributed to significantly improve it.

References

[BBR14] Krzysztof Barański, Balázs Bárány, and Julia Romanowska, *On the dimension of the graph of the classical Weierstrass function*, Adv. Math. **265** (2014), 32–59, DOI 10.1016/j.aim.2014.07.033. MR3255455

[Bot82] Raoul Bott, *Lectures on Morse theory, old and new*, Bull. Amer. Math. Soc. (N.S.) **7** (1982), no. 2, 331–358, DOI 10.1090/S0273-0979-1982-15038-8. MR663786

[Bot88] Raoul Bott, *Morse theory indomitable*, Inst. Hautes Études Sci. Publ. Math. **68** (1988), 99–114 (1989). MR1001450

[Bou04] Nicolas Bourbaki, *Theory of sets*, Elements of Mathematics (Berlin), Springer-Verlag, Berlin, 2004. Reprint of the 1968 English translation [Hermann, Paris; MR0237342], DOI 10.1007/978-3-642-59309-3. MR2102219

[Cal77] Alberto-P. Calderón, *An atomic decomposition of distributions in parabolic H^p spaces*, Advances in Math. **25** (1977), no. 3, 216–225, DOI 10.1016/0001-8708(77)90074-3. MR448066

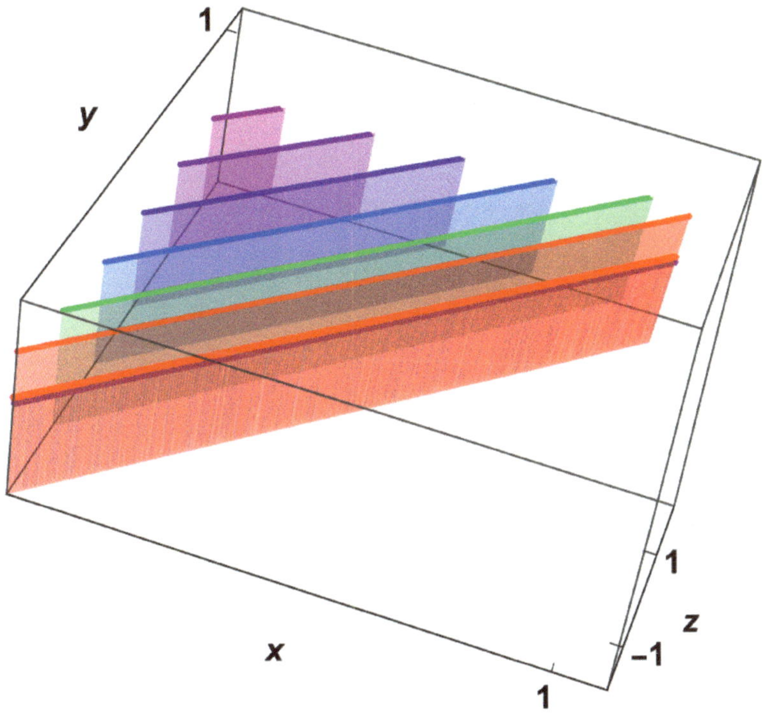

FIGURE 14. The vertical comb, before projection, for $m = 5$ – side view

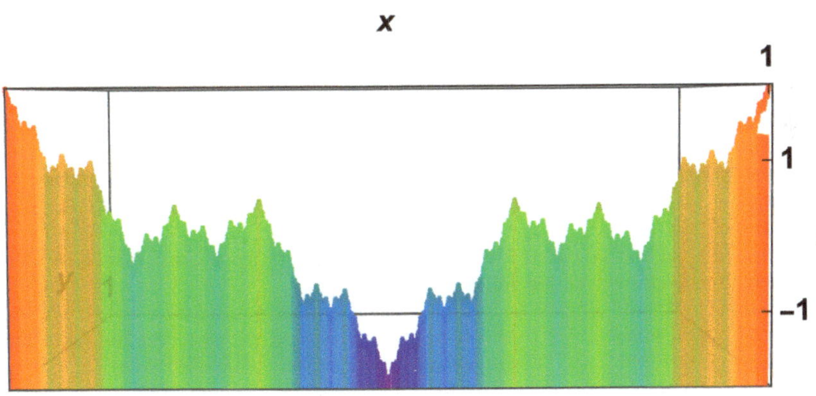

FIGURE 15. The resulting vertical comb, after projection – face view

[Coi74] Ronald R. Coifman, *A real variable characterization of H^p*, Studia Math. **51** (1974), 269–274, DOI 10.4064/sm-51-3-269-274. MR358318

[Dau92] Ingrid Daubechies, *Ten lectures on wavelets*, CBMS-NSF Regional Conference Series in Applied Mathematics, vol. 61, Society for Industrial and Applied Mathematics (SIAM), Philadelphia, PA, 1992, DOI 10.1137/1.9781611970104. MR1162107

[Dav18] Claire David, *Bypassing dynamical systems: a simple way to get the box-counting dimension of the graph of the Weierstrass function* (English, with English and Ukrainian summaries), Proc. Int. Geom. Cent. **11** (2018), no. 2, 53–68, DOI 10.15673/tmgc.v11i2.1028. MR3855920

FIGURE 16. The top view of the vertical comb. Each horizontal row coincides with a prefractal level set $\varepsilon_k^{k(2-D_{\mathcal{W}})}$, $0 \leqslant k \leqslant m$.

[Dav19] Claire David, *On fractal properties of Weierstrass-type functions* (English, with English and Ukrainian summaries), Proc. Int. Geom. Cent. **12** (2019), no. 2, 43–61, DOI 10.15673/tmgc.v12i2.1485. MR4033034

[Dav20] Claire David, *Laplacian, on the arrowhead curve* (English, with English and Ukrainian summaries), Proc. Int. Geom. Cent. **13** (2020), no. 2, 19–49, DOI 10.15673/tmgc.v13i2.1746. MR4157913

[Dav22] Claire David, *Wandering across the Weierstrass function, while revisiting its properties*, Lectures on the analysis of nonlinear partial differential equations. Part 6, Morningside Lect. Math., vol. 6, Int. Press, Somerville, MA, [2024] ©2024, pp. 91–124. MR4862074

[DL23a] Claire David and Michel L. Lapidus. New insights for fractal zeta functions: Polyhedral neighborhoods vs tubular neighborhoods, 2023.

[DL25a] Claire David and Michel L. Lapidus, *Polyhedral neighborhoods vs. tubular neighborhoods: new insights for fractal zeta functions and complex dimensions*, Ramanujan Journal, **67**(2025), no. 73, url: `https://hal.science/hal-04153049`.

[DL25b] Claire David and Michel L. Lapidus, *Understanding fractality: A polyhedral approach to the Koch curve and its complex dimensions*, Asymptotic Analysis, 2025; 0(0), . URL: `https://hal.sorbonne-universite.fr/hal-04348346`.

[DL25c] Claire David and Michel L. Lapidus, *Weierstrass fractal drums - I - A glimpse of complex dimensions*, Advances in Mathematics, in press (2025). doi: `https://doi.org/10.1016/j.aim.2025.110545`. `https://hal.sorbonne-universite.fr/hal-03642326`.

[DL24a] Claire David and Michel L. Lapidus, *Fractal complex dimensions and cohomology of the Weierstrass curve*, In Patricia Alonso Ruiz, Michael Hinz, Kasso A. Okoudjou, Luke G. Rogers, and Alexander Teplyaev, editors, *From Classical Analysis to Analysis on Fractals: A Tribute to Robert Strichartz*, volume 2 of *Applied and Numerical Harmonic Analysis*. Birkhäuser, Boston, in press, 2024. URL: `https://hal.science/hal-03797595v2`.

[DL24b] Claire David and Michel L. Lapidus, *From Weierstrass to Riemann: The Frobenius pass – When fractal flows come into play with zeta functions*, In John Friedlander,

Carl Pomerance, and Michael Th. Rassias, editors, *Essays in Analytic Number Theory – in Honor of Helmut Maier's 70th Birthday*. Springer Nature, Cham, 2024. URL: https://hal.sorbonne-universite.fr/hal-04614665.

[DL24c] Claire David and Michel L. Lapidus, *Weierstrass fractal drums: II towards a fractal cohomology*, Math. Z. **308** (2024), no. 2, Paper No. 35, 56, DOI 10.1007/s00209-024-03547-z. MR4800949

[FJ85] Michael Frazier and Björn Jawerth, *Decomposition of Besov spaces*, Indiana Univ. Math. J. **34** (1985), no. 4, 777–799, DOI 10.1512/iumj.1985.34.34041. MR808825

[Fol99] Gerald B. Folland, *Real analysis: Modern techniques and their applications*, 2nd ed., Pure and Applied Mathematics (New York), John Wiley & Sons, Inc., New York, 1999. A Wiley-Interscience Publication. MR1681462

[Har16] G. H. Hardy, *Weierstrass's non-differentiable function*, Trans. Amer. Math. Soc. **17** (1916), no. 3, 301–325, DOI 10.2307/1989005. MR1501044

[Jon09] Alf Jonsson, *Besov spaces on closed sets by means of atomic decomposition*, Complex Var. Elliptic Equ. **54** (2009), no. 6, 585–611, DOI 10.1080/17476930802669678. MR2537257

[JW84] Alf Jonsson and Hans Wallin, *Function spaces on subsets of* $\mathbf{R}^n$, Math. Rep. **2** (1984), no. 1, xiv+221. MR820626

[Kab12] Maryia Kabanava, *Characterization of Besov spaces on nested fractals by piecewise harmonic functions*, Z. Anal. Anwend. **31** (2012), no. 2, 183–201, DOI 10.4171/ZAA/1454. MR2914970

[Kel17] Gerhard Keller, *A simpler proof for the dimension of the graph of the classical Weierstrass function* (English, with English and French summaries), Ann. Inst. Henri Poincaré Probab. Stat. **53** (2017), no. 1, 169–181, DOI 10.1214/15-AIHP711. MR3606737

[Kig01] Jun Kigami, *Analysis on fractals*, Cambridge Tracts in Mathematics, vol. 143, Cambridge University Press, Cambridge, 2001, DOI 10.1017/CBO9780511470943. MR1840042

[KMPY84] James L. Kaplan, John Mallet-Paret, and James A. Yorke, *The Lyapunov dimension of a nowhere differentiable attracting torus*, Ergodic Theory Dynam. Systems **4** (1984), no. 2, 261–281, DOI 10.1017/S0143385700002431. MR766105

[Lap19] Michel L. Lapidus, *An overview of complex fractal dimensions: from fractal strings to fractal drums, and back*, Horizons of fractal geometry and complex dimensions, Contemp. Math., vol. 731, Amer. Math. Soc., [Providence], RI, [2019] ©2019, pp. 143–265, DOI 10.1090/conm/731/14677. MR3989820

[Lat78] Robert H. Latter, *A characterization of* $H^p(\mathbf{R}^n)$ *in terms of atoms*, Studia Math. **62** (1978), no. 1, 93–101, DOI 10.4064/sm-62-1-93-101. MR482111

[LP06] Michel L. Lapidus and Erin P. J. Pearse, *A tube formula for the Koch snowflake curve, with applications to complex dimensions*, J. London Math. Soc. (2) **74** (2006), no. 2, 397–414, DOI 10.1112/S0024610706022988. MR2269586

[LRŽ17a] Michel L. Lapidus, Goran Radunović, and Darko Žubrinić, *Fractal Zeta Functions and Fractal Drums: Higher-Dimensional Theory of Complex Dimensions*, Springer Monographs in Mathematics. Springer, New York, 2017.

[LRŽ17b] Michel L. Lapidus, Goran Radunović, and Darko Žubrinić, *Zeta functions and complex dimensions of relative fractal drums: theory, examples and applications*, Dissertationes Math. **526** (2017), 105, DOI 10.4064/dm757-4-2017. MR3764036

[LRŽ18] Michel L. Lapidus, Goran Radunović, and Darko Žubrinić, *Fractal tube formulas for compact sets and relative fractal drums: oscillations, complex dimensions and fractality*, J. Fractal Geom. **5** (2018), no. 1, 1–119, DOI 10.4171/JFG/57. MR3760302

[LvF13] Michel L. Lapidus and Machiel van Frankenhuijsen, *Fractal geometry, complex dimensions and zeta functions*, 2nd ed., Springer Monographs in Mathematics, Springer, New York, 2013. Geometry and spectra of fractal strings, DOI 10.1007/978-1-4614-2176-4. MR2977849

[Mar48] Andreï A. Markov. *Selected Works (in Russian*. Moscow-Leningrad, 1948.

[Mil63] J. Milnor, *Morse theory*, Annals of Mathematics Studies, No. 51, Princeton University Press, Princeton, NJ, 1963. Based on lecture notes by M. Spivak and R. Wells. MR163331

[Rud87] Walter Rudin, *Real and complex analysis*, 3rd ed., McGraw-Hill Book Co., New York, 1987. MR924157

[Rud91] Walter Rudin, *Functional analysis*, 2nd ed., International Series in Pure and Applied Mathematics, McGraw-Hill, Inc., New York, 1991. MR1157815

[She18] Weixiao Shen, *Hausdorff dimension of the graphs of the classical Weierstrass functions*, Math. Z. **289** (2018), no. 1-2, 223–266, DOI 10.1007/s00209-017-1949-1. MR3803788

[Str03] Robert S. Strichartz, *Function spaces on fractals*, J. Funct. Anal. **198** (2003), no. 1, 43–83, DOI 10.1016/S0022-1236(02)00035-6. MR1962353

[Str06] Robert S. Strichartz, *Differential equations on fractals: A tutorial*, Princeton University Press, Princeton, NJ, 2006. MR2246975

[Tri97] Hans Triebel, *Fractals and spectra*, Monographs in Mathematics, vol. 91, Birkhäuser Verlag, Basel, 1997. Related to Fourier analysis and function spaces, DOI 10.1007/978-3-0348-0034-1. MR1484417

[Tri06] Hans Triebel, *Theory of function spaces. III*, Monographs in Mathematics, vol. 100, Birkhäuser Verlag, Basel, 2006. MR2250142

[Wal89] Hans Wallin, *Interpolating and orthogonal polynomials on fractals*, Constr. Approx. **5** (1989), no. 1, 137–150, DOI 10.1007/BF01889603. Fractal approximation. MR982729

[Wal91] Hans Wallin, *The trace to the boundary of Sobolev spaces on a snowflake*, Manuscripta Math. **73** (1991), no. 2, 117–125, DOI 10.1007/BF02567633. MR1128682

[Wei75] Karl Weierstrass. Über continuirliche Funktionen eines reellen Arguments, die für keinen Werth des letzteren einen bestimmten Differential quotienten besitzen. *Journal für die reine und angewandte Mathematik*, 79:29–31, 1875.

[Zyg02] A. Zygmund, *Trigonometric series. Vol. I, II*, 3rd ed., Cambridge Mathematical Library, Cambridge University Press, Cambridge, 2002. With a foreword by Robert A. Fefferman. MR1963498

Sorbonne Université, CNRS, UMR 7598, Laboratoire Jacques-Louis Lions, 4, place Jussieu, BC 187 75252 Paris cedex 05, France

Email address: `Claire.David@sorbonne-universite.fr`

University of California, Riverside, Department of Mathematics, Skye Hall, 900 University Ave., Riverside, California 92521-0135

Email address: `lapidus@math.ucr.edu`

Contemporary Mathematics
Volume **828**, 2025
https://doi.org/10.1090/conm/828/16595

Green's function and eigenfunctions on random Sierpinski gaskets

Daniel Fontaine, Daniel J. Kelleher, and Alexander Teplyaev

ABSTRACT. We study Green's function and eigenfunctions on random Sierpinski gaskets. We use a randomized algorithm that can produce all possible Dirichlet forms on the Sierpinski gasket. We prove that the limiting local regular Dirichlet form exists, which implies the existence of a self-adjoint Laplacian. Numerically we demonstrate that for low randomness, Green's function is bounded and continuous, but for high randomness it is unbounded and discontinuous. We also study localization of eigenfunctions, as well as distribution and spacing of eigenvalues.

1. Introduction

This paper studies random resistance networks or, equivalently, energy (Dirichlet) forms, on random Sierpiński gaskets, see Figures 1 and 8. Such resistance networks can be described as certain limits of quantum graphs, as demonstrated in [**90**]. This is an active area of current research [**3–7, 9, 18, 20, 23–30, 37, 43, 46, 51–57, 80–82, 91**]. More precisely, we analyze the approximating sequences of quantum graphs which satisfy appropriate compatibility conditions. We use a specific algorithm, introduced in [**76**], that constructs all compatible Dirichlet forms based on a random choice of parameters. Defining an energy form on the Sierpiński gasket allows one to introduce such objects as Laplacian, diffusion processes etc.

It was proved in [**76**] that the harmonic functions are continuous with probability one. We prove that with probability one the random Sierpiński gasket has a homeomorphic harmonic embedding into the two dimensional Euclidean space, i.e. one can use a pair of harmonic functions as coordinates. Then it is possible to write the energy of a smooth function as the integral, with respect to a reference energy measure, of the squared norm of the gradient. Then one can represent the energy measure Laplacian as a second derivative in a sense. In addition, we show that, in harmonic coordinates, the angles of approximating triangles tend to zero with probability one at every junction point.

2020 *Mathematics Subject Classification.* Primary 60K37, 28A80; Secondary 05C81, 31E05, 35R60, 60J45, 60J46, 62R20, 81Q35.

Key words and phrases. Dirichlet form, resistance network, random fractal, Sierpinski gasket, quantum graphs, Green's function, Laplacian, eigenvalues, eigenfunctions, localization. PACS 02.30.Sa, 02.20.Bb, 02.50.Ga, 02.60.Lj, 02.70.Hm.

Supported in part by NSF grant DMS-0505622, DMS-2349433 and the Simons Foundation.

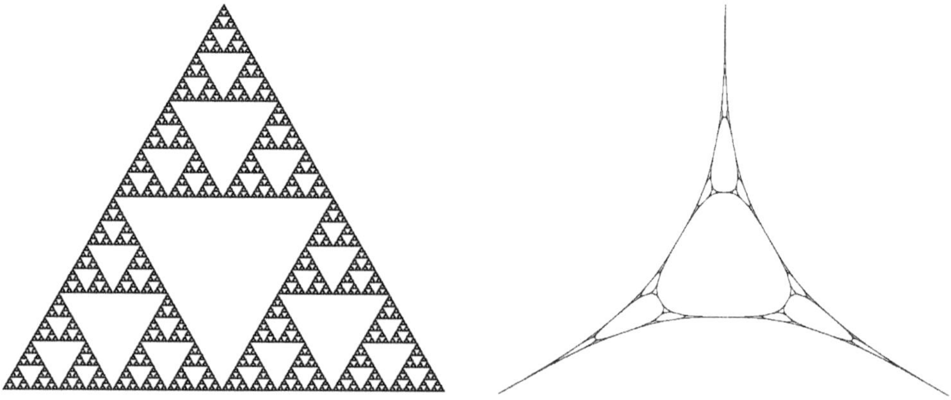

FIGURE 1. Sierpiński gasket with the usual self-similar embedding (left) and in the standard, nonrandom, harmonic coordinates (right).

A computer program was written that executes the randomization algorithm and calculates resistances which define Dirichlet forms on the Sierpiński gasket. Certain questions that arise naturally in our examination: Do the calculated resistances form a log normal distribution? How does the data change when we alter the domain of the parameters? We demonstrate that individual resistance tend to zero if the disorder is small, but for larger disorders "most" of the resistances tend to zero, but there are some resistances that approach infinity. We conjecture that for small disorder, with probability one, the effective resistance topology coincides with the standard topology on the Sierpiński gasket. However, as the disorder gets larger, with probability one there are points in the Sierpiński gasket which are at infinite effective resistance distance from the boundary. Equivalently, for small disorder the diffusion process is point recurrent on the Sierpiński gasket, and for larger disorder the diffusion process is point recurrent on a proper subset of the Sierpiński gasket.

The analysis on self-similar fractals was first developed in the physics and engineering literature, see [**2, 31, 35, 47, 83, 84, 86**] and references therein. There are three mathematical books, [**10, 62, 88**], that provide background to the analysis on self-similar fractals. The following papers deal with the analysis on the Sierpiński gasket and other fractals which is relevant to our work: [**16, 17, 44, 45, 48, 58–61, 63, 74, 76, 78, 85, 87, 89, 90**]. Also, there are many probabilistic works on the diffusions and random walks on self-similar fractals and graphs, see for instance [**14, 33, 36, 72, 73**] and references therein. It was recently demonstrated by Kigami, using the new theory of heat kernel estimates on metric measure spaces (see [**11–13, 40, 64, 71**] and references therein), that the energy measure diffusion on the standard Sierpiński gasket has Gaussian asymptotics in harmonic coordinates. Recently there has been renewed interest in studying the Sierpinski Gasket in Harmonic coordinates [**50, 51**]. Spectral analysis on (deterministic) non-self similar fractals has attracted some attention [**6, 8, 15, 49**]. Random fractals, primarily various modifications of random Sierpiński gaskets, were considered in [**38, 39, 41**], although the randomization procedure was significantly different from our work.

2. Energy forms and effective resistance

In this section we recall some basic facts from [**62, 63**] about limits of resistance networks. Although we state the results of this section for the Sierpiński gasket, they can be applied for general p.c.f. fractals with only minor changes.

If V is a finite set then an energy form $\mathcal{E}$ on V can be defined by

$$\mathcal{E}(f,f) = \sum_{x,y \in V} c(x,y)(f(x) - f(y))^2$$

where $c(x,y) \geqslant 0$ is called the conductance between x and y. Then the set V with the energy form $\mathcal{E}$ is often called a finite resistance network where the resistance between x and y is defined as $1/c(x,y)$ if $c(x,y) > 0$ and infinity otherwise. It is assumed that $c(x,x) = 0$ for all $x \in V$, and that the network is connected in the sense that any pair of points can be connected by a sequence of positive conductances.

Suppose $V_0 \subsetneq V_1 \subsetneq V_2 \subsetneq \ldots$ is an increasing sequence of finite sets, and an energy form $\mathcal{E}_n$ is defined on each V_n. Then this sequence of resistance networks is called compatible if

$$\mathcal{E}_n(f\big|_{V_n}, f\big|_{V_n}) \leqslant \mathcal{E}_{n+1}(f,f)$$

for any function f on V_{n+1} and for any function f_n on V_n there is a function f_{n+1} on V_{n+1} such that

$$\mathcal{E}_n(f_n, f_n) = \mathcal{E}_{n+1}(f_{n+1}, f_{n+1})$$

and f_n is the restriction of f_{n+1} to V_n. Then it is easy to see that such f_{n+1} is unique. In this case each one can see that $\mathcal{E}_n$ is equal to the so called trace on V_n of the energy form $\mathcal{E}_{n+1}$. Moreover, $\mathcal{E}_n$ is equal to the so called trace on V_n of the energy form $\mathcal{E}_{n+k}$ for any $k \geqslant 0$.

The limiting energy (Dirichlet) form $\mathcal{E}$ on $V_* = \bigcup_{n=0}^{\infty} V_n$ is defined by

$$\mathcal{E}(f,f) = \lim_{n \to \infty} \mathcal{E}_n(f\big|_{V_n}, f\big|_{V_n}).$$

By definition the domain of $\mathcal{E}$, denoted by $\operatorname{Dom}\mathcal{E}$, consists of all function for which this increasing limit is finite. It is not hard to see that $\mathcal{E}_n$ is the trace on V_n of $\mathcal{E}$ (see [**42, 75**] and references therein). It is also easy to see that every point of $V_* = \bigcup_{n \geq 0} V_n$ has positive capacity, in the sense of [**34**].

3. Parametrization of energy forms on the Sierpiński gasket

In this section we provide brief background information about the construction of non self-similar Dirichlet forms on the Sierpiński gasket. More detailed description is given in [**76**].

Suppose we start with a triangle with vertices v_0, v_1, and v_2 and some initial resistance values between these points. In our notation resistance r_j connects points v_{j-1} and v_{j+1}, where $j = 0, 1, 2$, considering the indices $mod(3)$, that is v_3 is the same as v_0 etc. Then this network is transformed by the $\Delta - Y$ transformation into the upside down Y shape, which is our initial resistance network Γ_0. The formulas for $\Delta - Y$ and $Y - \Delta$ transformations are

$$R_j = \frac{r_{j-1} r_{j+1}}{r_0 + r_1 + r_2}$$

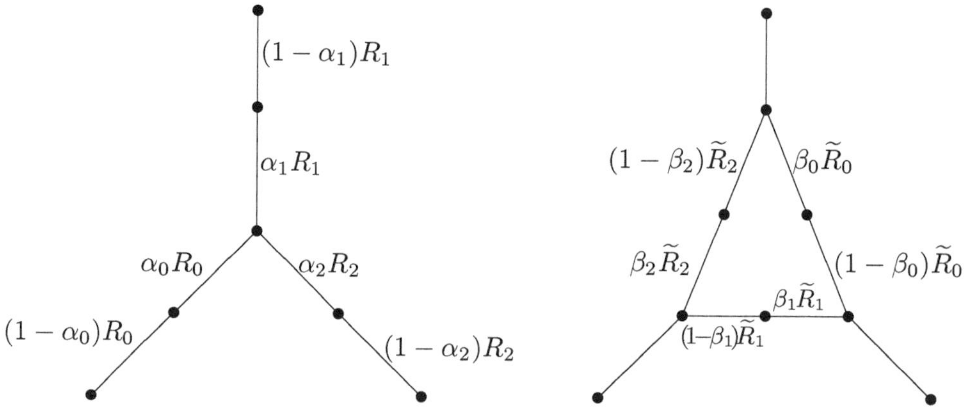

FIGURE 2. Networks Γ_0 and Γ_1.

and

$$r_j = \frac{R_0R_1 + R_0R_2 + R_1R_2}{R_j} \tag{3.1}$$

respectively.

We choose parameters $(\alpha_0, \alpha_1, \alpha_2) \in (0,1)$ to split the resistances R_j, as in the left hand side of Figure 2. Then create, using the $Y-\Delta$ transform (3.1), an inner triangle with new resistance values $\widetilde{R}_0$, $\widetilde{R}_1$, and $\widetilde{R}_2$, as seen in the right hand side of Figure 2. Then we choose parameters $(\beta_0, \beta_1, \beta_2) \in (0,1)$ to split these new resistances $\widetilde{R}_0$, $\widetilde{R}_1$, and $\widetilde{R}_2$. After that we have three Y-shaped networks, and we apply these steps to each of them independently. The effective resistances between vertices v_0, v_1, and v_2 remain the same under all these transformations. Moreover, the compatibility conditions of Section 2 hold in this case.

Note that it is convenient to label the 3^k triangles of depth k by words of length k of the alphabet $0, 1, 2$. For example, the whole Sierpiński gasket has an empty label; it is subdivided into three triangles with labels 0, 1 and 2. Each of these triangle, say labeled j, is subdivided into triangles $j0$, $j1$ and $j2$, and so on. The rule is that a depth k triangle labeled by word w is subdivided into three depth $k+1$ triangles labeled $w0$, $w1$ and $w2$ in the natural order.

To obtain a parametric description of all compatible Dirichlet forms on Γ_k we just use the algorithm iteratively on each of the triangles on each depth up to k. To transform our networks, for each word w we chose a 6-dimensional vector $\xi_w = (\alpha_0^w, \alpha_1^w, \alpha_2^w, \beta_0^w, \beta_1^w, \beta_2^w) \in (0,1)^6$; the choices of the 6 parameters can be made independently for each w.

LEMMA 3.1 ([**76**]). The space of all Dirichlet forms on Γ_1 compatible with a fixed Dirichlet form on Γ_0 is a manifold of dimension 6. More generally, the space of all Dirichlet forms on Γ_k compatible with a fixed Dirichlet form on Γ_0 is a manifold of dimension $6(1+3+\cdots+3^{k-1}) = 3(3^k-1)$.

THEOREM 1 ([**76**]). All the local regular resistance Dirichlet forms, in the sense of [**63**], on the Sierpiński gasket are in one to one correspondence, via the algorithm defined above, with the set of vectors $\xi_w = (\alpha_0^w, \alpha_1^w, \alpha_2^w, \beta_0^w, \beta_1^w, \beta_2^w) \in (0,1)^6$ where $w \in \bigcup_{k=0}^{\infty}\{0,1,2\}^k$.

EXAMPLE 3.2. It is an easy exercise to show that if the resistances in the initial network Γ_0 are equal to one, and for all i and w we set $\alpha_i^w = \frac{2}{5}, \beta_i^w = \frac{1}{2}$, then in the network Γ_k all resistances are equal to $\left(\frac{3}{5}\right)^k$. This corresponds to the so called standard energy form on the Sierpiński gasket. According to the results of [**62, 63, 76**] this is the only, up to a constant multiple, local regular Dirichlet form on the Sierpiński gasket which satisfies the following two assumptions: points have positive capacity and the effective resistance topology coincides with the standard topology; locally the Dirichlet form is preserved by the local symmetries of the Sierpiński gasket. The standard Dirichlet form on the Sierpiński gasket is self-similar with weights equal to $\frac{3}{5}$. Note that in this case the sub triangles have various energies, and so one can consider the distribution of energies even though this is a non random case. The histogram of this distribution can be found in [**32**]. This distribution is very interesting and has not been studied theoretically although papers [**17, 79**] contain some related results.

A computer program was written that executes the randomization of the algorithm of Section 3, and calculates resistances that define Dirichlet forms on the Sierpiński gasket. In particular, we are interested in how the data changes when we alter the domain of the parameters.

In our study we ran the program several times, and within each run the parameters $(\alpha_0, \alpha_1, \alpha_2, \beta_0, \beta_1, \beta_2)$ were chosen at random uniformly from the following:

$$\begin{array}{ll} \text{Nonrandom self-similar case:} & \alpha = .4, \beta = .5 \\ \text{Low randomness case:} & \alpha, \beta \in (.4, .6) \\ \text{High randomness case:} & \alpha, \beta \in (.1, .9) \end{array} \qquad (*)$$

As we increased the domains of random parameters, the resulting resistance networks and data changed significantly.

Each run of the program was done to a depth of 13, meaning one run consisted of executing the above three steps 3^{13} times. Thus, the program produced data for the resistance networks on $\Gamma_1, \ldots, \Gamma_{13}$. In [**32**] we computed the mean, standard deviation, fourth moment, and 5th and 95th percentiles for each depth. One can see that individual resistance tend to zero if the disorder is small, but for larger disorders "most" of the resistances tend to zero, there are some resistances that approach infinity. The statistical distribution of resistances resembles log normal in Figure 13.

4. Distribution of eigenvalues and localization of eigenfunctions

When dealing with random media, one of the natural questions is if the is Anderson type localization. Much work has focused on such localization, for example [**77**]. Our numerical results show that there is indeed very strong localization even with low randomness. Figure 3 demonstrates this most clearly by showing density estimates of the variance of eigenfunctions in nonrandom, low randomness and high randomness cases at depth 13. Figure 4 and Figure 5 also illustrate that eigenfunctions with numbers #50, #51 are localized even in the low randomness case. The Anderson type localization is further confirmed by Figure 6 which shows Poisson type sample densities of the logarithmic spacing of eigenvalues in low randomness and high randomness cases. Figure 7 shows sample densities of eigenvalues in nonrandom, low randomness and high randomness cases at depths 6, 7, 8.

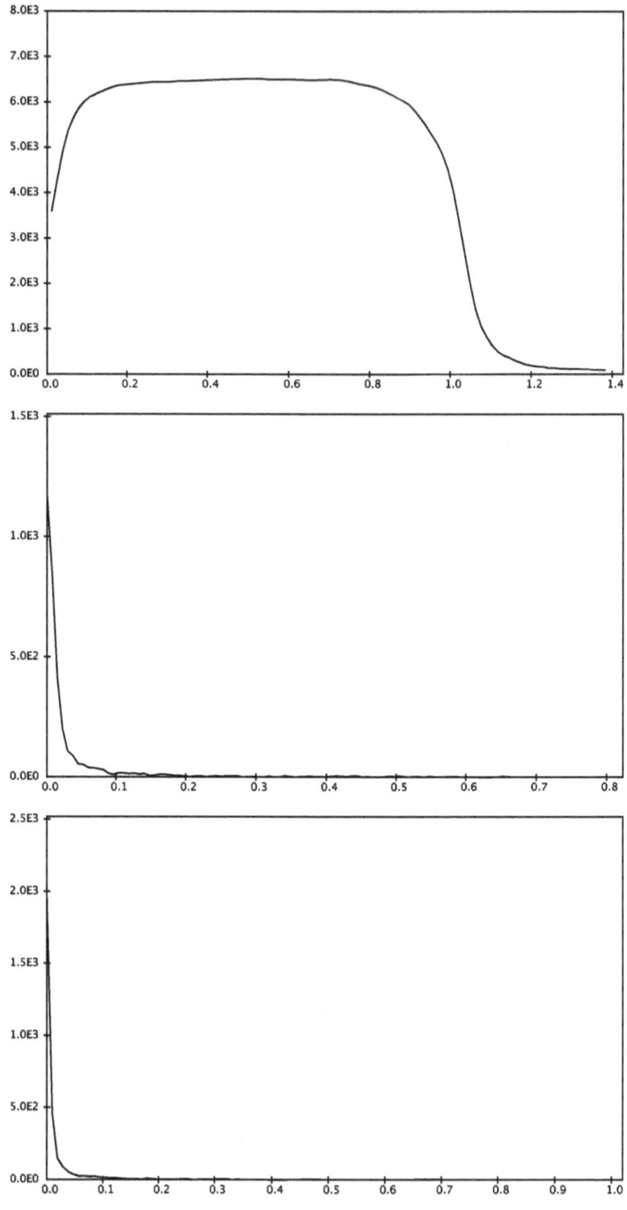

FIGURE 3. Density estimates of the variance of (from top to bottom) eigenfunctions in nonrandom, low randomness and high randomness cases at depth 13.

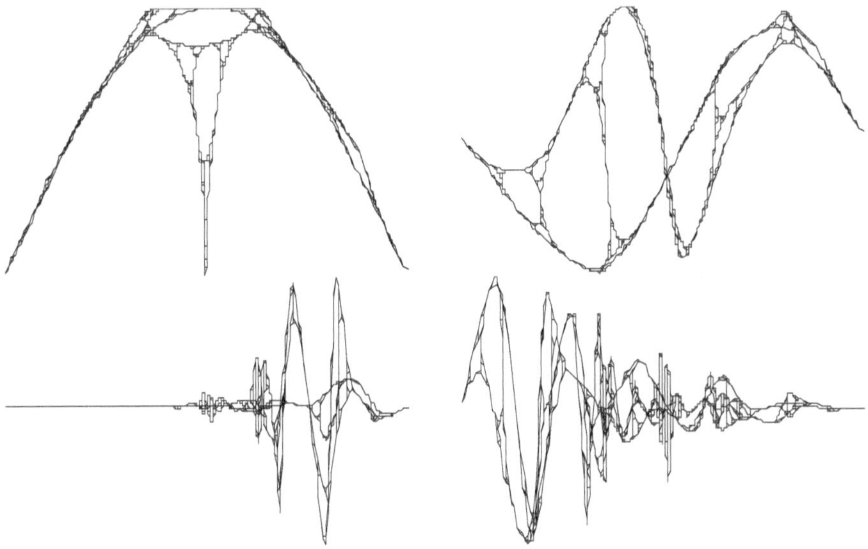

FIGURE 4. Pictures of the random eigenfunctions #1, #5, #50, #51 with low randomness.

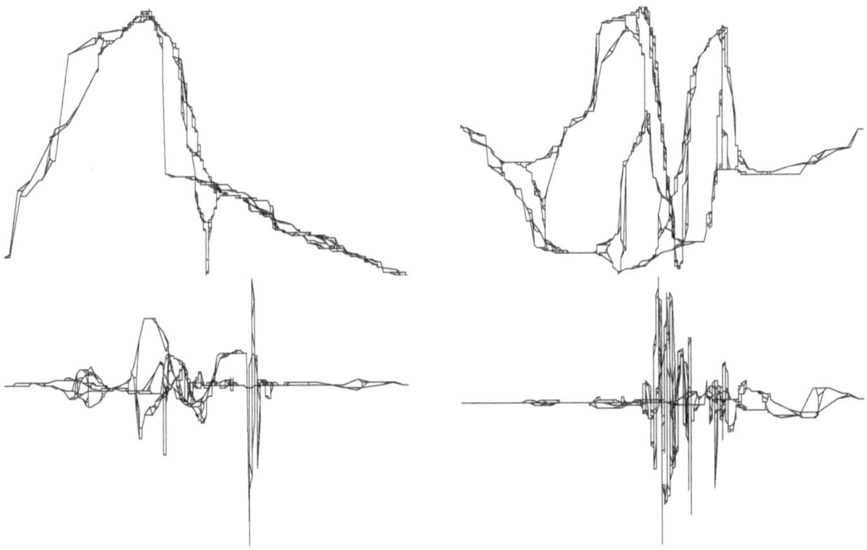

FIGURE 5. Pictures of the random eigenfunctions #1, #5, #50, #51 with high randomness.

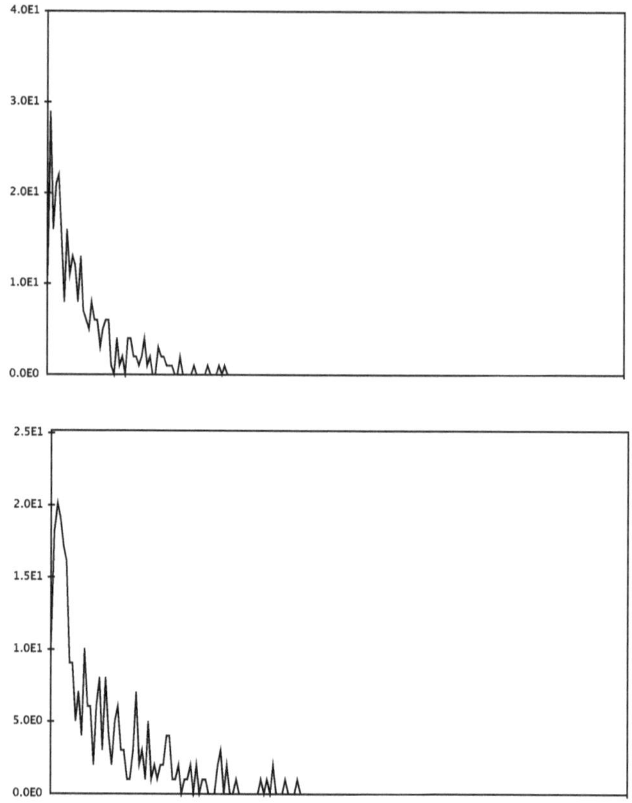

FIGURE 6. Sample densities of the logarithmic spacing of eigenvalues in low randomness and high randomness cases at depth 8

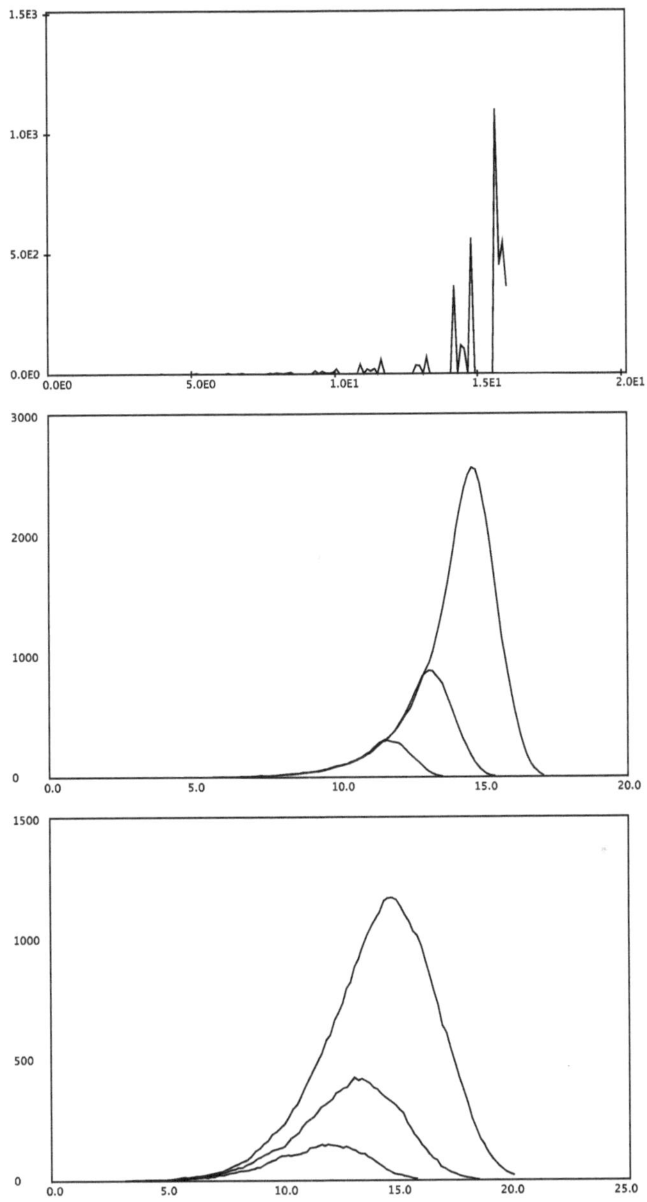

FIGURE 7. Sample densities of eigenvalues in nonrandom, low randomness and high randomness cases at depths 6, 7, 8.

5. Harmonic functions on random Sierpiński gaskets

Harmonic functions play an important role in the analysis on fractals. A function h defined on a finite set V is said to be harmonic if

$$\sum_{y\in V}(h(x)-h(y))c(x,y)=0$$

for every $x \in V\backslash\partial V$, where $c(x,y)$ is the conductance between vertices x and y. Conductance is the reciprocal of resistance, so for any resistance R, $c=\frac{1}{R}$. If $c(x,y)>0$, then x and y are said to be connected. On a connected resistance network a harmonic function is uniquely determined by it boundary values. In the case of the Sierpiński gasket it is the most natural and convienient to choose the boundary $\partial V=\{v_0,v_1,v_2\}$ to be the three corners of the largest triangle.

The next proposition is a consequence of general results on traces of resistance forms, see for instance [**22, 62, 63, 76**] and Section 2.

PROPOSITION 5.1. *Suppose x is a vertex of a triangle of depth k and h is a harmonic function in the network Γ_m. Then $h(x)$ is uniquely determined by the boundary values of h on $\partial V=\{v_0,v_1,v_2\}$, and is independent of m provided $m\geqslant k$.*

THEOREM 2 ([**76**]).

(1) *Suppose that $\xi_w=(\alpha_0^w,\alpha_1^w,\alpha_2^w,\beta_0^w,\beta_1^w,\beta_2^w)\in(0,1)^6$ are independent identically distributed random 6-dimensional vectors indexed by the words w of finite length. Then with probability one harmonic functions are continuous.*

(2) *Suppose that there is $\varepsilon>0$ such that $\alpha_j^w,\beta_j^w\in[\varepsilon,1-\varepsilon]$ for all w,j. Then harmonic functions are Hölder continuous with Hölder exponent $1-\varepsilon^2$.*

To define harmonic coordinates on the Sierpiński gasket S, we fix two harmonic functions h_1 and h_2, which are linearly independent over constants. Then for each point $x\in S$ we define

$$\psi : S\to\mathbb{R}^2,$$
$$\psi(x)=(h_1(x),h_2(x)).$$

For convenience we choose the harmonic functions h_1 and h_2 in such way that the boundary points of the Sierpiński gasket in harmonic coordinates are vertices of an isosceles triangle, that is $h_1(v_0)=-1$, $h_2(v_0)=0$, $h_1(v_1)=0$, $h_2(v_1)=\sqrt{3}$, $h_1(v_2)=1$, $h_2(v_2)=0$.

In [**60**] ψ is proved to be a homeomorphism in the case of the standard Dirichlet form on the Sierpiński gasket (see Example 3.2). This Kigami's result can be generalized as follows

THEOREM 3. *Suppose that $\xi_w=(\alpha_0^w,\alpha_1^w,\alpha_2^w,\beta_0^w,\beta_1^w,\beta_2^w)\in(0,1)^6$ are independent identically distributed random 6-dimensional vectors indexed by the words w of finite length. Then with probability one the coordinate map $\psi : S\to S_\psi := \psi(S)$ is a homeomorphism.*

PROOF. By Theorem 2 harmonic functions are continuous with probability one, and they separate points by [**76**, Lemma 5.3 and Proposition 6.1]. ☐

The next theorem says, essentially, that the angles of the curvilinear triangles that make the Sierpiński gasket in harmonic coordinates are zero.

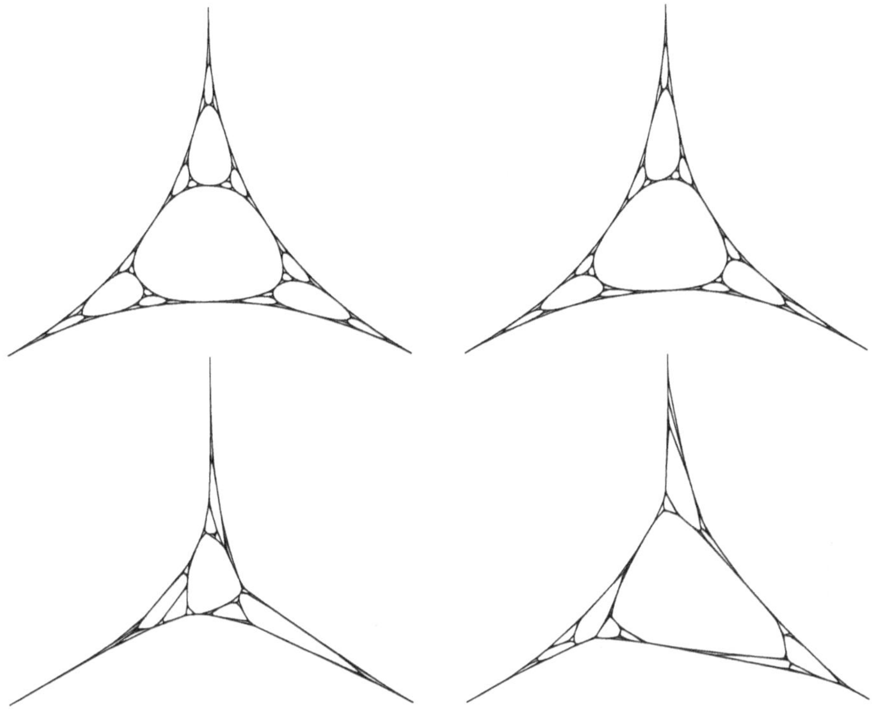

FIGURE 8. Samples of the random gaskets with low randomness (top) and with high randomness (bottom).

THEOREM 4. Suppose that $\xi_w = (\alpha_0^w, \alpha_1^w, \alpha_2^w, \beta_0^w, \beta_1^w, \beta_2^w) \in (0,1)^6$ are independent identically distributed random 6-dimensional vectors indexed by the words w of finite length. Then with probability one the following is true. Let $\{T_k\}_{k=0}^{\infty}$ be a sequence of triangles in approximation to the harmonic gasket with $T_{k+1} \subsetneq T_k$ for all k and which all share a corner. Then the the angle of the shared corner tends to zero as k tend to infinity.

PROOF. The result follows from [**76**, Lemma 4.1, Theorem 4.3 and Theorem 5.5]. □

We conjecture that some of the results of Kusuoka proved in [**17, 72, 73**] and recent results by Hino [**72, 73**] hold also with probability one for our random Sierpiński gaskets. In particular, we conjecture that the energy measures are singular with probability one with respect to all the product (Bernoulli) measures, and that the random Sierpiński gasket is one dimensional in any generic point. The latter means that the matrix Z (in Theorem 6 below) has rank one ν-almost everywhere.

6. Green's function and its local computation

The Green's function of the nonrandom Sierpiński gasket, see Figures 9, is investigated in [**67**] in detail. In this section we present experimental evidence that if the resistance form on the Sierpiński gasket is random, then there is a certain phase transition in the behavior of the Green's function. The nature of this phase transition is that for low randomness the Green's function is continuous (and hence

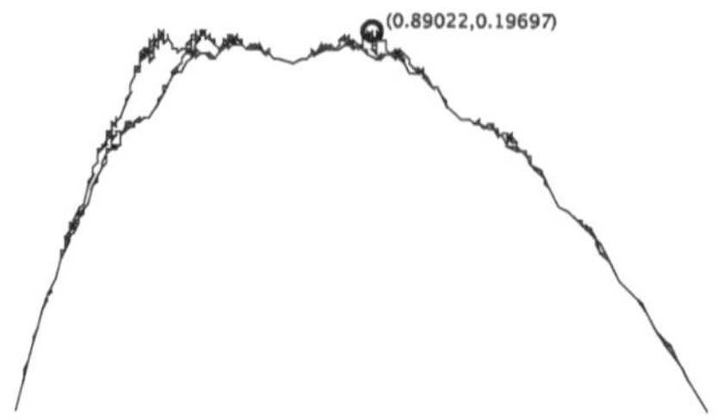

FIGURE 9. Green's functions of the nonrandom gasket.

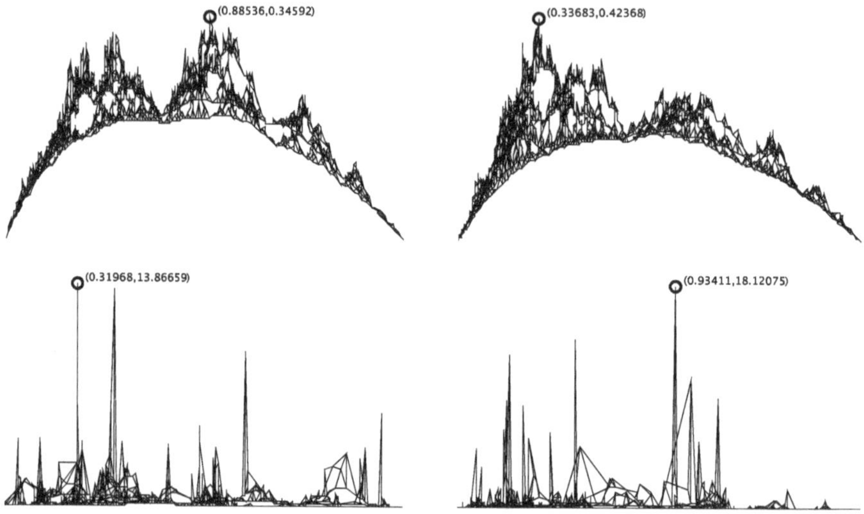

FIGURE 10. Samples of the diagonal of the Green's functions with low randomness (top) and with high randomness (bottom).

bounded), but for high randomness the Green's function is not bounded (and not continuous on the diagonal).
Figure 10 illustrates this experimental data by the pictures of the random sample Green's functions, where the difference between high and low randomness is apparent.

In Figure 11 we plot the sample of the logarithm of the maximum of the diagonal of the Green's functions depending on the depth of approximation. In the low randomness cases the sample curves grow slowly, if at all, but in the high randomness cases the curves grow fast (meaning exponential increase if a non-logarithmic plot).

The most definite experimental evidence for the phase transition is presented in 12, which shows that the median of the logarithm of the maximum of the diagonal of the Green's functions in the case of low randomness does not grow with the depth of approximation, but in the case of high randomness the median of the logarithm appears to grow linearly. This is compared with the density plot Figure 13.

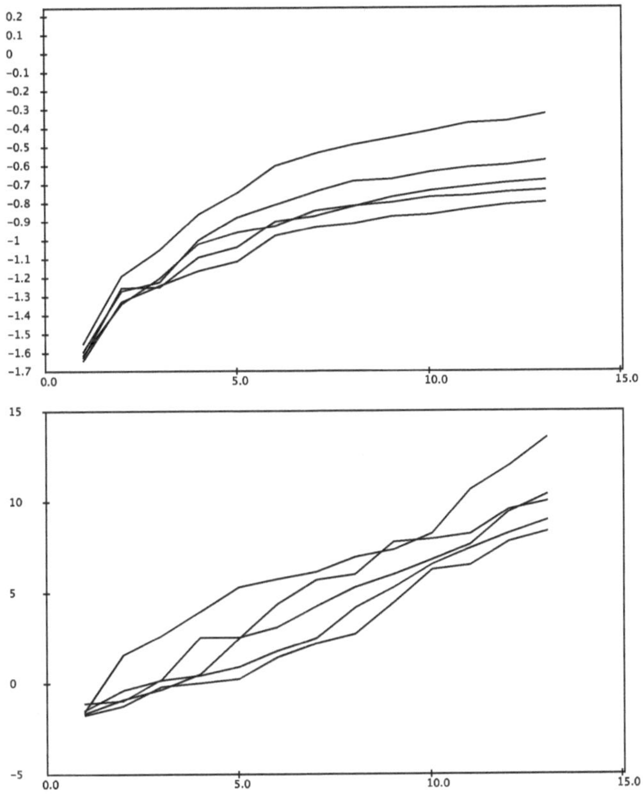

FIGURE 11. Samples of the logarithm of the maximum of the diagonal of the Green's functions with low randomness (top) and with high randomness (bottom) for depths 1–13.

Many questions related to such sequences of Dirichlet forms $\mathcal{E}_n$ are studied in [**62**,**63**] in detail. An important tool in this study is the so called effective resistance R, which is defined for any $x, y \in V_*$ by

$$R(x,y) = \Big(\min_u \{\mathcal{E}(u,u) | u(x) = 1, u(y) = 0\}\Big)^{-1}.$$

Here minimum is taken over all functions on V_*. Note that $x, y \in V_n$ for large enough n and $R(x, y)$ does not change if $\mathcal{E}$ is replaced by $\mathcal{E}_n$ because of the compatibility condition. By [**62**, Theorem 2.1.14], $R(x, y)$ is a metric on V_*. To avoid ambiguity, we may write, for example, R-continuity for continuity with respect to the effective resistance metric R. If $\mathcal{E}(u,u) < \infty$ then u is R-continuous by [**62**, Theorem 2.2.6(1)]. The main ingredient of the proof of this fact is the following inequality, which follows directly from the definition of R,

$$|u(x) - u(y)|^2 \le R(x,y)\mathcal{E}(u,u).$$

It implies, in particular, that any function of finite energy is R–Hölder continuous with respect to the effective resistance metric.

If Ω is the R–completion of V_*, then any function in $\operatorname{Dom} \mathcal{E}$ is a restriction of an R–continuous function on Ω. In other words, if u is a function on V_* such that

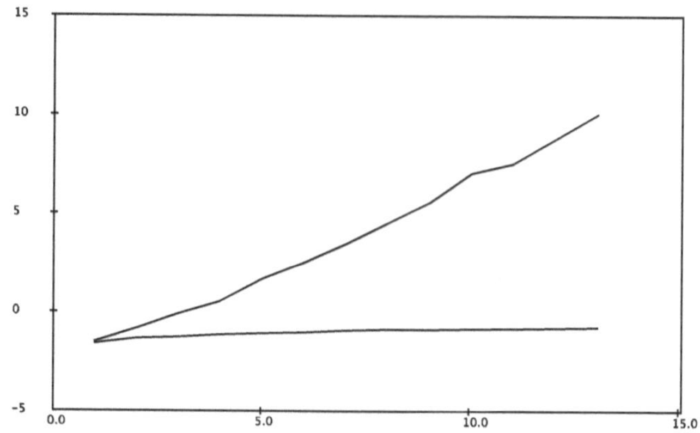

FIGURE 12. Median of the logarithm of the maximum of the diagonal of the Green's functions with low randomness (bottom curve) and with high randomness (top curve) over 50 runs for depths 1–13.

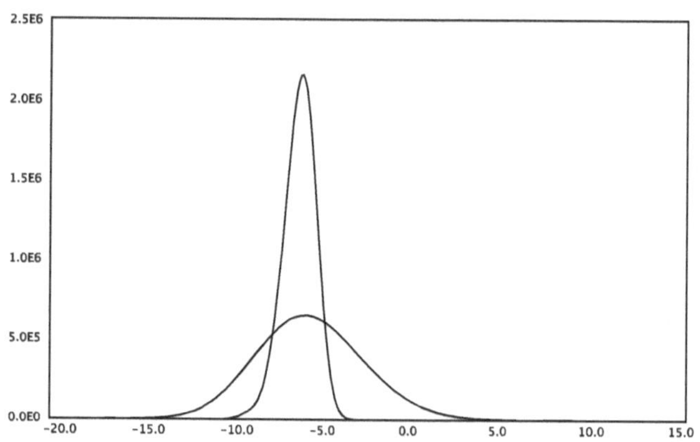

FIGURE 13. Density plot of resistances in low randomness (narrow curve) and with high randomness (wide curve) gaskets for depth 13.

$\mathcal{E}(u,u) < \infty$ then u has a unique continuation to Ω that is R–continuous. We will denote this continuation by the same symbol u and the set of such functions by $\operatorname{Dom}\mathcal{E}$.

An important question is whether Ω is equal to the Sierpiński gasket S. The answer is positive if all the conductances tend to infinity. This happens, for example, in the case of a so called regular self-similar harmonic structure (see [**59**, **62**]). Thus it is natural to say that a harmonic structure is regular if $\Omega = S$ and nonregular otherwise. It is easy to see that a harmonic structure is regular if all the conductances tend to infinity, but the converse is not true. It is proved in [**62**, Proposition 3.3.2] that if harmonic functions are continuous then there is a continuous injective map $\theta : \Omega \to S$ which is the identity on V_*. Therefore in this case we can (and will) consider Ω as a subset of S. Then Ω is the R-closure of V_*. In a sense, Ω is the set where the energy form $\mathcal{E}$ "lives". If Ω is not just an abstract completion then the name "energy form on the Sierpiński gasket S" is more justified. Strictly speaking [**62**, Proposition 3.3.2] is formulated for self-similar harmonic structures, but self-similarity is not used in the proof.

It is proved in [**62**, Theorem 3.5.6] that if $x \in \Omega$ then $\{x\}$ has positive capacity. The converse of this statement is proved in [**62**] for any self-similar harmonic structure.

To define the Green's function we use the construction invented by Kigami for the self-similar harmonic structures. Let Green's function on $V_n \backslash \partial V$ be defined as $G_n = (X_n)^{-1}$ where X_n is the matrix of the energy form $\mathcal{E}_n$, and the inverse defined only for functions with zero boundary conditions. Then the compatibility condition implies that the restriction of Green's function on V_{n+1} to V_n is Green's function on V_n. Naturally, this allows us to define Green's function on V_*, which is denoted by $g(\cdot, \cdot)$.

THEOREM 5 ([**62**, **63**]).
(1) $g(x, y) = g(y, x)$ for all $x, y \in V_*$;
(2) $g(x, y) > 0$ for all $x, y \in V_* \backslash \partial V$;
(3) $g(x, y) = 0$ for all $x \in \partial V$;
(4) $g(x, \cdot) \in \operatorname{Dom} \mathcal{E}$, in particular it is R-continuous;
(5) $\mathcal{E}(g(x, \cdot), h) = 0$ for any harmonic function h;
(6) $\mathcal{E}(g(x, \cdot), f) = f(x)$ for any $f \in \operatorname{Dom} \mathcal{E}$ which vanishes on the boundary;
(7) $g(\cdot, \cdot)$ has a continuation from $V_* \times V_*$ to $\Omega \times \Omega$, in particular if $x \in \Omega$ then $g(x, x) < \infty$.
(8) For any probability measure μ on Ω, which is non zero on any nonempty open set, the energy form $\mathcal{E}$ is a local regular Dirichlet form on L^2_μ.

Note, again, that if harmonic functions are continuous then this theorem holds for Ω which is the R-closure of V_* in S. For discussion related to the last item of this theorem see also [**90**].

7. Energy measure, eigenfunctions and approximation by quantum graphs

In this section we assume that a random Sierpiński gasket is homeomorphically embedded into $\mathbb{R}^2$ using harmonic coordinates, which is possible by Theorem 3. In our exposition we follow [**90**]. For the background on quantum graphs and the interest they hold to physics see [**1**, **69**, **70**] and references therein.

We start with defining a different sequence of approximating energy forms. In various situations these forms are associated with so called quantum graphs, photonic crystals and cable systems. If $f \in C^1(\mathbb{R}^2)$ then we define

$$\mathcal{E}_n^Q(f, g) = \sum_{x, y \in V_n} c_{n,x,y} \mathcal{E}_{x,y}^Q(f, f)$$

where

$$\mathcal{E}_{x,y}^Q(f, f) = \int_0^1 \left(\tfrac{d}{dt} f\big(x(1-t) + ty\big)\right)^2 dt$$

is the integral of the square of the derivative

$$\tfrac{d}{dt} f\big(x(1-t) + ty\big) = \langle \nabla f\big(x(1-t) + ty\big), y - x \rangle$$

of f along the straight line segment connecting x and y. Thus $\mathcal{E}_{x,y}^Q(f, f)$ is the usual one dimensional energy of a function on a straight line segment. If f is

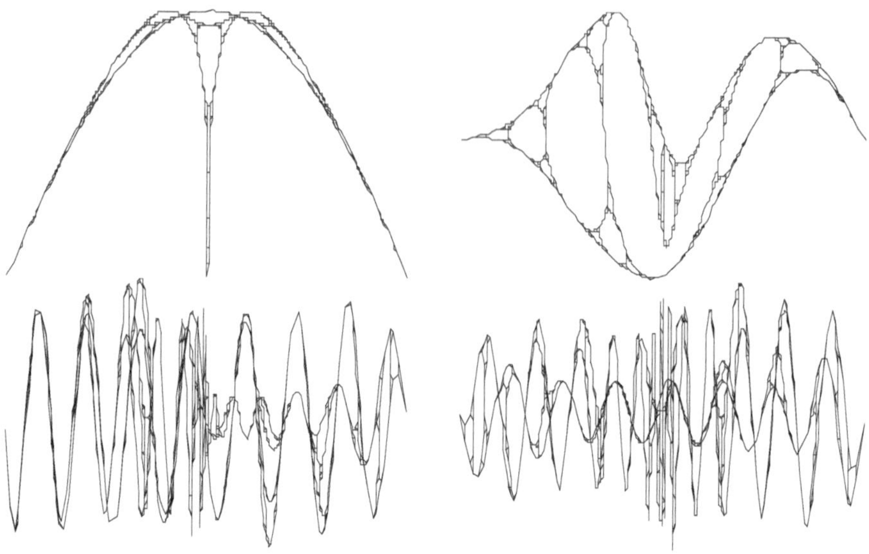

FIGURE 14. Pictures of the non-random energy measure eigenfunctions #1, #5, #50, #51.

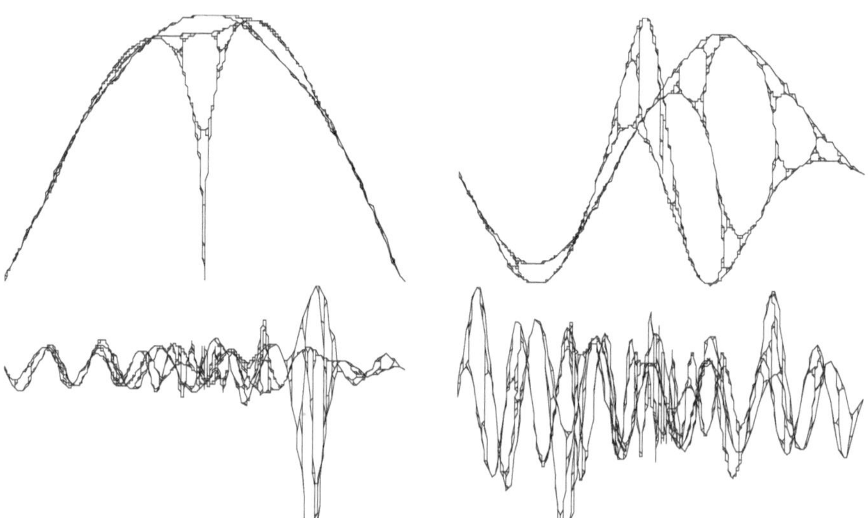

FIGURE 15. Pictures of the random energy measure eigenfunctions #1, #5, #50, #51 with low randomness.

linear then $\mathcal{E}^{Q}_{x,y}(f,f) = \big(f(x) - f(y)\big)^2$. Therefore if f is piecewise harmonic then $\mathcal{E}^{Q}_{n}(f,f) = \mathcal{E}_n(f,f)$ for all large enough n. Also $\mathcal{E}^{Q}_{n}$ satisfies estimate

$$\mathcal{E}_n(f,f) \leqslant \|f\|^2_{C^1(\mathbb{R}^m)} \nu(S). \tag{7.1}$$

Therefore for any $C^1(R^2)$-function we have

$$\lim_{n\to\infty} \mathcal{E}^{Q}_{n}(f,f) = \mathcal{E}(f,f)$$

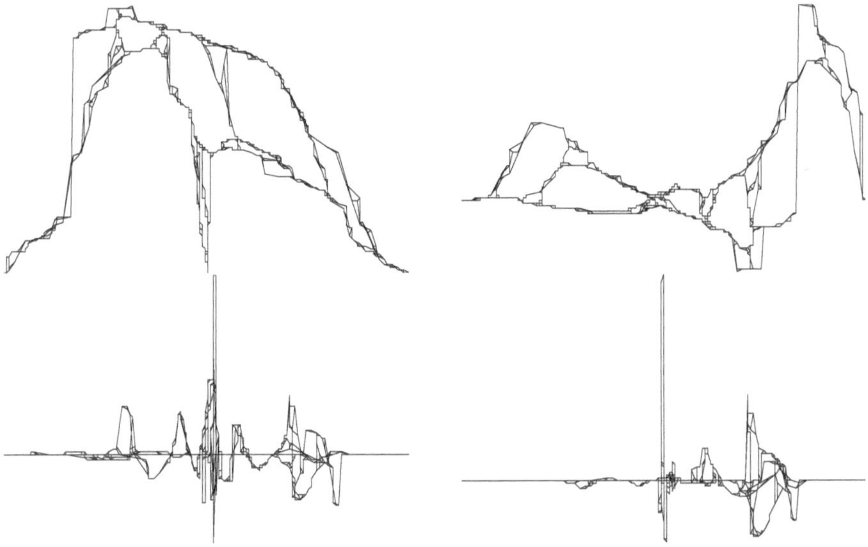

FIGURE 16. Pictures of the random energy measure eigenfunctions #1, #5, #50, #51 with high randomness.

by [**90**, Theorem 5].

THEOREM 6 ([**90**]). If f is the restriction to S of a $C^1(\mathbb{R}^m)$ function then $f \in \operatorname{Dom}\mathcal{E}$, and such functions are dense in $\operatorname{Dom}\mathcal{E}$. In particular we have the Kigami formula

$$\mathcal{E}(f,f) = \int_S \langle \nabla f, Z\nabla f\rangle d\nu$$

for any $f \in C^1(\mathbb{R}^m)$, where Z is a positive trace one matrix defined ν-almost everywhere.

It is easy to see that if g is a $C^1(\mathbb{R}^2)$-function vanishing on V_0 and f is a $C^2(\mathbb{R}^2)$-function then

$$\mathcal{E}_n^Q(f,g) = \sum_{x,y\in V_n} c_{n,x,y} \int_0^1 g\big(x(1-t)+ty\big)\Big(\tfrac{d^2}{dt^2} f\big(x(1-t)+ty\big)\Big)dt$$

because after integration by parts all the boundary terms are canceled.

By [**63**] there is a densely defined operator Δ_ν, called the energy Laplacian (which is self-adjoint with Dirichlet or Neumann boundary conditions), such that for any function $g \in \operatorname{Dom}\mathcal{E}$, vanishing on the boundary V_0, and any function $f \in \operatorname{Dom}\Delta_\nu$, we have the analog of the Gauss-Green formula:

$$\mathcal{E}(f,g) = -\int_S g\Delta_\nu f d\nu,$$

see [**21**, **34**].

LEMMA 7.1 ([**90**]). If f is the restriction to S of a $C^2(\mathbb{R}^2)$ function, and g is the restriction to S of a $C^1(\mathbb{R}^2)$ function, then

$$|\mathcal{E}_n(f,g)| \leqslant const\|g\|_{C^1(\mathbb{R}^2)}\|f\|_{C^2(\mathbb{R}^2)}\nu(S).$$

THEOREM 7 ([**90**]). If f is the restriction to S of a $C^2(\mathbb{R}^2)$ function then $f \in \operatorname{Dom}\Delta_\nu$, and such functions are dense in $\operatorname{Dom}\Delta_\nu$. Moreover, ν-almost everywhere

$$\Delta_\nu f = \operatorname{Tr}(ZD^2 f)$$

where $D^2 f$ is the matrix of the second derivatives of f.

COROLLARY 7.2. If $f(x) = \|x\|^2$ then $\Delta_\nu f = 1$. Moreover, $\Delta_\nu f \in L^\infty(S)$ for any $f \in C^2(\mathbb{R}^2)$.

One can also obtain Theorem 7 using the general theory of Dirichlet forms in [**21, 34**]. However there is a different constructive proof using the approximation by quantum graphs (see [**90**]). Energy measure eigenfunctions are illustrated in Figures 14, 15 and 16.

Acknowledgment

The last author thanks Stanislav Molchanov and Ofer Zeitouni for helpful discussions.

References

[1] Eric Akkermans, Alain Comtet, Jean Desbois, Gilles Montambaux, and Christophe Texier, *Spectral determinant on quantum graphs*, Ann. Physics **284** (2000), no. 1, 10–51, DOI 10.1006/aphy.2000.6056. MR1780489
Phys. Rev. B **29** (1984), 5504–5508.

[2] Shlomo Alexander, *Some properties of the spectrum of the Sierpiński gasket in a magnetic field*, Phys. Rev. B (3) **29** (1984), no. 10, 5504–5508, DOI 10.1103/physrevb.29.5504. MR743875

[3] Patricia Alonso Ruiz, *Explicit formulas for heat kernels on diamond fractals*, Comm. Math. Phys. **364** (2018), no. 3, 1305–1326, DOI 10.1007/s00220-018-3221-x. MR3875827

[4] Patricia Alonso-Ruiz, Fabrice Baudoin, Li Chen, Luke Rogers, Nageswari Shanmugalingam, and Alexander Teplyaev, *Besov class via heat semigroup on Dirichlet spaces III: BV functions and sub-Gaussian heat kernel estimates*, Calc. Var. Partial Differential Equations **60** (2021), no. 5, Paper No. 170, 38, DOI 10.1007/s00526-021-02041-2. MR4290383

[5] P. Alonso Ruiz, Y. Chen, H. Gu, R. S. Strichartz, and Z. Zhou, *Analysis on hybrid fractals*, Commun. Pure Appl. Anal. **19** (2020), no. 1, 47–84, DOI 10.3934/cpaa.2020004. MR4025934

[6] P. Alonso Ruiz and U. R. Freiberg: *Hanoi attractors and the Sierpinski gasket.* Special issue of IJMMNO on Fractals, Fractal-based Methods and Applications, **3(4)**: 251–265, 2012.

[7] Patricia Alonso Ruiz and Uta R. Freiberg, *Weyl asymptotics for Hanoi attractors*, Forum Math. **29** (2017), no. 5, 1003–1021, DOI 10.1515/forum-2015-0179. MR3692024

[8] N. Bajorin, T. Chen, A. Dagan, C. Emmons, M. Hussein, M. Khalil, P. Mody, B. Steinhurst, and A. Teplyaev, *Vibration modes of 3n-gaskets and other fractals*, J. Phys. A **41** (2008), no. 1, 015101, 21, DOI 10.1088/1751-8113/41/1/015101. MR2450694

[9] Sebastian Andres, David Croydon, and Takashi Kumagai, *Heat kernel fluctuations for stochastic processes on fractals and random media*, From classical analysis to analysis on fractals. Vol. 1. A tribute to Robert Strichartz, Appl. Numer. Harmon. Anal., Birkhäuser/Springer, Cham, [2023] ©2023, pp. 265–281, DOI 10.1007/978-3-031-37800-3_12. MR4676391

[10] Martin T. Barlow, *Diffusions on fractals*, Lectures on probability theory and statistics (Saint-Flour, 1995), Lecture Notes in Math., vol. 1690, Springer, Berlin, 1998, pp. 1–121, DOI 10.1007/BFb0092537. MR1668115

[11] Martin T. Barlow, Richard F. Bass, and Takashi Kumagai, *Stability of parabolic Harnack inequalities on metric measure spaces*, J. Math. Soc. Japan **58** (2006), no. 2, 485–519. MR2228569

[12] Martin Barlow, Thierry Coulhon, and Alexander Grigor'yan, *Manifolds and graphs with slow heat kernel decay*, Invent. Math. **144** (2001), no. 3, 609–649, DOI 10.1007/s002220100139. MR1833895

[13] M. T. Barlow and B. M. Hambly, *Transition density estimates for Brownian motion on scale irregular Sierpinski gaskets* (English, with English and French summaries), Ann. Inst. H. Poincaré Probab. Statist. **33** (1997), no. 5, 531–557, DOI 10.1016/S0246-0203(97)80104-5. MR1473565

[14] Martin T. Barlow and Edwin A. Perkins, *Brownian motion on the Sierpiński gasket*, Probab. Theory Related Fields **79** (1988), no. 4, 543–623, DOI 10.1007/BF00318785. MR966175

[15] Matthew Begué, Levi Devalve, David Miller, and Benjamin Steinhurst, *Spectrum and heat kernel asymptotics on general Laakso spaces*, Fractals **20** (2012), no. 2, 149–162, DOI 10.1142/S0218348X12500144. MR2950203

[16] Matthew Begue, Daniel J. Kelleher, Aaron Nelson, Hugo Panzo, Ryan Pellico, and Alexander Teplyaev, *Random walks on barycentric subdivisions and the Strichartz hexacarpet*, Exp. Math. **21** (2012), no. 4, 402–417, DOI 10.1080/10586458.2012.715542. MR3004256

[17] Oren Ben-Bassat, Robert S. Strichartz, and Alexander Teplyaev, *What is not in the domain of the Laplacian on Sierpinski gasket type fractals*, J. Funct. Anal. **166** (1999), no. 2, 197–217, DOI 10.1006/jfan.1999.3431. MR1707752

[18] Ugo Bessi, *Another point of view on Kusuoka's measure*, Discrete Contin. Dyn. Syst. **41** (2021), no. 7, 3241–3271, DOI 10.3934/dcds.2020404. MR4247139

[19] Erik J. Bird, Sze-Man Ngai, and Alexander Teplyaev, *Fractal Laplacians on the unit interval* (English, with English and French summaries), Ann. Sci. Math. Québec **27** (2003), no. 2, 135–168. MR2103098

[20] Adam M. Bowditch and David A. Croydon, *Biased random walk on supercritical percolation: anomalous fluctuations in the ballistic regime*, Electron. J. Probab. **27** (2022), Paper No. 68, 22, DOI 10.1214/22-ejp794. MR4422966

[21] Nicolas Bouleau and Francis Hirsch, *Dirichlet forms and analysis on Wiener space*, De Gruyter Studies in Mathematics, vol. 14, Walter de Gruyter & Co., Berlin, 1991, DOI 10.1515/9783110858389. MR1133391

[22] Brighid Boyle, Kristin Cekala, David Ferrone, Neil Rifkin, and Alexander Teplyaev, *Electrical resistance of N-gasket fractal networks*, Pacific J. Math. **233** (2007), no. 1, 15–40, DOI 10.2140/pjm.2007.233.15. MR2366367

[23] Shiping Cao, *Convergence of energy forms on Sierpinski gaskets with added rotated triangle*, Potential Anal. **59** (2023), no. 4, 1793–1825, DOI 10.1007/s11118-022-10034-9. MR4684376

[24] Shiping Cao and Hua Qiu, *Resistance forms on self-similar sets with finite ramification of finite type*, Potential Anal. **54** (2021), no. 4, 581–606, DOI 10.1007/s11118-020-09840-w. MR4230706

[25] D. A. Croydon, *Scaling limits of stochastic processes associated with resistance forms* (English, with English and French summaries), Ann. Inst. Henri Poincaré Probab. Stat. **54** (2018), no. 4, 1939–1968, DOI 10.1214/17-AIHP861. MR3865663

[26] D. A. Croydon, B. M. Hambly, and T. Kumagai, *Heat kernel estimates for FIN processes associated with resistance forms*, Stochastic Process. Appl. **129** (2019), no. 9, 2991–3017, DOI 10.1016/j.spa.2018.08.011. MR3985553

[27] D. A. Croydon and D. Shiraishi, *Scaling limit for random walk on the range of random walk in four dimensions* (English, with English and French summaries), Ann. Inst. Henri Poincaré Probab. Stat. **59** (2023), no. 1, 166–184, DOI 10.1214/22-aihp1243. MR4533724

[28] David Croydon, Ben Hambly, and Takashi Kumagai, *Time-changes of stochastic processes associated with resistance forms*, Electron. J. Probab. **22** (2017), Paper No. 82, 41, DOI 10.1214/17-EJP99. MR3718710

[29] David A. Croydon, *The random conductance model with heavy tails on nested fractal graphs*, Fractal geometry and stochastics VI, Progr. Probab., vol. 76, Birkhäuser/Springer, Cham, [2021] ©2021, pp. 239–254. MR4237256

[30] David A. Croydon, Ryoki Fukushima, and Stefan Junk, *Anomalous scaling regime for one-dimensional Mott variable-range hopping*, Ann. Appl. Probab. **33** (2023), no. 5, 4044–4090, DOI 10.1214/22-aap1915. MR4663504

[31] T. Reese, D. Yott, A. Brzoska, and D. Kelleher: *Analysing self-similar and fractal properties of the C. Elegans neural network.* PLoS ONE **7(10)** (2012). doi:10.1371/journal.pone.0040483

[32] D. Fontaine, T. Smith and A. Teplyaev, *Random Sierpiński gasket*, Proceedings of the Joint Summer Research Conference on Quantum Graphs and Their Applications, Contemporary Mathematics, AMS, 2006.

[33] Masatoshi Fukushima, *Dirichlet forms, diffusion processes and spectral dimensions for nested fractals*, Ideas and methods in mathematical analysis, stochastics, and applications (Oslo, 1988), Cambridge Univ. Press, Cambridge, 1992, pp. 151–161. MR1190496

[34] Masatoshi Fukushima, Yōichi Ōshima, and Masayoshi Takeda, *Dirichlet forms and symmetric Markov processes*, De Gruyter Studies in Mathematics, vol. 19, Walter de Gruyter & Co., Berlin, 1994, DOI 10.1515/9783110889741. MR1303354

[35] Yuval Gefen, Amnon Aharony, and Benoit B. Mandelbrot, *Phase transitions on fractals. I. Quasilinear lattices, II. Sierpiński gaskets, III. Infinitely ramified lattices*, J. Phys. A **16** (1983), no. 6, 1267–1278; **17** (1984), 435–444 and 1277–1289. MR0706699, MR0734134, MR0755653

[36] Rostislav I. Grigorchuk and Andrzej Żuk, *On the asymptotic spectrum of random walks on infinite families of graphs*, Random walks and discrete potential theory (Cortona, 1997), Sympos. Math., XXXIX, Cambridge Univ. Press, Cambridge, 1999, pp. 188–204. MR1802431

[37] Alexander Grigor'yan and Naotaka Kajino, *Localized upper bounds of heat kernels for diffusions via a multiple Dynkin-Hunt formula*, Trans. Amer. Math. Soc. **369** (2017), no. 2, 1025–1060, DOI 10.1090/tran/6784. MR3572263

[38] B. M. Hambly, *Heat kernels and spectral asymptotics for some random Sierpinski gaskets*, Fractal geometry and stochastics, II (Greifswald/Koserow, 1998), Progr. Probab., vol. 46, Birkhäuser, Basel, 2000, pp. 239–267. MR1786351

[39] B. M. Hambly, *On the asymptotics of the eigenvalue counting function for random recursive Sierpinski gaskets*, Probab. Theory Related Fields **117** (2000), no. 2, 221–247, DOI 10.1007/s004400050005. MR1771662

[40] Ben M. Hambly and Takashi Kumagai, *Heat kernel estimates for symmetric random walks on a class of fractal graphs and stability under rough isometries*, Fractal geometry and applications: a jubilee of Benoît Mandelbrot, Part 2, Proc. Sympos. Pure Math., vol. 72, Amer. Math. Soc., Providence, RI, 2004, pp. 233–259, DOI 10.1090/pspum/072.2/2112125. MR2112125

[41] B. M. Hambly and Michel L. Lapidus, *Random fractal strings: their zeta functions, complex dimensions and spectral asymptotics*, Trans. Amer. Math. Soc. **358** (2006), no. 1, 285–314, DOI 10.1090/S0002-9947-05-03646-9. MR2171234

[42] B. M. Hambly, V. Metz and A. Teplyaev, *Admissible refinements of energy on finitely ramified fractals*, Proc. London Math. Soc. 74, 93–112.

[43] Ben Hambly and Weiye Yang, *Degenerate limits for one-parameter families of non-fixed-point diffusions on fractals*, J. Fractal Geom. **6** (2019), no. 1, 1–51, DOI 10.4171/JFG/67. MR3910540

[44] Masanori Hino, *On singularity of energy measures on self-similar sets*, Probab. Theory Related Fields **132** (2005), no. 2, 265–290, DOI 10.1007/s00440-004-0396-1. MR2199293

[45] Masanori Hino and Kenji Nakahara, *On singularity of energy measures on self-similar sets. II*, Bull. London Math. Soc. **38** (2006), no. 6, 1019–1032, DOI 10.1112/S0024609306019072. MR2285256

[46] Michael Hinz and Melissa Meinert, *Approximation of partial differential equations on compact resistance spaces*, Calc. Var. Partial Differential Equations **61** (2022), no. 1, Paper No. 19, 47, DOI 10.1007/s00526-021-02119-x. MR4350193

[47] R. G. Hohlfeld and N. Cohen: *Self–Similarity and the Geometric Requirements for Frequency Independence in Antennae.* Fractals **7** (1999), 79–84.

[48] Michael J. Ignatowich, Daniel J. Kelleher, Catherine E. Maloney, David J. Miller, and Khrystyna Serhiyenko, *Resistance scaling factor of the pillow and fractalina fractals*, Fractals **23** (2015), no. 2, 1550018, 9, DOI 10.1142/S0218348X15500188. MR3351925

[49] Christopher J. Kauffman, Robert M. Kesler, Amanda G. Parshall, Evelyn A. Stamey, and Benjamin A. Steinhurst, *Quantum mechanics on Laakso spaces*, J. Math. Phys. **53** (2012), no. 4, 042102, 18, DOI 10.1063/1.3702099. MR2953261

[50] Naotaka Kajino, *Heat kernel asymptotics for the measurable Riemannian structure on the Sierpinski gasket*, Potential Anal. **36** (2012), no. 1, 67–115, DOI 10.1007/s11118-011-9221-5. MR2886454

[51] Naotaka Kajino, *Time changes of local Dirichlet spaces by energy measures of harmonic functions*, Forum Math. **24** (2012), no. 2, 339–363, DOI 10.1515/form.2011.065. MR2900010

[52] Naotaka Kajino, *Heat kernel asymptotics for the measurable Riemannian structure on the Sierpinski gasket*, Potential Anal. **36** (2012), no. 1, 67–115, DOI 10.1007/s11118-011-9221-5. MR2886454

[53] Naotaka Kajino, *Analysis and geometry of the measurable Riemannian structure on the Sierpiński gasket*, Fractal geometry and dynamical systems in pure and applied mathematics. I. Fractals in pure mathematics, Contemp. Math., vol. 600, Amer. Math. Soc., Providence, RI, 2013, pp. 91–133, DOI 10.1090/conm/600/11932. MR3203400

[54] Naotaka Kajino, *Equivalence of recurrence and Liouville property for symmetric Dirichlet forms* (English, with English and Russian summaries), Mat. Fiz. Komp′yut. Model. **3(40)** (2017), 89–98, DOI 10.15688/mpcm.jvolsu.2017.3.7. MR3706131

[55] Naotaka Kajino, *Energy measures for diffusions on fractals: a survey*, Analysis and partial differential equations on manifolds, fractals and graphs, Adv. Anal. Geom., vol. 3, De Gruyter, Berlin, [2021] ©2021, pp. 119–142, DOI 10.1515/9783110700763-005. MR4320088

[56] Naotaka Kajino, *On singularity of energy measures for symmetric diffusions with full off-diagonal heat kernel estimates II: some borderline examples*, Dirichlet forms and related topics, Springer Proc. Math. Stat., vol. 394, Springer, Singapore, [2022] ©2022, pp. 223–251, DOI 10.1007/978-981-19-4672-1_13. MR4508848

[57] Daniel J. Kelleher, Hugo Panzo, Antoni Brzoska, and Alexander Teplyaev, *Dual graphs and modified Barlow-Bass resistance estimates for repeated barycentric subdivisions*, Discrete Contin. Dyn. Syst. Ser. S **12** (2019), no. 1, 27–42, DOI 10.3934/dcdss.2019002. MR3836590

[58] Jun Kigami, *A harmonic calculus on the Sierpiński spaces*, Japan J. Appl. Math. **6** (1989), no. 2, 259–290, DOI 10.1007/BF03167882. MR1001286

[59] Jun Kigami, *Harmonic calculus on p.c.f. self-similar sets*, Trans. Amer. Math. Soc. **335** (1993), no. 2, 721–755, DOI 10.2307/2154402. MR1076617

[60] J. Kigami, *Harmonic metric and Dirichlet form on the Sierpiński gasket*, Asymptotic problems in probability theory: stochastic models and diffusions on fractals (Sanda/Kyoto, 1990), Pitman Res. Notes Math. Ser., vol. 283, Longman Sci. Tech., Harlow, 1993, pp. 201–218. MR1354156

[61] Jun Kigami, *Effective resistances for harmonic structures on p.c.f. self-similar sets*, Math. Proc. Cambridge Philos. Soc. **115** (1994), no. 2, 291–303, DOI 10.1017/S0305004100072091. MR1277061

[62] Jun Kigami, *Analysis on fractals*, Cambridge Tracts in Mathematics, vol. 143, Cambridge University Press, Cambridge, 2001, DOI 10.1017/CBO9780511470943. MR1840042

[63] Jun Kigami, *Harmonic analysis for resistance forms*, J. Funct. Anal. **204** (2003), no. 2, 399–444, DOI 10.1016/S0022-1236(02)00149-0. MR2017320

[64] Jun Kigami, *Local Nash inequality and inhomogeneity of heat kernels*, Proc. London Math. Soc. (3) **89** (2004), no. 2, 525–544, DOI 10.1112/S0024611504014807. MR2078700

[65] Jun Kigami and Michel L. Lapidus, *Weyl's problem for the spectral distribution of Laplacians on p.c.f. self-similar fractals*, Comm. Math. Phys. **158** (1993), no. 1, 93–125. MR1243717

[66] Jun Kigami and Michel L. Lapidus, *Self-similarity of volume measures for Laplacians on p.c.f. self-similar fractals*, Comm. Math. Phys. **217** (2001), no. 1, 165–180, DOI 10.1007/s002200000326. MR1815029

[67] Jun Kigami, Daniel R. Sheldon, and Robert S. Strichartz, *Green's functions on fractals*, Fractals **8** (2000), no. 4, 385–402, DOI 10.1142/S0218348X00000421. MR1810881

[68] Serguei M. Kozlov, *Harmonization and homogenization on fractals*, Comm. Math. Phys. **153** (1993), no. 2, 339–357. MR1218305

[69] Peter Kuchment, *Quantum graphs. I. Some basic structures*, Waves Random Media **14** (2004), no. 1, S107–S128, DOI 10.1088/0959-7174/14/1/014. Special section on quantum graphs. MR2042548

[70] Peter Kuchment, *Quantum graphs. II. Some spectral properties of quantum and combinatorial graphs*, J. Phys. A **38** (2005), no. 22, 4887–4900, DOI 10.1088/0305-4470/38/22/013. MR2148631

[71] Takashi Kumagai, *Heat kernel estimates and parabolic Harnack inequalities on graphs and resistance forms*, Publ. Res. Inst. Math. Sci. **40** (2004), no. 3, 793–818. MR2074701

[72] Shigeo Kusuoka, *Dirichlet forms on fractals and products of random matrices*, Publ. Res. Inst. Math. Sci. **25** (1989), no. 4, 659–680, DOI 10.2977/prims/1195173187. MR1025071

[73] S. Kusuoka: *Lecture on diffusion process on nested fractals.* Lecture Notes in Math. **1567** 39–98, Springer-Verlag, Berlin, 1993.

[74] Shigeo Kusuoka and Xian Yin Zhou, *Waves on fractal-like manifolds and effective energy propagation*, Probab. Theory Related Fields **110** (1998), no. 4, 473–495, DOI 10.1007/s004400050156. MR1626955

[75] Volker Metz, *The cone of diffusions on finitely ramified fractals*, Nonlinear Anal. **55** (2003), no. 6, 723–738, DOI 10.1016/j.na.2003.08.007. MR2060526
[76] Robert Meyers, Robert S. Strichartz, and Alexander Teplyaev, *Dirichlet forms on the Sierpiński gasket*, Pacific J. Math. **217** (2004), no. 1, 149–174, DOI 10.2140/pjm.2004.217.149. MR2105771
[77] S. Molchanov, *Lectures on random media*, Lectures on probability theory (Saint-Flour, 1992), Lecture Notes in Math., vol. 1581, Springer, Berlin, 1994, pp. 242–411, DOI 10.1007/BFb0073874. MR1307415
[78] Jonathan Needleman, Robert S. Strichartz, Alexander Teplyaev, and Po-Lam Yung, *Calculus on the Sierpinski gasket. I. Polynomials, exponentials and power series*, J. Funct. Anal. **215** (2004), no. 2, 290–340, DOI 10.1016/j.jfa.2003.11.011. MR2150975
[79] Anders Öberg, Robert S. Strichartz, and Andrew Q. Yingst, *Level sets of harmonic functions on the Sierpiński gasket*, Ark. Mat. **40** (2002), no. 2, 335–362, DOI 10.1007/BF02384540. MR1948069
[80] Anders Öberg and Konstantinos Tsougkas, *The Kusuoka measure and the energy Laplacian on level-k Sierpiński gaskets*, Rocky Mountain J. Math. **49** (2019), no. 3, 945–961, DOI 10.1216/RMJ-2019-49-3-945. MR3983309
[81] Olaf Post and Jan Simmer, *Approximation of fractals by discrete graphs: norm resolvent and spectral convergence*, Integral Equations Operator Theory **90** (2018), no. 6, Paper No. 68, 30, DOI 10.1007/s00020-018-2492-0. MR3861286
[82] Olaf Post and Jan Simmer, *Graph-like spaces approximated by discrete graphs and applications*, Math. Nachr. **294** (2021), no. 11, 2237–2278, DOI 10.1002/mana.201900108. MR4371297
[83] R. Rammal, *Spectrum of harmonic excitations on fractals*, J. Physique **45** (1984), no. 2, 191–206, DOI 10.1051/jphys:01984004502019100. MR737523
[84] R. Rammal and G. Toulouse: *Random walks on fractal structures and percolation clusters.*: J. Physique Letters **44** (1983), L13–L22.
[85] Jeremy Stanley, Robert S. Strichartz, and Alexander Teplyaev, *Energy partition on fractals*, Indiana Univ. Math. J. **52** (2003), no. 1, 133–156, DOI 10.1512/iumj.2003.52.2115. MR1970024
[86] R. B. Stinchcombe, *Fractals, phase transitions and criticality*, Proc. Roy. Soc. London Ser. A **423** (1989), no. 1864, 17–33. Fractals in the natural sciences. MR1059263
[87] Robert S. Strichartz, *Analysis on fractals*, Notices Amer. Math. Soc. **46** (1999), no. 10, 1199–1208. MR1715511
[88] Robert S. Strichartz, *Differential equations on fractals: A tutorial*, Princeton University Press, Princeton, NJ, 2006. MR2246975
[89] Alexander Teplyaev, *Energy and Laplacian on the Sierpiński gasket*, Fractal geometry and applications: a jubilee of Benoît Mandelbrot. Part 1, Proc. Sympos. Pure Math., vol. 72, Amer. Math. Soc., Providence, RI, 2004, pp. 131–154, DOI 10.1090/pspum/072.1/2112105. MR2112105
[90] Alexander Teplyaev, *Harmonic coordinates on fractals with finitely ramified cell structure*, Canad. J. Math. **60** (2008), no. 2, 457–480, DOI 10.4153/CJM-2008-022-3. MR2398758
[91] Konstantinos Tsougkas, *Non-degeneracy of the harmonic structure on Sierpiński gaskets*, J. Fractal Geom. **6** (2019), no. 2, 143–156, DOI 10.4171/JFG/73. MR3950670

Department of Mathematics, University of Connecticut, Storrs, Connecticut 06269
Email address: daniel.fontaine@uconn.edu

Department of Mathematics, University of Connecticut, Storrs, Connecticut 06269
Email address: daniel.kelleher@uconn.edu

Department of Mathematics, University of Connecticut, Storrs, Connecticut 06269
Email address: alexander.teplyaev@uconn.edu
URL: http://www.math.uconn.edu/~teplyaev

Contemporary Mathematics
Volume **828**, 2025
https://doi.org/10.1090/conm/828/16596

Spectra of three-peg Hanoi towers graphs

Brett Hungar, Gamal Mograby, Madison Phelps, Luke G. Rogers, and Jonathan Wheeler

ABSTRACT. We consider the relationship between the Laplacians on two sequences of planar graphs, one from the theory of self-similar groups and one from analysis on fractals. By establishing spectral decimation maps between these sequences we provide an alternative to the method of Grigorchuk and Šunić for computing the spectrum of the first sequence of graphs, and an alternative to Quint's computation of the eigenfunctions. Our goal in treating this example is to illustrate the applicability of spectral decimation techniques for computing spectral features for graphs arising from self-similar groups.

1. Introduction

We consider the spectra of the Laplacians for the sequence of graphs H_n in Figure 1, which are the Schreier graphs of the action of the Hanoi towers group on three pegs on the 3-regular rooted tree, and the spectra of the Laplacians for the sequence of Sierpinski gasket graphs G_n in Figure 2. Both sequences limit to the Sierpinski gasket in a natural manner, see [**9, 13**] for H_n and [**7**] for G_n.

On the graphs G_n the spectrum of the Laplacian Δ_n^G may be obtained from that of Δ_{n-1}^G by a method called spectral decimation [**4, 8, 12**]; the decimation map involved is independent of n, so iteration of it provides a dynamical system for computing $\sigma(\Delta_n^G)$, see Theorem 5.1 below. This technique also gives access to spectra of structures based on these graphs [**14, 17**]. An independently discovered approach applies to the graphs H_n arising from their realization as the Schreier graphs of the self-similar Hanoi towers group, see [**5**]. In this latter setting one first obtains a two-dimensional dynamics for the spectrum and then proves it is semi-conjugate to a one-dimensional dynamical system; again the spectrum $\sigma(\Delta_n^H)$ can be recovered from the resulting simple dynamics.

2020 *Mathematics Subject Classification.* Primary 05C50; Secondary 05C25, 20E08, 28A80.

Key words and phrases. graph spectrum, self-similar graph, fractal graph, substitution graph, Laplacian, spectral decimation, Hanoi Towers.

The authors were supported by grants NSF DMS-1659643 and DMS-1613025.

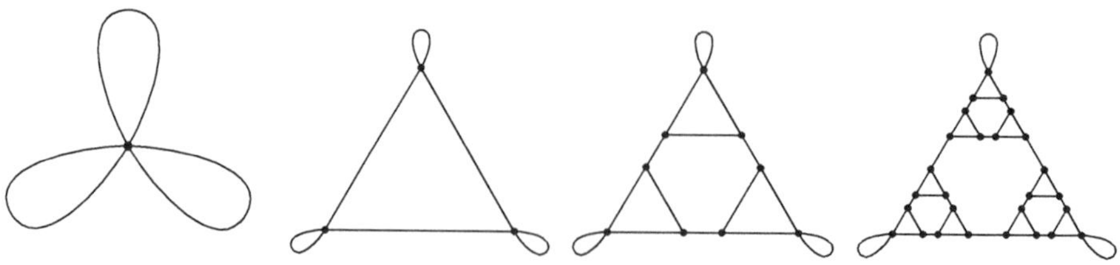

FIGURE 1. The graphs H_n, $n = 0, 1, 2, 3$

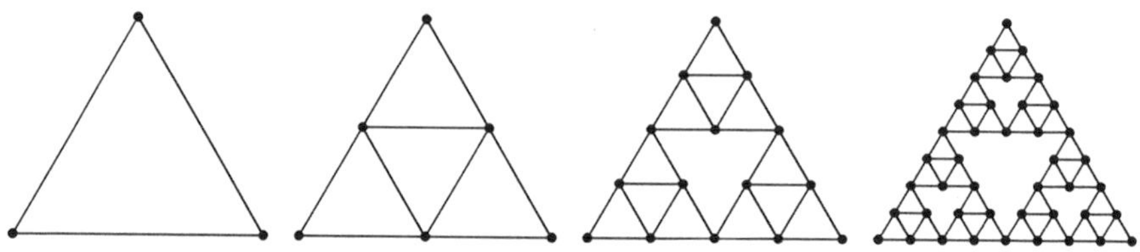

FIGURE 2. The graphs G_n, $n = 0, 1, 2, 3$

In [**5**] it is suggested that their main result might be derived by spectral decimation, and indeed it is easy to check that the dynamical systems for G_n and H_n are the same. Our purpose here is to examine the connection between the graphs G_n and H_n in more detail, use this to explain a connection between the methods used to obtain the dynamics for the spectra, and determine the structure of eigenfunctions on the graphs H_n. From this we obtain the following result; note that definitions of the Laplacian operators are in Section 2.

THEOREM 1.1. *The spectrum of Δ_n^H is the set*

$$\sigma(\Delta_n^H) = \{0\} \cup \left(\bigcup_{i=0}^{n-1} R_H^{-i}(-1) \right) \cup \left(\bigcup_{j=0}^{n-2} R_H^{-j}\left(-\frac{5}{3}\right) \right)$$

where $R_H(z) = z(3z + 5)$. The multiplicities and eigenspaces are as follows:

- *0 has multiplicity 1 with constant eigenfunction*
- *$R_H^{-i}(-1)$ contains 2^i eigenvalues when $0 \le i \le n-1$, each with multiplicity $\frac{1}{2}(3^{n-1-i} + 3)$. For each eigenvalue there is a basis for the eigenspace in which basis elements are localized to the union of two adjacent copies of H_{i+2} in H_n, including copies that are self-adjacent via a loop. (In the case $i = n - 1$ this description is vacuous and the support is all of H_n.) The basis elements can be derived from a specific set of values on H_{n-i}, see Figure 6, by i iterations of an explicit algorithm (5.2).*
- *$R_H^{-j}(-\frac{5}{3})$ contains 2^j eigenvalues when $0 \le j \le n - 2$, each with multiplicity $\frac{1}{2}(3^{n-1-j} - 1)$. Eigenfunctions reflect the homology of H_{n-1-j}: a basis may be obtained by taking alternating values around the corresponding cycles in H_{n-j} and extending to H_k for $n - j + 1 \le k \le n$ by iteration of an explicit algorithm (5.2).*

The above description of the spectrum is equivalent to that given by Grigorchuk and Šunić in Theorem 1.1 of [**5**], and the description of the eigenfunctions is

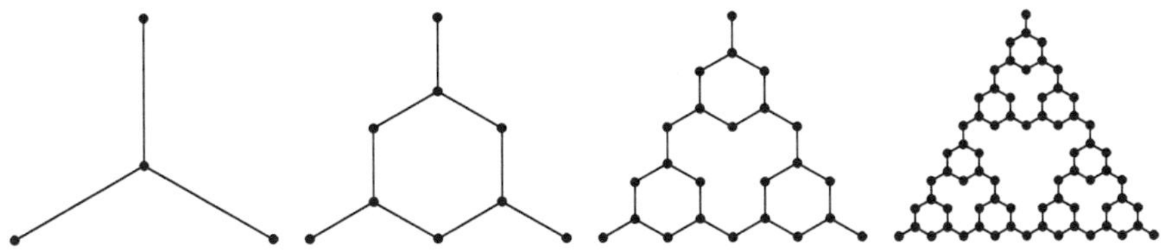

FIGURE 3. The graphs J_n, $n = 0, 1, 2, 3$

equivalent to that obtained by Quint, Theorem 1.1 of [**11**]. What is new here is the method, which involves a simpler dynamical system that that appearing in [**5**] and applies the existing spectral decimation method rather than the functional equation approach of [**11**], allowing a more elementary and potentially more flexible analysis to determine eigenfunctions. We compare our method to that in [**5**] in Section 7 and to that of [**11**] in Section 8. The method used in [**5**]has also been used to treat other important examples of Schreier graphs from self-similar groups [**2**], and our hope is that the flexibility of the spectral decimation approach illustrated on the simple example treated here may permit it to generalize to some of these examples.

The structure of the paper is as follows. In Section 2 we give formal definitions of our graphs and Laplacians. Section 3 contains the definition and some known results about spectral similarity, which is our main tool. Our results are in Section 4, which establishes spectral similarities between the Laplacians on the graphs G_n and H_n, Section 5, where we prove the Laplacian on H_n is spectrally self-similar and establish the iteration (5.2) for obtaining eigenfunctions of H_{n+1} from those of H_n, and Section 6, where we discuss the construction of eigenfunctions and prove Theorem 1.1.

2. Graphs and Laplacians

Graphs. A rapid way to formally define the graphs G_n in Figure 2 is to take $V_0^G = \{p_1, p_2, p_3\}$ to be the vertices of a triangle in the plane and let G_0 be the complete graph on V_0^G. Then set $F_j(x) = \frac{1}{2}(x + p_j)$ for $j = 1, 2, 3$ and $V_n^G = \cup_{j=1}^3 F_j(V_{n-1})$ for $n \geq 1$. G_n has vertex set V_n^G and inductively defined edge relation $x \sim_{G_n} y$ if and only if $x = F_j(x')$, $y = F_j(y')$ for some $j \in \{1, 2, 3\}$ and points $x' \sim_{G_{n-1}} y'$ from V_{n-1}^G. Figure 2 is then derived by iterative application of the three maps F_j to the graph G_0. Note that the three copies $F_j(V_{n-1}^G)$ overlap at three points, so there are $\frac{1}{2}(3^n + 3)$ points in V_n^G.

We define the graphs H_n in Figure 1 indirectly, obtaining along the way a third series of graphs J_n, shown in Figure 3 that will be useful in their own right. Let $p_0 = \frac{1}{3}\sum_1^3 p_j$ be the midpoint of the previously-defined triangle and $V_0^J = \{p_0, p_1, p_2, p_3\}$. Define J_0 to be the graph on V_0^J with edges from p_0 to each p_j, $j = 1, 2, 3$. Then define J_n to have vertex set $V_n^J = \cup_1^3 F_j(V_{n-1})$ for $n \geq 1$ and edges $x \sim_{J_n} y$ precisely when $x = F_j(x')$, $y = F_j(y')$ for some $j \in \{1, 2, 3\}$ and points $x' \sim_{J_{n-1}} y'$ from V_{n-1}^J.

Now we can define H_n on the vertex set obtained inductively from $V_0^H = \{p_0\}$ and $V_n^H = \cup_1^3 F_j(V_{n-1}^H)$ for $n \geq 1$ by providing an edge $x \sim_{H_n} x'$ for $x \neq x' \in V_n^H$ when there is a vertex $y \in V_n^J$ so both $y \sim_{J_n} x$ and $y \sim_{J_n} x'$. Evidently $V_n^J = V_n^G \cup V_n^H$ and this point y is actually from V_n^G. Moreover for each $y \in V_n^G \setminus V_0^G$

there is a uniquely determined pair $x \neq x'$ with this property. In the case $y \in V_0^G$ we follow [**5**] and define a special case, placing a loop at the unique neighboring $x \in V_n^H$. The purpose of this loop will become more clear later; for now we note that it will make H_n a 3-regular graph.

The reader might find this to be a convenient point to check that H_n is the level n Schreier graph of the action of the Hanoi Towers group on the rooted 3-tree, as described in [**5**]. Indeed, if we introduce a word space $W_n = \{1,2,3\}^n$ and for $w = w_1 \cdots w_n \in W_n$ write $F_w = F_{w_1} \circ \cdots \circ F_{w_n}$ then $w \mapsto F_w(p_0)$ is a bijection $W_n \to V_n^H$ and there is an edge from $F_w(p_0)$ to $F_{w'}(p_0)$ exactly when there are $j, j' \in \{1,2,3\}$ so $F_w(p_j) = F_{w'}(p_{j'})$, which is precisely when the Hanoi Towers group contains an element taking w to w'. A characterization of the connection between sequences of self-similar graphs and sequences of Schreier graphs of self-similar groups is in [**6**]; other connections between the methods used to analyze these sequences are in [**10**].

Laplacian. There are several standard notions of Laplacian associated to a graph. In order to simplify the use of some results from [**8**] we use the graph Laplacian, which is defined for a function f on the vertex set by

$$\Delta_n f(x) = \frac{1}{\deg(x)} \sum_{x \sim_n y} (f(y) - f(x))$$

where the degree $\deg(x)$ is the number of edges incident at the vertex x. However the case of a loop at x is slightly different: then $x = y$ and $f(x) - f(y) = 0$, and we count the loop as contributing 1 to the degree. It is easy to check that this is consistent with the usual simple random walk in which a mass at x is divided by $\deg(x)$ and propagated to the neighbors, one of which neighbors is x itself if there is a loop at x.

Observe that on G_n we have $\deg(x) = 4$ for $x \in V_n^G \setminus V_0^G$ and $\deg(x) = 2$ on V_0^G. The graph H_n is regular, with $\deg(x) = 3$ at all vertices, because the loops contribute 1 to the degree of the three vertices at the corners of the triangle. The graphs J_n have $\deg(x) = 3$ at those vertices in V_n^J that come from V_n^H, $\deg(x) = 2$ on $V_n^G \setminus V_0^G$ and $\deg(x) = 1$ on V_0^G.

Rather than giving a recursion for the entries of the Laplacian matrices, which is possible but would not be used later, we illustrate our definition with the following explicit examples. For clarity we have made blocks of vertices with the same degree and ordered them by decreasing degree. Some other ordering choices have been used to emphasize the symmetry of the matrices. Setting

$$M = \begin{bmatrix} 0 & 1 & 1 \\ 1 & 0 & 1 \\ 1 & 1 & 0 \end{bmatrix}$$

and writing I for the 2×3 identity matrix and 0 for the zero matrix we have $\Delta_1^H = \frac{1}{3}M - I$ and in block form:

$$\Delta_1^G = \begin{bmatrix} \frac{1}{4}M - I & \frac{1}{4}M \\ \frac{1}{2}M & -I \end{bmatrix} \qquad \Delta_1^J = \begin{bmatrix} -I & \frac{1}{3}M & \frac{1}{3}M \\ \frac{1}{2}M & -I & 0 \\ I & 0 & -I \end{bmatrix}.$$

3. Spectral similarity and decimation

We present the elementary features of spectral similarity and spectral decimation as they apply to the circumstances in the problem at hand. These include the connection between the classical Schur complement formula for relating the spectrum of a matrix to that of submatrices in a block decomposition and the notion of spectral similarity, and a construction from [**8**] which permits one to glue spectrally similar graph Laplacians in a manner that preserves spectral similarity.

Let $\mathcal{H}$ be a finite dimensional complex vector space and $\tilde{\mathcal{H}}$ be a subspace. Write $P : \mathcal{H} \to \tilde{\mathcal{H}}$ for the orthogonal projection (so P^* is the inclusion) and Q for the projection to the orthogonal complement. Let D be an operator on $\mathcal{H}$ and $\tilde{D}$ be an operator on $\tilde{\mathcal{H}}$.

In accordance with Definition 2.1 in [**8**] we say D is spectrally similar to $\tilde{D}$ with functions ϕ and $\tilde{\phi}$ if $P^*(D-z)^{-1}P = \left(\tilde{\phi}(z)\tilde{D} - \phi(z)\right)^{-1}$ as meromorphic functions. This is related to the Schur complement in the following manner (Lemma 3.3 in [**8**]). Writing the symmetric matrix D in the block form on $\tilde{\mathcal{H}}^{\perp} \otimes \tilde{\mathcal{H}}$

$$D = \begin{pmatrix} QDQ^* & PDQ^* \\ QDP^* & PDP^* \end{pmatrix}$$

we find that D is spectrally similar to $\tilde{D}$ if and only if the Schur complement from $D - z$ satisfies

$$PDP^* - z - QDP^*(QDQ^* - z)^{-1}PDQ^* = \tilde{\phi}(z)\tilde{D} - \phi(z).$$

In particular, we can compute the spectrum of D from that of $\tilde{D}$ using the associated functions. Writing $\sigma(D)$ for the spectrum we have the following.

THEOREM 3.1 (Theorem 3.6 in [**8**]). *If D is spectrally similar to $\tilde{D}$ with functions ϕ and $\tilde{\phi}$, let the exceptional set be $E = \sigma(QDQ^*) \cup \{z : \tilde{\phi}(z) = 0\}$ and the (spectral) decimation function be $R(z) = \phi(z)/\tilde{\phi}(z)$. Suppose $z \notin E$. Then $R(z)$ is an eigenvalue of $\tilde{D}$ if and only if z is an eigenvalue of D, and there is a bijection $\tilde{f} \mapsto f = \tilde{f} - (QDQ^* - z)^{-1}PDQ^*\tilde{f}$ between the corresponding eigenspaces.*

Some general theory regarding the exceptional set is summarized in [**1**].

EXAMPLE 3.2. A case of special interest in the problem at hand occurs when D_N is the Laplacian of the graph Γ_N in which one vertex p_0 is connected by a single edge to each of N other vertices p_j, $j = 1, \dots N$, so $\mathcal{H} = \mathbb{C}^{N+1}$. Take $\tilde{D}_N$ to be the Laplacian of the complete graph on the vertices p_j, $j = 1, \dots N$ and let $\tilde{\mathcal{H}}$ be the corresponding subspace. Writing the block form

$$D_N - z = \begin{pmatrix} -(1+z) & (\frac{1}{N})1_{1\times N} \\ 1_{N\times 1} & -(1+z)I_N \end{pmatrix}$$

where I_N is the $N \times N$ identity and $1_{m\times n}$ is the $m \times n$ matrix with all entries equal 1 we can compute the Schur complement S_z to be

$$S_z = -(1+z)I_N + \frac{1}{N(1+z)}1_{N\times N} = \frac{N-1}{N(1+z)}\tilde{D}_N - \Big(1 + z - \frac{1}{1+z}\Big)I_N$$

so that D_N is spectrally similar to $\tilde{D}_N$ with $\phi_N(z) = 1 + z - \frac{1}{1+z}$ and $\tilde{\phi}_N(z) = \frac{N-1}{N(1+z)}$. This relation is valid provided z is not in the exceptional set E. However, the exceptional set is small: the spectrum of the Schur block is simply $\{-1\}$ and

$\tilde{\phi}_N(z) = \frac{N-1}{N(1+z)}$ is always non-zero, so $E = \{-1\}$ independent of N. The reader will note that this Schur complement reduction is the spectral version of the star-mesh transform that is well known in electrical network theory; the case $N = 3$ of the latter is often called the $\Delta - Y$ transform.

It is worth noting what happens to eigenfunctions in this context. If $z \neq -1$, $R(z) \in \sigma(\tilde{D}_N)$ and $\tilde{f}$ is an eigenfunction of $\tilde{D}_N$ with eigenvalue $R(z)$, then the corresponding eigenfunction of D_N with eigenvalue z is

$$f = \tilde{f} - (QD_NQ^*)^{-1}\tilde{f} = \tilde{f} + \frac{1}{1+z}(\frac{1}{N})1_{1\times N}\tilde{f}_0$$

so that $f = f_0$ on $\tilde{V}$ and $f(p_0) = \frac{1}{N(1+z)}\sum_{j=1}^N f_0(p_j)$. The remaining possibility is $z = -1$, in which case the spectral similarity cannot be used but

$$D_N + I = \begin{pmatrix} 0 & (\frac{1}{N})1_{1\times N} \\ 1_{N\times 1} & 0_N \end{pmatrix}$$

so -1 is an eigenvalue with an $N-1$ dimensional eigenspace consisting of those functions f with $f(p_0) = 0$ and $\sum_{j=1}^N f(p_j) = 0$.

4. Spectral similarities for G_n, H_n and J_n

The significance of Example 3.2 is that the graph J_n can be built by replacing the edges of H_n by copies of Γ_2, and also by replacing all copies of the graph V_0^G (the complete graph on 3 vertices) in V_n^G with copies of $V_0^J = \Gamma_3$. The effect of this type of graph replacement operation on the spectra of graphs was previously studied in [**16**]. The appropriate substitution operation preserves spectral similarity of graph Laplacians according to the following result from [**8**]

LEMMA 4.1 (Lemma 4.7 in [**8**]). *If Γ is a graph with vertex set V and is symmetric with respect to $\tilde{V} \subset V$ in the sense that every bijection of $\tilde{V}$ extends to a graph automorphism of Γ then the graph Laplacian Δ^Γ of Γ is spectrally similar to $\Delta^{\tilde{\Gamma}}$ for the complete graph $\tilde{\Gamma}$ on $\tilde{V}$, with some functions ϕ and $\tilde{\phi}$. Now take disjoint copies of Γ over some index set $\mathcal{A}$ and form a quotient graph Γ' by identifying vertices in $V \times \mathcal{A}$ via an equivalence relation on $\tilde{V} \times \mathcal{A}$ for which equivalence classes are finite. Repeat the same quotient construction on $\tilde{\Gamma} \times \mathcal{A}$ to get a quotient graph $\tilde{\Gamma}'$. Then $\Delta^{\Gamma'}$ is spectrally similar to $\Delta^{\tilde{\Gamma}'}$ with the same functions ϕ and $\tilde{\phi}$.*

PROPOSITION 4.2. *The Laplacian Δ_n^J on J_n is spectrally similar to Δ_n^G on G_n with exceptional set $E = \{-1\}$ and decimation function $R_3(z) = \frac{3}{2}(z^2 + 2z)$. If $z \neq -1$ and $R_3(z)$ is an eigenvalue of Δ_n^G with eigenfunction $\tilde{f}$ then the corresponding eigenfunction of Δ_n^J with eigenvalue z is obtained by extending $\tilde{f}$ to each point $x \in V_n^J \setminus V_n^G = V_n^H$ so that $f(x) = \frac{1}{3(1+z)}\sum_{y\sim_{J_n}x}\tilde{f}(y)$. Moreover, -1 is an eigenvalue of Δ_n^J with a $\frac{1}{2}(3^n+3)$ dimensional eigenspace consisting of functions g that vanish at each $x \in V_n^H \subset V_n^J$ and such that $\sum_{y\sim_{J_n}x} g(y) = 0$ for all such x.*

PROOF. Recall we wrote $F_w = F_{w_1} \circ \dots F_{w_n}$ where $w_1 \cdots w_n$ is a length n word with letters $w_j \in \{1,2,3\}$. The vertices V_n^J are then all points of the form $F_w(p_j)$ as w ranges over n-words and $j \in \{0,1,2,3\}$, and the inductive definition of the edges of J_n implies there is an edge between $F_w(p_j)$ and $F_{w'}(p_k)$ if and only if $w = w'$ and $j \neq k$. This exhibits J_n as a quotient as in Lemma 4.1 of copies of J_0 indexed by the length n-words $\{1,2,3\}^n$.

Adopting the notation of the lemma, $\Gamma = J_0$, $\mathcal{A} = \{1,2,3\}^n$, and the equivalence relation is that (p_j, w) and (p_k, w') are equivalent when $F_w(p_j) = F_{w'}(p_k)$. Observe that this latter can never occur if j or k is zero, so the equivalence relation is really on $\tilde{V} \times \mathcal{A}$, where $\tilde{V} = \{p_1, p_2, p_3\} = V_0^G$, and it is easily seen that V_0^J is symmetric with respect to V_0^G. According to Lemma 4.1, then Γ is spectrally similar to the complete graph $\tilde{\Gamma}$ on $\tilde{V}$, which we observe is actually G_0. In fact this is the case $N = 3$ of the spectral similarity in Example 3.2, so we know the functions ϕ_3 and $\tilde{\phi}_3$ explicitly.

We have seen that $J_n = \Gamma'$, the quotient in Lemma 4.1. At the same time, the quotient $\tilde{\Gamma}'$ is that obtained by gluing the corresponding copies of $\tilde{\Gamma} = V_0^G$, and the same reasoning as used on J_n shows the result is $\tilde{\Gamma}' = G_n$. Now the result of the lemma says that J_n and G_n are spectrally similar with functions $\phi(z) = \phi_3(z) = 1 + z - \frac{1}{1+z}$ and $\tilde{\phi}(z) = \tilde{\phi}_3(z) = \frac{2}{3(1+z)}$, giving the formula for $R_3(z)$. Moreover, the discussion in Example 3.2 tells us that if $z \neq -1$ and $R_3(z) \in \sigma(\Delta_n^G)$ with eigenfunction $\tilde{f}$ then we can extend to an eigenfunction f of Δ_n^J with eigenvalue z by taking $f(F_w(p_0)) = \frac{1}{3(1+z)} \sum_{j=1}^3 \tilde{f}(F_w(p_j))$. At the same time, $z = -1$ is an eigenvalue of Δ_n^J with eigenspace consisting of functions that vanish at all points $F_w(p_0)$ and have mean zero over the neighbors of such points. Evidently this space involves the values on the $\frac{1}{2}(3^{n+1} + 3)$ points in V_n^G subject to 3^n constraints (because the mean is zero on each of the 3^n cells), so is dimension $\frac{1}{2}(3^n + 3)$. □

PROPOSITION 4.3. *The Laplacian Δ_n^J on J_n is spectrally similar to Δ_n^H on H_n with exceptional set $E = \{-1\}$ and decimation function $R_2(z) = 2(z^2 + 2z)$. If $z \neq -1$ and $R_2(z)$ is an eigenvalue of Δ_n^H with eigenfunction $\tilde{f}$ then the corresponding eigenfunction of Δ_n^J with eigenvalue z is obtained by extending $\tilde{f}$ to each point $x \in V_n^J \setminus V_n^G = V_n^H$ so that $f(x) = \frac{1}{2(1+z)} \sum_{y \sim_{J_n} x} \tilde{f}(y)$.*

PROOF. We again use Lemma 4.1, in particular following its notation. This time Γ is the graph called Γ_2 in Example 3.2, so the vertex set V has three elements which we label $\{q_0, q_1, q_2\}$, with q_0 being the one incident on two edges. Set $\tilde{V} = \{q_1, q_2\}$. Evidently V is symmetric with respect to $\tilde{V}$ and the complete graph $\tilde{\Gamma}$ on $\tilde{V}$ is a single edge. The spectral symmetry functions are ϕ_2 and $\tilde{\phi}_2$ from Example 3.2.

Our index set $\mathbb{A}$ will be the edge set of H_n, including the loops at the boundary vertices. The disjoint union of copies of $\tilde{\Gamma}$ indexed by $\mathcal{A}$ has vertex set $\tilde{V} \times \mathcal{A}$. On it we define an equivalence relation such that the quotient $\tilde{\Gamma}'$ is precisely H_n. This is easy to do: whenever $(q_1, e), (q_2, e)$ are the two points in $\tilde{V} \times \{e\}$ for some $e \in \mathcal{A}$ we simply identify them with the vertices of V_n^H that are the endpoints of e. The identification is a bijection unless e is a loop, in which case it is two-to-one. Evidently this is a surjection $T : \tilde{V} \times \mathcal{A}$ to V_n^H, so we obtain an equivalence relation $(q_1, e_1) \sim (q_2, e_2)$ if $T(q_1, e_1) = T(q_2, e_2)$. This equivalence is not uniquely defined because we must choose one of the two possible bijections at each e, except in the case that e is a loop. However, every edge in the quotient comes from an edge in a copy of $\tilde{\Gamma}$, so for any set of choices of the bijections we see that there is an edge between the equivalence classes of (q_1, e_1) and (q_2, e_2) in $\tilde{\Gamma}'$ if and only if we can choose representatives such that $e_1 = e_2 = e$, which happens precisely when $T(q_1, e)$ and $T(q_2, e)$ are joined by the edge e in H_n. Moreoever this is a loop exactly when

$(q_1, e) \sim (q_2, e)$, meaning $T(q_1, e) = T(q_2, e)$, which occurs exactly when e is a loop in H_n. Thus $\tilde{\Gamma}' = H_n$.

Now we wish to identify the quotient Γ'. We had $T : \tilde{V} \times \mathcal{A} \to V_n^H$ and we extend it to $V \times \mathcal{A} \to V_n^J$ by defining it on $(V \setminus \tilde{V}) \times \mathcal{A} = \{q_0\} \times \mathcal{A}$ as follows. Given (q_0, e) we have defined $T(q_1, e)$ and $T(q_2, e)$ to be endpoints of the edge e from H_n. If e is a loop then $T(q_1, e) = T(q_2, e) \in V_n^H \subset V_n^J$ and this vertex is joined in J_n to a unique vertex $p \in V_0^G \subset V_n^J$; in this case let $T(q_0, e) = p$. Evidently this extension is a bijection from $\{(q_0, e) : e \text{ is a loop in } H_n\}$ to V_0^G.

If e is an edge in H_n with endpoints $T(q_1, e) = F_w(p_0)$ and $T(q_2, e) = F_{w'}(p_0)$ from V_n^H, where w, w' are words of length n, then our definition of H_n says there is a unique point $x \in V_n^G$ so that $F_w(p_0)$ and $F_{w'}(p_0)$ are both joined to x by edges in J_n. We define $T(q_0, e) = x$. Since every point of $V_n^G \setminus V_0^G$ is of the form $F_w(p_j) = F_{w'}(p_k)$ for some length n words $w \neq w'$ and $j \neq k \in \{1, 2, 3\}$ we see there is a corresponding edge of H_n from $F_w(p_0)$ to $F_{w'}(p_0)$ and thus T is bijective from $\{(q_0, e) : e \text{ is an edge of } H_n\} \subset V \times \mathcal{A}$ to $V_n^G \setminus V_0^G$.

Our extension is a surjection $T : V \times \mathcal{A} \to V_n^J$ that is bijective from $(V \setminus \tilde{V}) \times \mathcal{A}$ to V_n^G. Taking the quotient Γ' by the previously defined equivalence on $\tilde{V} \times \mathcal{A}$ we see that T identifies the vertex set of Γ' with V_n^J. To complete the proof we need to show that pairs of equivalence classes from Γ' that are joined by an edge correspond under T to pairs of points in V_n^J that are joined by an edge of J_n. To this end, observe that edges in $V \times \mathcal{A}$ are between pairs (q_0, e) and (q_j, e) for $j = 1, 2$ so edges in Γ' are between pairs of equivalence classes represented by (q_0, e) and (q_j, e), $j = 1, 2$. We will deal with the case where e is a loop of H_n separately from the case where it is an edge.

Fix equivalence classes in Γ' that are joined by an edge and choose representatives as described above. Bijectivity of $T : (V \setminus \tilde{V}) \times \mathcal{A} \to V_n^G$ says the equivalence class of (q_0, e) is a one-element set corresponding to the point $x = T(q_0, e) \in V_n^G$. If e is an edge (not a loop) of H_n then $x \notin V_0^G$ so $x = F_w(p_j) = F_{w'}(p_k)$ for exactly two distinct length n words $w \neq w'$ and corresponding $j \neq k \in \{1, 2, 3\}$. Our construction ensures e is the edge of H_n between $F_w(p_0)$ and $F_{w'}(p_0)$ and thus $T(q_j, e)$ is one of $F_w(p_0)$ or $F_{w'}(p_0)$. However, the edges in J_n that are incident at x connect to precisely the points $F_w(p_0)$ and $F_{w'}(p_0)$, so we conclude that $T(q_0, e)$ and $T(q_j, e)$ are connected by an edge in J_n. A similar argument applies if e is a loop of H_n, as then $x \in V_0^G$ is $p_k = F_{k^n}(p_k)$ and $T(q_j, e) = F_{k^n}(p_0)$ for both $j = 1, 2$. This latter is exactly the neighbor of p_k in J_n.

Conversely, fix an edge of J_n which connects $F_w(p_0)$ to $x \in V_n^G$. If $x \notin V_0$ we can write $x = F_w(p_k) = F_{w'}(p_l)$ for some length n words $w \neq w'$ and some $k \neq l \in \{1, 2, 3\}$. The definition of H_n gives an edge e between $F_w(p_0)$ and $F_{w'}(p_0)$. Our construction provides $T(q_0, e) = x$ and $\{T(q_j, e) : j = 1, 2\} = \{F_w(p_0), F_{w'}(p_0)\}$, so $T^{-1}(x) = \{(q_0, e)\}$ and $T^{-1}(F_w(p_0))$ is the equivalence class of either (q_1, e) or (q_2, e), both of which neighbor $\{(q_0, e)\}$ in Γ'. If, on the other hand, $x \in V_0^G$ then $x = p_k$ neighbors $F_{k^n}(p_0)$ in J_n, and the definition of T gives $T^{-1}(x) = (q_0, e)$ for the loop of H_n attached at $F_{k^n}(p_0)$. Thus $(q_j, e) \in T^{-1}(F_{k^n}(p_0))$ for both $j = 1, 2$ and this equivalence class is a neighbor of $T^{-1}(x)$ in Γ'.

The preceding proves that T induces a graph isomorphism between Γ' and J_n, so we can apply Lemma 4.1 to conclude that J_n is spectrally similar to H_n with functions (from Example 3.2) $\phi = \phi_2 = 1 + z - \frac{1}{1+z}$ and $\tilde{\phi}_2 = \frac{1}{2(1+z)}$. The stated expression for $R_2(z)$ follows, as does the fact that the exceptional set is

$E = \{-1\}$. Moreover, we find from the discussion in Example 3.2 that if $z \neq -1$ and $\tilde{f}$ is an eigenfunction of Δ_n^H with eigenvalue $R_2(z)$ then we can extend $\tilde{f}$ to an eigenfunction f of Δ_n^J with eigenvalue z by setting $f(x) = \frac{1}{2(1+z)} \sum_{y \sim_{J_n} x} \tilde{f}(y)$ at points of $V_n^G \setminus V_0^G$, while if $x \in V_0^G$ and $y \in V_n^H \subset V_n^J$ is its unique neighbor in J_n then $f(x) = \tilde{f}(y)$.

The discussion in Example 3.2 also suggests that we could construct eigenfunctions g of Δ_n^J with eigenvalue -1 by requiring that g vanish at each $x \in V_n^G \subset V_n^J$ and have $\sum_{y \sim_{J_n} x} g(y) = 0$ for all such x. However these requirements imply g is identically zero, so there are no such eigenfunctions. To see this, take points $y, y; \in V_n^H$ and let $x \in V_n^G$ be their common neighbor in J_n. The constraint equation says that if $g(y) = 0$ then also $g(y') = 0$, so by the obvious path connectedness of J_n we determine that if g satisfying the constraint equation vanishes at some $y \in V_n^H$ then it vanishes on all of V_n^H. Since the other constraint says g vanishes on V_n^G, we conclude that if g vanishes at some $y \in V_n^H$ it is identically zero on J_n. However, if we take $x = p_1$ in the constraint equation then the fact that p_1 has only one neighbor in J_n ensures g will vanish at that neighbor, which is a point of V_n^H. □

REMARK 4.4. There is a subtlety about the application of Lemma 4.1 in the proof of Theorem 4.3 that may warrant additional explanation. It is clear that the identification of two endpoints of an edge to a single point gives a loop, but when we identify the two endpoints q_1, q_2 of a copy of $\Gamma = \Gamma_2$ to the same single point it appears to give a double edge rather than the single edge we expect to see in J_n. The resolution of this apparent difficulty is embedded in the proof of Lemma 4.1, so is not visible in the proof of Theorem 4.3. However, we can see what happens by looking at the weights for the connections from $p_j \in V_0^G$ to the adjacent point $y = F_{j^n}(p_0)$. In J_n we have a single edge to y with weight $1 = 1/\deg(p_j)$. In Γ' our construction appears to provide two edges from p_j to y, but since each has adjacency weight $1/\deg(p_j) = 1/2$ this is equivalent to having a single edge of weight 1.

THEOREM 4.5. *The sets $\sigma(\Delta_n^G) \setminus \{-\frac{3}{2}\}$ and $\sigma(\Delta_n^H) \setminus \{-2\}$ are bijective under the map $\zeta \mapsto \frac{4}{3}\zeta$. Moreover, if we define for $f : V_n^G \to \mathbb{C}$ and $g : V_n^H \to \mathbb{C}$*

$$\Phi_3 f(y) = \frac{1}{3} \sum_{x \sim_{J_n} y} f(x) \quad \text{for } y \in V_n^H, \text{ and}$$

$$\Phi_2 g(x) = \begin{cases} \frac{1}{2} \sum_{y \sim_{J_n} x} g(y) & \text{if } x \in V_n^G \setminus V_0^G \\ \sum_{y \sim_{J_n} x} g(y) & \text{if } x \in V_0^G \end{cases}$$

then Φ_3 is a bijection from the eigenspace of Δ_n^G with eigenvalue $\zeta \in \sigma(\Delta_n^G) \setminus \{-\frac{3}{2}\}$ to the eigenspace of Δ_n^H with eigenvalue $\frac{4}{3}\zeta \in \sigma(\Delta_n^H) \setminus \{-2\}$, and Φ_2 is its inverse.

PROOF. Propositions 4.2 and 4.3 tell us that the following maps are surjections

$$R_3 : \sigma(\Delta_n^J) \setminus E \to \sigma(\Delta_n^G) \setminus \{-\frac{3}{2}\}$$

$$R_2 : \sigma(\Delta_n^J) \setminus E \to \sigma(\Delta_n^H) \setminus \{-2\}$$

and it is easily seen that -1 is the only critical point of both R_3 and R_2, so these maps are actually two-to-one covers. By inspection of the formulas for R_3 and R_2

we conclude that

$$\sigma(\Delta_n^G) \setminus \{-\frac{3}{2}\} \xrightarrow{\zeta \mapsto \frac{4}{3}\zeta} \sigma(\Delta_n^H) \setminus \{-2\}$$

is a bijection.

Now comparing the definition of the maps Φ_2 and Φ_3 with the extensions given in Propositions 4.2 and 4.3 we see that if $z \in \sigma(\Delta_n^J) \setminus E$ with eigenfunction f then $f|_{V_n^G} = (1+z)^{-1}\Phi_2(f|_{V_n^H})$ and $f|_{V_n^H} = (1+z)^{-1}\Phi_3(f|_{V_n^G})$. Since $f|_{V_n^G}$ is an eigenfunction of Δ_n^G with eigenvalue $R_3(z)$ and $f|_{V_n^H}$ is an eigenfunction of Δ_n^H with eigenvalue $R_2(z) = \frac{4}{3}R_3(z)$, we conclude that Φ_3 maps the eigenspace of Δ_n^G with eigenvalue $\zeta \in \sigma(\Delta_n^G) \setminus \{-\frac{3}{2}\}$ bijectively to the eigenspace of Δ_n^H with eigenvalue $\frac{4}{3}$. Repeating the argument for Φ_2 shows that it is inverse to Φ_3 on these spaces. This last point may seem counterintuitive; it is more readily apparent when one recalls that these eigenfunctions on J_n are orthogonal to the functions having mean zero on all sets $\{F_w(p_j) : j = 1, 2, 3\}$ where w is a length n word, because the latter are eigenfunctions with eigenvalue -1. □

5. Spectral self-similarity of Δ_n^G and Δ_n^H

Theorem 4.5 allows us to get the spectrum of H_n from that of G_n or vice-versa. The spectrum for the graphs G_n has been known for a long time [**4**, **12**] as a consequence of the spectral self-similarity of the Laplacians Δ_n^G given in Theorem 5.1 below. While the spectrum for H_n was given in [**5**], we obtain it by transferring this spectral self-similarity to the Laplacians Δ_n^H via Theorem 4.5.

THEOREM 5.1 ([**3**, **4**, **12**]). *For $n \geq 0$ the Laplacian Δ_{n+1}^G is spectrally similar to Δ_n^G with exceptional set $E_G = \{-\frac{3}{2}, -\frac{5}{4}, -\frac{1}{2}\}$ and decimation function $R_G(z) = z(4z+5)$. If $z \notin E_G$ and $R_G(z)$ is an eigenvalue of Δ_n^G with eigenfunction $\tilde{f}$ then the corresponding eigenfunction of Δ_{n+1}^G with eigenvalue z is obtained by extending $\tilde{f}$ to each point of $V_{n+1}^G \setminus V_n^G$ in the following manner: for w a length n word we have $\tilde{f}$ at $F_w(p_j)$, $j = 1, 2, 3$; write $m_j = F_{j+1}(p_{j-1})$ for the vertex opposite p_j in V_1 (all subscripts are taken modulo 3 so as to lie in $\{1, 2, 3\}$), and set*

$$f(F_w(m_j)) = \frac{f(F_w(p_j)) + 2(z+1)\big(f(F_w(p_{j+1})) + f(F_w(p_{j-1}))\big)}{(4z+5)(2z+1)}. \tag{5.1}$$

REMARK 5.2. The interested reader may care to prove this theorem herself by checking that $D = \Delta_1^G$ is spectrally similar to $\tilde{D} = \Delta_0^G$ with $\phi = z(2z+3)/(2z+1)$ and $\tilde{\phi} = (2z+3)/(4z+5)(2z+1)$ and exceptional set E_G, then working as in Proposition 4.2 to write G_n as a quotient of copies of G_0 indexed by words in $\{1, 2, 3\}^n$ and checking that the same quotient of copies of G_1 is G_{n+1}. Alternatively, an elementary and very readable treatment appears in [**15**, Chapter 3], with a different normalization for the Laplacian.

The bijection in Theorem 4.5 allows us to transfer this spectral similarity to the graphs H_n.

THEOREM 5.3. *The Laplacian Δ_{n+1}^H is spectrally similar to Δ_n^H with exceptional set $E_H = \{-2, -\frac{5}{3}, -1, -\frac{2}{3}\}$ and decimation function $R_H(z) = z(3z+5)$. If $z \notin E_H$ and $R_H(z)$ is an eigenvalue of Δ_n^H with eigenfunction f then the corresponding eigenfunction g of Δ_{n+1}^H with eigenvalue z is obtained as follows. A point $x \in V_{n+1}^H$ is $F_{wj}(p_0)$ for some w of length n and $j \in \{1, 2, 3\}$. We have $y_0 = F_w(p_0) \in V_n^H$;*

write y_k, $k = 1,2,3$ *for the neighbors of* y_0 *in* H_n, *choosing the labels so that* y_0 *is joined to* y_k *via* $F_w(p_k)$ *in* J_n. *(Note that* $y_k = y_0$ *if* $F_w(p_k) \in V_0^G$, *meaning there is a loop at* y_0.*) Then*

$$g(x) = \frac{2z+3}{6(4z+5)(2z+1)}\Big((4z+5)f(y_0)+(4z+3)f(y_j)+f(y_{j+1})+f(y(j-1))\Big) \tag{5.2}$$

where, as usual, $j \pm 1$ *are taken modulo 3 so as to give values in* $\{1,2,3\}$.

PROOF. From Theorem 4.5 we have a bijection between $\sigma(\Delta_m^G) \setminus \{-\frac{3}{2}\}$ and $\sigma(\Delta_m^H) \setminus \{-2\}$, simply by $\zeta \mapsto \frac{4}{3}\zeta$. Suppose $z \in \sigma(\Delta_{n+1}^H) \setminus E_H$. Then this bijection gives $\frac{3}{4}z \in \sigma(\Delta_m^G) \setminus (E_G \cup \{-\frac{3}{4}\})$, and therefore Theorem 5.1 says that $R_G(\frac{3}{4}z) \in \sigma(\Delta_n^G)$. Moreover we have $R_G^{-1}(-\frac{3}{2}) = \{-\frac{3}{4}, -\frac{1}{2}\} \subset (E_G \cup \{-\frac{3}{4}\})$, so in fact $R_G(\frac{3}{4}z) \in \sigma(\Delta_n^G) \setminus \{-\frac{3}{2}\}$ and we can again use the bijection from Theorem 4.5, this time with $m = n$, to get that $\frac{4}{3}R_G(\frac{3}{4}z) \in \sigma(\Delta_n^H) \setminus \{-2\}$. Hence $R_H(z) = \frac{4}{3}R_G(\frac{3}{4}z) = z(3z+5)$ is a spectral decimation map from Δ_{n+1}^H to Δ_n^H with the stated exceptional set.

Given an eigenfunction as in the statement of the theorem we apply Φ_2 to get an eigenfunction of Δ_n^G, then the map in (5.1), and finally Φ_3 to get an eigenfunction of Δ_{n+1}^H. For the reader's convenience we illustrate with the sequence of diagrams in Figure 4. One is given the values a, b, c, d on a piece of H_n shown on the left (dashed lines). Treating these as values on J_n (dotted lines), apply Φ_2 to obtain values at the vertices in G_n. Then (5.1) gives

$$2\alpha = \mu(2a+b+d)) + \nu(a+c)$$
$$2\beta = \mu(2a+b+c)) + \nu(a+d)$$
$$2\gamma = \mu(2a+c+d)) + \nu(a+b)$$

where $\mu = \frac{2(z+1)}{(4z+5)(2z+1)}$ and $\nu = \frac{1}{(4z+5)(2z+1)}$. Apply Φ_3 to get

$$6t = 2\alpha + 2\gamma + a + d = \mu(4a+b+c+2d)) + \nu(2a+b+c) + a + d$$
$$= \frac{2z+3}{(4z+5)(2z+1)}\Big((4z+5)a + (b+c) + (4z+3)d\Big) \tag{5.3}$$

This is precisely (5.2). To see this, write $t = g(F_{wj}(p_0))$. Then $f(F_w(p_0)) = a$ is apparent, and $y_0 = F_w(p_0)$. There are edges in H_n from y_0 to each y_k, and these correspond to paths in J_n via points $F_w(p_k) \in V_n^G$. Since d has a different weight to b and c in (5.3) it is important that the edge between the corresponding vertices of H_n goes via $F_w(p_j)$, and thus $d = f(y_j)$. The other two values $f(y_k)$, $k \in \{1,2,3\} \setminus \{j\}$ are b and c. □

6. Eigenfunctions of G_n and H_n

It is well-known that one can give a complete description of all eigenfunctions of Δ_n^G using the results in the previous section. This is because eigenfunctions on G_{n+1} either have eigenvalue in E_G or are obtained by decimation from Theorem 5.1: one simply makes a direct construction of eigenfunctions with eigenvalues in E_G and counts to ensure they generate the whole spectrum. In this section we use the same idea to construct the eigenfunctions of Δ_n^H and determine multiplicities in the spectrum. A convenient approach is to state known results for Δ_n^G as propositions and derive the implications for Δ_n^H sequentially.

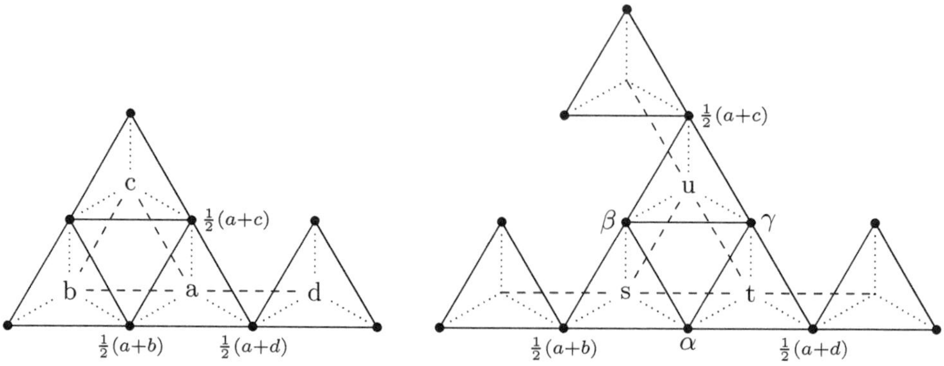

FIGURE 4. Computation of (5.2): values a, b, c, d on H_n (dashed lines) are extended to G_n (solid lines) via Φ_2, then to α, β, γ in G_{n+1} by (5.1) and thus to s, t, u on H_{n+1} via Φ_3. Dotted lines show J_n and J_{n+1}.

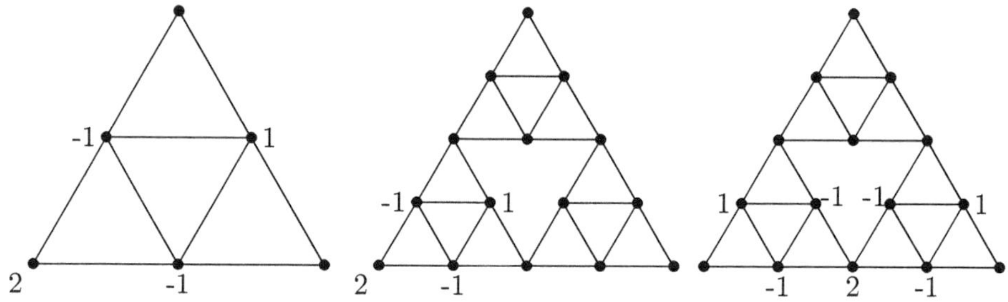

FIGURE 5. Eigenfunctions of V_1^G (left) and V_2^G (center and right) with eigenvalue $-\frac{3}{2}$.

PROPOSITION 6.1 ([**3**], see also [**15**, Chapter 3]). *The exceptional value $-\frac{3}{2}$ is an eigenvalue of G_n for all $n \geq 0$; the corresponding eigenfunctions are exactly those functions satisfying $\sum_{j=1}^{3} f(F_w(p_j)) = 0$ for all words w of length n and form a $\frac{1}{2}(3^n + 3)$ dimensional eigenspace.*

A basis for this eigenspace can be chosen so as to be localized around points of V_{n-1}^G. For $n \geq 1$ it is generated by placing copies of the eigenfunction on V_1^G shown on the left in Figure 5 on $n-1$ cells in V_n^G, so that the value 2 occurs at a point of $x \in V_{n-1}^G$. If $x \in V_0^G$ one copy suffices, as shown in the center of the figure. If $x \in V_{n-1}^G \setminus V_0^G$ two copies are needed, as shown on the right. The functions are zero at all unlableled vertices in the figure.

Observe that the functions in Proposition 6.1 exactly satisfy the condition given in Proposition 4.2 for the restriction of an eigenfunction f of Δ_n^J with eigenvalue -1 to the set V_n^G. Moreover, if $f|_{V_n^G}$ is such a function then $\Phi_3 f$ vanishes on V_n^H, so coincides with $f_{V_n^H}$, also by Proposition 4.2. We have therefore extended the bijective correspondence of eigenspaces in Proposition 4.2 to the exceptional value.

COROLLARY 6.2. $R_3 : \sigma(\Delta_n^J) \to \sigma(\Delta_n^G)$ *with a corresponding bijection of eigenspaces. From J_n to G_n the bijection is simply restriction of an eigenfunction f*

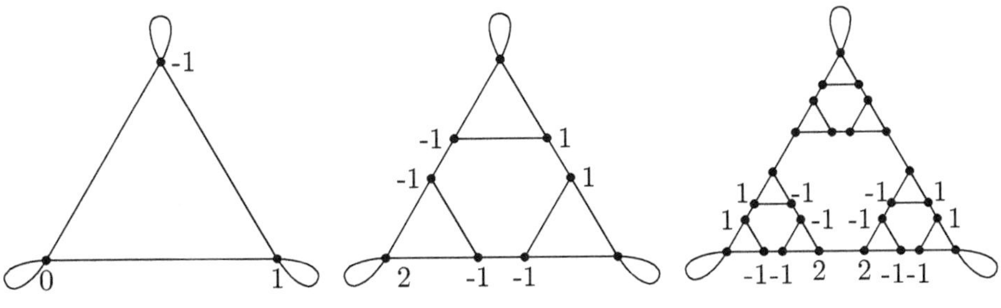

FIGURE 6. Eigenfunctions of Δ_n^H with eigenvalue -1 for $n = 1, 2, 3$

of Δ_n^J to V_n^G. From G_n to J_n the map is as follows. Suppose g is an eigenfunction of Δ_n^G with eigenvalue ζ. If $\zeta = -\frac{3}{2}$ extend g to V_n^J so it vanishes on V_n^H. If $\zeta \neq -\frac{3}{2}$ and $z \in R_3^{-1}(\zeta)$ is one of the corresponding elements of $\sigma(\Delta_n^J) \setminus \{-1\}$ then define $f = g$ on V_n^G and $f = (1+z)^{-1}\Phi_3 g$ on V_n^H.

COROLLARY 6.3. *The kernel of Φ_3 is the eigenspace of Δ_n^G with eigenvalue $-\frac{3}{2}$.*

Since our method for obtaining eigenfunctions of Δ_n^H from those of Δ_n^G (in Theorem 4.5) is to apply Φ_3, Corollary 6.3 makes it clear that nothing of the kind is possible for the $-\frac{3}{2}$ eigenspace of Δ_n^G. However, we can use it to get eigenfunctions of Δ_n^H with eigenvalue -1.

PROPOSITION 6.4. *For $n \geq 1$ the exceptional value -1 of the decimation in Theorem 5.3 is an eigenvalue of Δ_n^H with multiplicity $\frac{1}{2}(3^{n-1}+3)$. If $n = 1$ a basis consists of the function on the left in Figure 6 and one rotate thereof. For $n \geq 2$ a basis can be obtained from copies of the function on H_2 shown at center in Figure 6 scaled to cells of scale $n-1$ and arranged as follows. A single copy may be placed so the value 2 occurs at the attachment of a loop, or two copies may be placed on adjacent cells such that both endpoints of the connecting edge carry the value 2, as shown at right in Figure 6.*

PROOF. By Proposition 6.1 we know $-\frac{3}{2}$ is an eigenvalue of Δ_{n-1}^G with multiplicity $\frac{1}{2}(3^{n-1}+3)$. Now $R_G(-\frac{3}{4}) = -1$, so by Theorem 5.1 we have that $-\frac{3}{4}$ is an eigenvalue of Δ_n^G, and there is a bijection between the eigenspaces. By Theorem 4.5 then -1 is an eigenvalue of Δ_n^H; again there is a bijection between the eigenspaces from which we deduce the stated multiplicity.

One can obtain the eigenfunctions by taking values on V_{n-1}^G as illustrated in Figure 5 and applying (5.1), calling the result f. Recall from Proposition 6.1 that then $\sum_{j=1}^3 f(F_w(p_j)) = 0$ for all words w of length n. We compute from this constraint on the values and the requirement that $z = -\frac{3}{4}$ that the value at a point in $V_n^G \setminus V_{n-1}^G$ is the average of the values at its neighbors.

When $n = 1$ we can reason as above for the eigenfunction of Δ_0^G having values $1, -1, 0$ on V_0. Multiplying by 2 gives the function on the left in Figure 6, and any two of the three rotates are linearly independent. When $n = 2$ we do the same for the left function in Figure 5. Then the above reasoning followed by the map $2\Phi_3$ produces the function on H_2 shown at the center in Figure 6. For any $n \geq 2$ copies of this can be rescaled to cells of size $n-1$ and placed so that the common value 2 occurs at the attachment point of a loop. If $n \geq 3$ it is also possible to place two

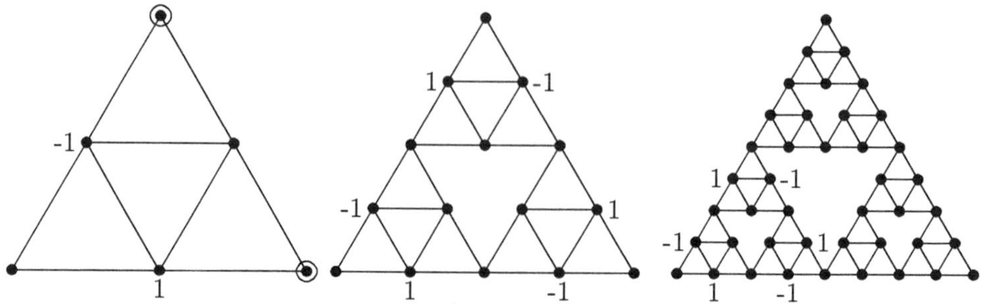

FIGURE 7. Eigenfunctions of Δ_n^G with eigenvalue $-\frac{5}{4}$ for $n = 2$ (center) and $n = 3$ (right) are constructed from the function at left, which satisfies the eigenfunction equation at all but the circled points, by chaining around holes in the graph.

copies so the common value 2 occurs at both ends of one of the $\frac{1}{2}(3^{n-1} - 3)$ edges in H_n that were already present in H_{n-1}, see at right in Figure 6. The fact that these are linearly independent follows from the previously stated bijectivity or can readily be verified by hand. □

REMARK 6.5. An easy consequence of the preceding is that the values of the -1 eigenfunctions of Δ_n^H sum to zero over all triangles of scale n in the graph, though this is not the only constraint on such eigenfunctions.

The eigenfunctions corresponding to the exceptional value $-\frac{5}{4}$ for the decimation of Δ_n^G are known to be in one-to-one correspondence with certain loops on the cell graph. The left image in Figure 7 shows a function on V_1^G by listing the values at vertices, with the vertices where the function value is zero left unmarked. This function satisfies the eigenfunction equation with eigenvalue $-\frac{5}{4}$ at all vertices except the two that are circled. By symmetry, one may join copies of this function along non-trivial cycles so as to obtain eigenfunctions of Δ_n^G with eigenvalue $-\frac{5}{4}$ for any $n \geq 2$. The cycles surround "holes" in the cell graph with size at least one scale larger than a cell, as illustrated on the center and right in Figure 7. Counting the holes gives the multiplicity of the eigenspace.

PROPOSITION 6.6 ([**3**], see also [**15**, Chapter 3]). *The exceptional value $-\frac{5}{4}$ is an eigenvalue for $n \geq 2$ with multiplicity $\frac{1}{2}(3^{n-1} - 1)$. An explicit basis for the eigenspace is obtained by chaining the function on the left in Figure 7 along cycles around holes of scale at least $n - 1$.*

Figure 8 shows some of these eigenfunctions. In particular, the functions at the left and center are obtained by applying $3\Phi_3$ to the functions at the center and right in Figure 7. Evidently each such function is supported on a cycle around a hole and has alternating values going around the cycle. Applying the bijection from Theorem 4.5 we obtain the following result.

COROLLARY 6.7. *For $n \geq 2$ the exceptional value $-\frac{5}{3} \in E_H$ is an eigenvalue of Δ_n^H with multiplicity $\frac{1}{2}(3^{n-1} - 1)$. A basis for the eigenspace consists of alternating functions around cycles in H_n of any size larger than the smallest cycles, so has the homology of H_{n-1}.*

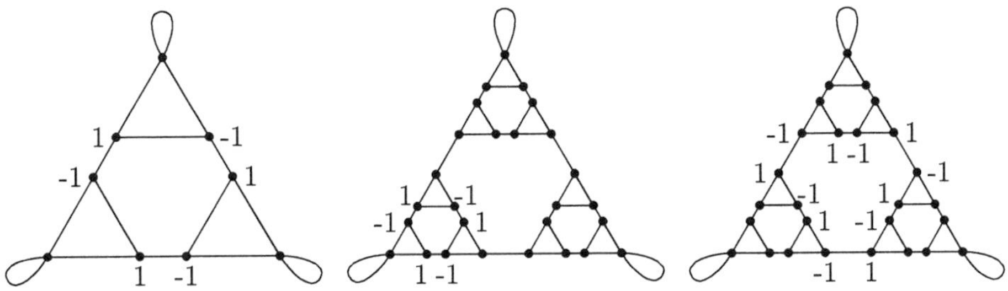

FIGURE 8. Eigenfunctions of Δ_n^H with eigenvalues $-\frac{5}{3}$ for $n = 2$ (left) and $n = 3$ (center and right) form loops around holes in the graph of scale at least $n - 1$.

The following well-known theorem is included only for completeness of the description of $\sigma(\Delta_n^G)$ and to emphasize the connection to Theorem 1.1. It is proved in the same manner as the latter by counting the eigenfunctions obtained in Propositions 6.1 and 6.6 and by decimation using Theorem 5.1.

THEOREM 6.8 ([**3**, **4**, **12**], see also [**15**, Chapter 3]). *The spectrum of Δ_n^G is the set*

$$\sigma(\Delta_n^G) = \{0\} \cup \left(\bigcup_{i=0}^{n} R_G^{-i}\left(-\frac{3}{2}\right)\right) \cup \left(\bigcup_{j=0}^{n-2} R_G^{-j}\left(-\frac{5}{4}\right)\right).$$

The multiplicities and eigenspaces are as follows:

- *0 has multiplicity 1 with constant eigenfunction*
- *$R_G^{-i}(-\frac{3}{2})$ contains 2^{i-1} eigenvalues when $1 \le i \le n$, each with multiplicity $\frac{1}{2}(3^{n-i}+3)$, while $-\frac{3}{2}$ has multiplicity $\frac{1}{2}(3^n+3)$. A basis for the eigenspace of any eigenvalue in $R_G^{-i}(-\frac{3}{2})$ is obtained by taking the basis for the -1 eigenspace of Δ_{n-i}^G described in Proposition 6.1 and extending to G_n by i iterations of (5.1).*
- *$R_G^{-j}(-\frac{5}{4})$ contains 2^j eigenvalues when $0 \le j \le n-2$, each with multiplicity $\frac{1}{2}(3^{n-j-1} - 1)$. A basis for the eigenspace is obtained by taking the basis for the $-\frac{5}{4}$ eigenspace for Δ_{n-j}^G from Proposition 6.6 and extending to G_n by j iterations of (5.1).*

The corresponding theorem for Δ_n^H is Theorem 1.1.

PROOF OF THEOREM 1.1. For $n = 0$ and $n = 1$ it is easy to check the description is correct using Proposition 6.4 and the fact that the number of vertices in V_n^H is 3^n. For $n \ge 2$ we induct.

Consider the eigenvalues and eigenfunctions obtained by the decimation in Theorem 5.3. Assuming for the induction that the spectrum for Δ_{n-1}^H is as stated we see that the forward orbit of any $\zeta \in \Delta_{n-1}^H$ under R_H hits the set $S = \{-\frac{5}{3}, -1, 0\}$ (indeed it hits the smaller set $\{-1, 0\}$, but we do not need this fact). We use this to check that ζ cannot be a critical value of R_H and to understand when $R_H^{-1}(\zeta)$ can be in E_H. A useful observation in this regard is that $z > 0$ implies $R_H(z) > 0$ and the orbit of z cannot be in S. Applying this to the critical value $-\frac{25}{12}$ of R_H we see $R_H\left(-\frac{25}{12}\right) = \frac{125}{12}$, so $\zeta \in \Delta_{n-1}^H$ implies ζ is not the critical value and thus $R_H^{-1}(\zeta)$ contains exactly two points. For the set E_H we compute the forward orbits $R_H(-\frac{5}{3}) = 0$ and $R_H(0) = 0$, so these fall onto the fixed point $0 \in S$, while $R_H(-1) = R_H(-\frac{2}{3}) = -2$ and also $R_H(-2) = 2$, after which the orbit never hits S.

Our inductive hypothesis says that for $0 \leq i \leq n-2$ there are 2^i eigenvalues of Δ_{n-1}^H of the form $R_H^{-i}(-1)$, each with multiplicity $\frac{1}{2}(3^{n-2-i}+3)$. For each of these eigenvalues ζ the above reasoning ensures $R_H^{-1}(\zeta)$ contains exactly two points, neither of which is in E_H. Then Theorem 5.3 shows both of these are eigenvalues of Δ_n^H with the same multiplicity as ζ and eigenfunctions obtained by the decimation formula (5.2). Accordingly, for $1 \leq i \leq n-1$ there are 2^i eigenvalues of Δ_n^H of the form $R_H^{-i}(-1)$; the case $i = 0$ is then obtained from Proposition 6.4.

In the case $n = 2$ we can count that the preceding gives $3^n - 1$ eigenfunctions and the remaining one comes from Corollary 6.7. At the same time we have inductively that for $n \geq 3$ and $0 \leq j \leq n-3$ there are 2^j eigenvalues of Δ_{n-1}^H each with multiplicity $\frac{1}{2}(3^{n-2-j}-1)$. For each of these eigenvalues ζ our earlier reasoning about the orbits of E_H and the critical value ensures $R_H^{-1}(\zeta)$ contains exactly two points, neither of which is in E_H, and thus by Theorem 5.3 we obtain two eigenvalues of Δ_n^H of the form $R_H^{-1}(\zeta)$ with eigenfunctions from (5.2). Thus for $1 \leq j \leq n-2$ we have 2^j eigenvalues of Δ_n^H with multiplicity $\frac{1}{2}(3^{n-1-j}-1)$, and this is also true for $j = 0$ by Corollary 6.7.

The remaining possible eigenvalue $\zeta \in \Delta_{n-1}^H$ is $\zeta = 0$, with multiplicity 1. In this case $R_H^{-1}(0) = \{-\frac{5}{3}, 0\}$ and Theorem 5.3 says only that this ensures 0 is an eigenvalue of Δ_n^H, also with multiplicity 1. Evidently the corresponding eigenfunction is constant.

The preceding does not yet ensure we have a full description of the spectrum, as it is a priori possible that the other values in E_H could be eigenvalues. This possibility is eliminated by counting. Supposing inductively that our description gives the whole spectrum for Δ_{n-1}^H, observe that this consists of 3^{n-1} eigenvalues counted with multiplicity. All but the zero eigenvalue gave two new eigenvalues of Δ_n^H with preservation of multiplicity by spectral decimation; adding back in the zero eigenvalue we thus obtain $2(3^{n-1})-1$ eigenvalues. We also have eigenvalues -1 and $-\frac{5}{4}$ from Proposition 6.4 and Corollary 6.7, with (respectively) multiplicities $\frac{1}{2}(3^{n-1}+3)$ and $\frac{1}{2}(3^{n-1}-1)$. Together these sum to 3^n, which is the number of vertices in V_n^H, so we have found all eigenvalues and eigenfunctions.

What remains is the topological interpretation of the eigenspaces, which simply corresponds to the fact that all eigenfunctions of the types we have described were constructed by iterated application of (5.2) to the functions described in Proposition 6.4 and Corollary 6.7. □

REMARK 6.9. In our proof of Theorem 1.1 we used the fact that we knew the eigenspaces and hence the multiplicity of eigenvalues of Δ_n^H for the exceptional values -1 and $-\frac{5}{3}$ exactly. However it is apparent in the proof that we needed only lower bounds for these multiplicities: once these lower bounds match the upper bound obtained by counting the vertices the proof is complete. This observation shows that once Theorem 5.3 is known it is perfectly possible to obtain Theorem 1.1 as follows. Construct the eigenfunctions of Δ_n^H by hand for small n and observe the topological structures described in Proposition 6.4 and Corollary 6.7 that allow one to produce some eigenfunctions of Δ_n^H for any n with these exceptional values. Without knowing whether this gives all eigenfunctions, one still obtains lower bounds on the dimensions of both the eigenspaces corresponding to the exceptional values -1 and $-\frac{5}{3}$ and the eigenspaces obtained from Theorem 5.3. The argument given in the proof of Theorem 1.1 gives matching upper bounds, completing the proof.

7. The method of Grigorchuk and Šunić

We have already mentioned that Grigorchuk and Šunić [**5**] computed the spectra of the sequence Δ_n^H by a different method. Our work shares many features with theirs, so it seems useful to identify the connections and most important differences. For this reason we sketch some aspects of their approach in the same language used for our results above. Note that although they consider the adjacency matrix rather than the Laplacian we phrase our discussion in terms of the latter; there is no essential difference because the graph is 3-regular.

Considering the sequence of graphs H_n in Figure 1 it is apparent that we could assign vertices $\tilde{V}_n^H$ so that $\tilde{V}_n^H \subset \tilde{V}_{n+1}^H$ and the graph H_{n+1} is obtained by subdividing edges of H_n at these new vertices and introducing new edges between certain pairs of the new vertices. From this perspective it seems natural to seek a spectral decimation from Δ_{n+1}^H to Δ_n^H by taking the Schur complement of the subspace of functions supported on $\tilde{V}_{n+1}^H \setminus \tilde{V}_n^H$. Although they come to this point by considering the permutation representation of the Hanoi towers group on the n-th level of a rooted 3-tree, this Schur complement is the first step in the spectral analysis in [**5**].

Unfortunately, direct computation immediately shows that this Schur complement does not provide a spectral similarity between Δ_{n+1}^H and Δ_n^H; indeed, the Schur complement corresponds to the Laplacian on H_n for a weighted graph in which the edges between pairs of vertices both of which are in $\tilde{V}_n^H$ are different than the weight on all other edges. This motivates the introduction of such a weighted Laplacian, which we illustrate in Figure 9. Grigorchuk and Šunić then show this weighted Laplacian on H_{n+1} is spectrally similar to the weighted Laplacian on H_n, only now the decimation function depends on both the spectral value x and the weight y. The remarkable thing is that they are then able to find an explicit semiconjugacy of this two dimensional dynamics for the weighted Laplacian to the one-dimensional dynamics we found in Theorem 5.3.

This brings us to the difference between our methods, which is that our vertex set V_n^H does not lie in V_{n+1}^H, and we therefore are not taking the Schur complement with respect to the subspace of functions supported on $\tilde{V}_{n+1}^H \setminus \tilde{V}_n^H$. Indeed, the subspace with respect to which we are taking the Schur complement does not correspond to any subset of the vertices of the graph H_{n+1}. As far as the authors are aware, this is the first example using such a spectral decimation on self-similar graphs, though the use of spectral decimation with respect to subspaces coming from vertex subsets has been in common use for many years. Given the increased flexibility obtained, it seems this variation of the spectral decimation method warrants further investigation.

8. The method of Quint

As previously mentioned, the structure of the Laplacian eigenfunctions exhibited in Theorem 1.1 was previously established by Quint [**11**]. Unlike the approach in [**5**], Quint's analysis depends primarily on functional equations for the graph

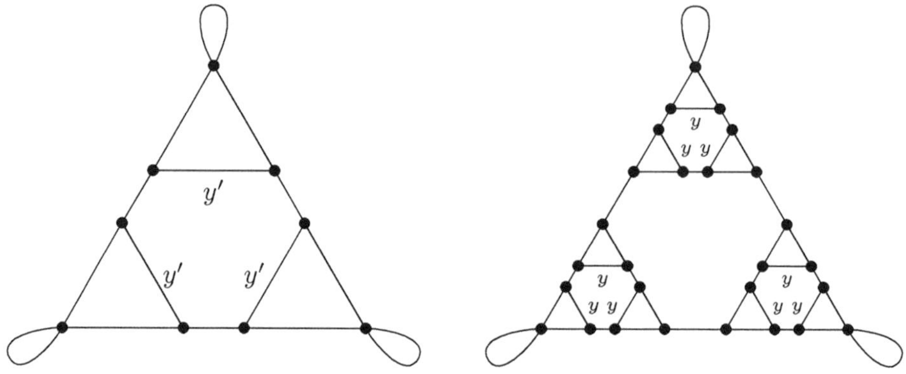

FIGURE 9. Weighted graphs on H_2 and H_3; unmarked edges have weight 1.

Laplacians rather than on Schur complement methods. Specifically, he associates to a 3-regular graph Θ a new graph $\tilde{\Theta}$ in which each point of Θ has been replaced by a triangle, establishes some elementary functional equations (Lemmas 3.1 and 4.1 in [**11**]) and uses these to relate the spectrum, spectral measure, eigenfunctions, and spectral decomposition of the l^2 space for these pair of graphs (Sections 2-6 in [**11**]).

The difference between our approach and that of Quint is that we use the established spectral decimation method, which is a hybrid of the Schur and functional equation approach for which the relationships between spectrum and eigenfunctions are known (Theorem 3.1), thus allowing a shorter and somewhat more elementary analysis. Specifically, spectral similarity is a functional equation for the Laplacian (much like that used by Quint) that can be established by computing a Schur complement. In Quint's approach the functional equations are established quickly by direct computation, but there is considerable work in determining the consequences for the spectral structure. In ours, most of the work is in establishing a specific type of functional equation, namely the spectral self-similarity for the sequence Δ_n^H, but then the spectrum can just be written down.

It should be noted that Quint's approach handles one major aspect of the theory that ours does not: he is able to determine the spectrum and spectral measure in the case of a Hanoi graph with boundary. At the same time, we note that his method depends on knowing an appropriate functional equation relating the Laplacians in a sequence of graphs, which he resolves only for the case of a pair Θ, $\tilde{\Theta}$ as described above. By contrast, our analysis is concerned with a specific type of functional equation (spectral similarity) but spectral similarity is not constrained to be associated to the l^2-spaces associated to the graphs and, as noted at the end of the previous section, it is therefore potentially applicable to a wider variety of graphs.

Acknowledgments

The authors thank Alexander Teplyaev for suggesting this enjoyable problem and the University of Connecticut for hosting the authors.

References

[1] N. Bajorin, T. Chen, A. Dagan, C. Emmons, M. Hussein, M. Khalil, P. Mody, B. Steinhurst, and A. Teplyaev, *Vibration modes of 3n-gaskets and other fractals*, J. Phys. A **41** (2008), no. 1, 015101, 21, DOI 10.1088/1751-8113/41/1/015101. MR2450694

[2] L. Bartholdi and R. I. Grigorchuk, *On the spectrum of Hecke type operators related to some fractal groups*, Tr. Mat. Inst. Steklova **231** (2000), no. Din. Sist., Avtom. i Beskon. Gruppy, 5–45; English transl., Proc. Steklov Inst. Math. **4(231)** (2000), 1–41. MR1841750

[3] Kyallee Dalrymple, Robert S. Strichartz, and Jade P. Vinson, *Fractal differential equations on the Sierpinski gasket*, J. Fourier Anal. Appl. **5** (1999), no. 2-3, 203–284, DOI 10.1007/BF01261610. MR1683211

[4] M. Fukushima and T. Shima, *On a spectral analysis for the Sierpiński gasket*, Potential Anal. **1** (1992), no. 1, 1–35, DOI 10.1007/BF00249784. MR1245223

[5] Rostislav Grigorchuk and Zoran Šunić, *Schreier spectrum of the Hanoi Towers group on three pegs*, Analysis on graphs and its applications, Proc. Sympos. Pure Math., vol. 77, Amer. Math. Soc., Providence, RI, 2008, pp. 183–198, DOI 10.1090/pspum/077/2459869. MR2459869

[6] Daniel J. Kelleher, Benjamin A. Steinhurst, and Chuen-Ming M. Wong, *From self-similar structures to self-similar groups*, Internat. J. Algebra Comput. **22** (2012), no. 7, 1250056, 16, DOI 10.1142/S0218196712500567. MR2999364

[7] Jun Kigami, *Analysis on fractals*, Cambridge Tracts in Mathematics, vol. 143, Cambridge University Press, Cambridge, 2001, DOI 10.1017/CBO9780511470943. MR1840042

[8] Leonid Malozemov and Alexander Teplyaev, *Self-similarity, operators and dynamics*, Math. Phys. Anal. Geom. **6** (2003), no. 3, 201–218, DOI 10.1023/A:1024931603110. MR1997913

[9] Volodymyr Nekrashevych, *Self-similar groups*, Mathematical Surveys and Monographs, vol. 117, American Mathematical Society, Providence, RI, 2005, DOI 10.1090/surv/117. MR2162164

[10] Volodymyr Nekrashevych and Alexander Teplyaev, *Groups and analysis on fractals*, Analysis on graphs and its applications, Proc. Sympos. Pure Math., vol. 77, Amer. Math. Soc., Providence, RI, 2008, pp. 143–180, DOI 10.1090/pspum/077/2459868. MR2459868

[11] J.-F. Quint, *Harmonic analysis on the Pascal graph*, J. Funct. Anal. **256** (2009), no. 10, 3409–3460, DOI 10.1016/j.jfa.2009.01.011. MR2504530

[12] R. Rammal and G. Toulouse, *Random walks on fractal structures and percolation clusters*, J. Phys. Lett., 44:L13–L22, 1983.

[13] Robert S. Strichartz, *The Laplacian on the Sierpinski gasket via the method of averages*, Pacific J. Math. **201** (2001), no. 1, 241–256, DOI 10.2140/pjm.2001.201.241. MR1867899

[14] Robert S. Strichartz, *Fractafolds based on the Sierpiński gasket and their spectra*, Trans. Amer. Math. Soc. **355** (2003), no. 10, 4019–4043, DOI 10.1090/S0002-9947-03-03171-4. MR1990573

[15] Robert S. Strichartz, *Differential equations on fractals: A tutorial*, Princeton University Press, Princeton, NJ, 2006. MR2246975

[16] Robert S. Strichartz, *Transformation of spectra of graph Laplacians*, Rocky Mountain J. Math. **40** (2010), no. 6, 2037–2062, DOI 10.1216/RMJ-2010-40-6-2037. MR2764237

[17] Robert S. Strichartz and Alexander Teplyaev, *Spectral analysis on infinite Sierpiński fractafolds*, J. Anal. Math. **116** (2012), 255–297, DOI 10.1007/s11854-012-0007-5. MR2892621

Department of Mathematics, Ohio State University, 231 W 18th Ave, Columbus, Ohio 43210
Email address: bretthungar@gmail.com

Department of Mathematics, University of Cincinnati, Cincinnati, Ohio 45221
Email address: mograbgl@ucmail.uc.edu

Department of Mathematics, Kidder Hall 368, Oregon State University, Corvallis, Oregon 97331-4605
Email address: phelpmad@oregonstate.edu

Department of Mathematics, University of Connecticut, Storrs, Connecticut 065269-1009
Email address: luke.rogers@uconn.edu

Department of Mathematics, Northern Arizona University, 801 S. Osborne Dr. PO Box: 5717, Flagstaff, Arizona 86011
Email address: jlw697@nau.edu

Contemporary Mathematics
Volume **828**, 2025
https://doi.org/10.1090/conm/828/16597

Homogeneous Newton-Sobolev spaces in metric measure spaces and their Banach space properties

Nageswari Shanmugalingam

ABSTRACT. In this note we prove the Banach space properties of the homogeneous Newton-Sobolev spaces $HN^{1,p}(X)$ of functions on an unbounded metric measure space X equipped with a doubling measure supporting a p-Poincaré inequality, and show that when $1 < p < \infty$, even with the lack of global L^p-integrability of functions in $HN^{1,p}(X)$, we have that every bounded sequence in $HN^{1,p}(X)$ has a strongly convergent convex-combination subsequence. The analogous properties for the inhomogeneous Newton-Sobolev classes $N^{1,p}(X)$ are proven elsewhere in existing literature (as for example in [*Sobolev spaces on metric measure spaces: An approach based on upper gradients*, Cambridge University Press, Cambridge, 2015] and [Rev. Mat. Iberoamericana 16 (2000), no. 2, 243–279]).

1. Introduction

The Sobolev spaces play a vital role in the study of partial differential equations and potential theory in the setting of Euclidean spaces, and more generally, Riemannian manifolds. The research carried out during the past two decades also developed function spaces, in the setting of more general metric measure spaces, that can play the role of Sobolev spaces in the setting where the metric space is non-smooth and carries no natural differential structure. One such class of function-spaces, called the Newton-Sobolev spaces, have turned out to be highly useful in developing potential theory and studying properties of energy-minimizers. The homogeneous versions of these spaces is the focus of the present paper.

Homogeneous versions of Newton-Sobolev spaces, where the energy of the function in that version is globally finite without the function itself being necessarily in the global L^p-class, arise naturally when dealing with large-scale potential theory of a metric measure space; see for example [**1, 4, 10, 11, 18**]. For this reason, the behavior of such functions has attracted great interest recently, and an interesting discussion about limits at infinity of such functions can be found in the metric space context in [**3, 13–15**].

As with classical Sobolev spaces, the Banach space property is vital in the use of Newton-Sobolev spaces. The normed vector-space property of these function spaces is clear. It was shown in [**19**] (see also [**9**]) that the Newton-Sobolev spaces $N^{1,p}(X)$, with $1 \le p < \infty$ and (X, d, μ) the non-smooth metric measure space on

2020 *Mathematics Subject Classification.* Primary 46E36; Secondary 31E05, 31L15.
The author's work was partially supported by the NSF (U.S.A.) grant DMS #2054960.

which functions from $N^{1,p}(X)$ act, are always Banach spaces, that is, as normed vector spaces, $N^{1,p}(X)$ are complete. The situation of homogeneous Newton-Sobolev spaces $D^{1,p}(X)$ is different; even the norm on $D^{1,p}(X)$ is only a seminorm. In this note we will establish the Banach space property for $HN^{1,p}(X) := D^{1,p}(X)/\sim$, where $\sim$ is an equivalence relationship on $D^{1,p}(X)$ that ensures that the seminorm on $D^{1,p}(X)$ transforms into a norm on $HN^{1,p}(X)$.

THEOREM 1.1. *If the measure μ of the metric measure space (X, d, μ) is doubling and supports a p-Poincaré inequality, then $HN^{1,p}(X)$ is a Banach space. If $1 < p < \infty$, then any bounded sequence in $HN^{1,p}(X)$ has a convex-combination subsequence that converges in $HN^{1,p}(X)$.*

Here by a convex combination subsequence of a sequence $(u_k)_k$ we mean a sequence of the form $(w_k)_k$ such that for each $k \in \mathbb{N}$ there are nonnegative real numbers $\lambda_{j,k}$, $j = 1, \cdots, N_k$ with $\sum_{j=0}^{N_k} \lambda_{j,k} = 1$ and $w_k = \sum_{j=0}^{N_k} \lambda_{j,k}\, u_{k+j}$. Should $HN^{1,p}(X)$ be reflexive, then by Mazur's lemma the second claim of the theorem would follow from the first part of the theorem; the proof that $HN^{1,p}(X)$ is reflexive follows along the same lines as the proof the fact that $N^{1,p}(X)$ is reflexive under the hypotheses of the above theorem, with added complications due to lack of global L^p-integrability of the functions in $HN^{1,p}(X)$; see for example the discussion in [**9**], which uses the tools developed in [**2**]. However, the consequence of Mazur's lemma does not need the reflexivity property of $HN^{1,p}(X)$, and we provide a more direct proof here.

2. Background

The classical Sobolev spaces are based on the notion of weak derivatives, where we expect to have a vector-valued function associated with a Sobolev function so that an integration-by-parts formula holds. The comprehensive text [**17**] gives a wide-ranging overview of the theory of Sobolev spaces in Euclidean setting, while the lecture note [**20**] gives a connection between classical Sobolev functions and functions that are absolutely continuous on almost every curve in the Euclidean space, and so for almost every curve, the function should be differentiable almost everywhere along that curve. This is the idea behind the notion of upper gradients, first proposed by Heinonen and Koskela in [**8**], see also [**7**].

In this note, (X, d, μ) is a metric measure space with μ a Radon measure such that for each ball $B \subset X$ we have that $0 < \mu(B) < \infty$. We say that μ is *doubling* if there is a constant $C \geq 1$ such that

$$\mu(B(x, 2r)) \leq C\, \mu(B(x, r)),$$

whenever $x \in X$ and $r > 0$. Here $B(x, r) := \{y \in X : d(x, y) < r\}$ denotes the ball centered at x and with radius r. We will also fix $1 \leq p < \infty$ here.

2.1. Construction of Newton-Sobolev spaces $N^{1,p}(X)$. A function $f : X \to [-\infty, \infty]$ is said to have a Borel function $g : X \to [0, \infty]$ as an *upper gradient* if we have

$$|f(\gamma(b)) - f(\gamma(a))| \leq \int_\gamma g\, ds$$

whenever $\gamma : [a, b] \to X$ is a rectifiable curve with $a < b$. In the above, we interpret the inequality as also meaning that $\int_\gamma g\, ds = \infty$ whenever at least one of $f(\gamma(b))$,

$f(\gamma(a))$ is not finite. We say that $f \in \widetilde{N^{1,p}}(X)$ for a fixed choice of $1 \le p < \infty$ if

$$\|f\|_{N^{1,p}(X)} := \left(\int_X |f|^p \, d\mu\right)^{1/p} + \inf_g \left(\int_X g^p \, d\mu\right)^{1/p}$$

is finite. In the above, the infimum is over all upper gradients g of f. It is not difficult to see that $\widetilde{N^{1,p}}(X)$ is a vector space, and that if g_1 is an upper gradient of f_1 and g_2 is an upper gradient of f_2, then $|\alpha|g_1 + |\beta|g_2$ is an upper gradient of $\alpha f_1 + \beta f_2$ when $\alpha, \beta \in \mathbb{R}$.

Note that it is possible for a function to be not identically zero on X but at the same time have $\|f\|_{N^{1,p}(X)} = 0$. This is seen, for example, in the context of Euclicean spaces, with $f = \chi_{\{0\}}$, see the discussion in [**9**]. Therefore, to set $\|\cdot\|_{N^{1,p}(X)}$ to be a norm, we deal with a quotient space. To this end, when $f_1, f_2 \in \widetilde{N^{1,p}}(X)$, we say that $f_1 \sim f_2$ if $\|f_1 - f_2\|_{N^{1,p}(X)} = 0$.

We say that a family Γ of rectifiable curves in X has *p-modulus zero*, that is, $\mathrm{Mod}_p(\Gamma) = 0$, if there is a non-negative Borel function ρ on X such that $\int_X \rho^p \, d\mu$ is finite, and at the same time, $\int_\gamma \rho \, ds = \infty$ for each $\gamma \in \Gamma$. It follows from [**9**] that the countable union of families of curves with zero p-modulus must have p-modulus zero.

With $E \subset X$, the collection Γ_E is the family of all non-constant compact rectifiable curves in X that intersect E. The collection Γ_E^+ is the family of all non-constant compact rectifiable curves γ in X for which $\mathcal{H}^1(\gamma_s^{-1}(E)) > 0$, where $\gamma_s : [0, \ell(\gamma)] \to X$ is the arc-length parametrization of γ. Here, $\ell(\gamma)$ is the length of the rectifiable curve γ. Note that when E is a Borel set with $\mu(E) = 0$, necessarily we must have that $\mathrm{Mod}_p(\Gamma_E^+) = 0$. However, it is still possible to have $\mathrm{Mod}_p(\Gamma_E)$ to be positive.

Following Ohtsuka, we say that a set $E \subset X$ is *p-exceptional* if both $\mu(E) = 0$ and $\mathrm{Mod}_p(\Gamma_E) = 0$ are satisfied.

LEMMA 2.1. *If $f_1, f_2 \in \widetilde{N^{1,p}}(X)$, then $f_1 \sim f_2$ if and only if the set $E := \{x \in X : f_1(x) \ne f_2(x)\}$ is p-exceptional.*

PROOF. Let $f_1, f_2 \in \widetilde{N^{1,p}}(X)$. If $E := \{x \in X : f_1(x) \ne f_2(x)\}$ satisfies $\mu(E) = 0$, then we already have that $\int_X |f_1 - f_2|^p \, d\mu = 0$. If in addition we have that $\mathrm{Mod}_p(\Gamma_E) = 0$, then with ρ the Borel function associated with this property for Γ_E, we see that for each $\varepsilon > 0$ the function $g_\varepsilon := \varepsilon\rho$ is an upper gradient of $f_1 - f_2$, and $\int_X g_\varepsilon^p \, d\mu = \varepsilon^p \int_X \rho^p \, d\mu$. As $\int_X \rho^p \, d\mu < \infty$, it follows that $\inf_g \int_X g^p \, d\mu = 0$ where the infimum is over all upper gradients g of $f_1 - f_2$.

Now suppose that $f_1 \sim f_2$. Then we must have that $\int_X |f_1 - f_2|^p \, d\mu = 0$, and so $\mu(E) = 0$. Moreover, for each positive integer n we can find an upper gradient g_n of $f_1 - f_2$ such that $\int_X g_n^p \, d\mu < 2^{-np}$. Set $\rho_0 = \sum_n g_n$. For each positive integer k let $E_k = \{x \in X : |f_1(x) - f_2(x)| > 1/k\}$. Then $E = \bigcup_k E_k$.

By the summability of $\sum_n 2^{-n}$, we know that $\int_X \rho_0^p \, d\mu < \infty$. Since μ is a Radon measure, there is a Borel set $E_0 \subset X$ with $E \subset E_0$ such that $\mu(E_0) = 0$; let $\rho = \rho_0 + \infty\chi_{E_0}$. Then again we have that $\int_X \rho^p \, d\mu = \int_X \rho_0^p \, d\mu < \infty$, and moreover, if $\gamma \in \Gamma_E^+$, then $\int_\gamma \rho \, ds = \infty$. On the other hand, if $\gamma \in \Gamma_{E_k} \setminus \Gamma_E^+$, then there is some point along the trajectory of γ at which $f_1 - f_2$ takes on the value of 0, and there is some point in this trajectory where $|f_1 - f_2|$ is at least $1/k$. It follows that for each positive integer n we have $\int_\gamma g_n \ge 1/k$, and so for each such γ we have that $\int_\gamma k\rho\, ds = \infty$. Thus $\mathrm{Mod}_p(\Gamma_{E_k} \cup \Gamma_E^+) = 0$. As the countable union of families of curves, each of zero p-modulus, must be of p-modulus zero, it follows that Γ_E is of zero p-modulus. This completes the proof of the lemma. $\square$

The *Newton-Sobolev class* is the collection of all equivalence classes, that is, $N^{1,p}(X) := \widetilde{N^{1,p}}(X)/\sim$. Given the above lemma, we see that when $f_1 \sim f_2$, we have $\|f_1\|_{N^{1,p}(X)} = \|f_2\|_{N^{1,p}(X)}$, and so the seminorm $\|\cdot\|_{N^{1,p}(X)}$ is inherited by $N^{1,p}(X)$ as a norm. Indeed, with E the set of points $x \in X$ for which $f_1(x) \neq f_2(x)$, we have that E is p-exceptional, and so there is some non-negative Borel function $\rho \in L^p(X)$ such that $\int_\gamma \rho\, ds = \infty$ for each γ that intersects E; hence, for each $\varepsilon > 0$ we have that $\varepsilon\rho$ is an upper gradient of $f_1 - f_2$, and so whenever g_1, g_2 are upper gradients of f_1, f_2 respectively, then $g_i + \varepsilon\rho$ is an upper gradient of f_j for $i, j \in \{1, 2\}$ with $i \neq j$. Thus $\|f_1\|_{N^{1,p}(X)} = \|f_2\|_{N^{1,p}(X)}$.

LEMMA 2.2 ([**19**, Theorem 3.7],[**9**, Proposition 7.3.1, Proposition 7.2.8]). *The space $N^{1,p}(X)$ is a Banach space. Suppose that $(u_k)_k$ is a Cauchy sequence of functions in $N^{1,p}(X)$, converging to $u \in N^{1,p}(X)$. Let $(u_{k_j})_j$ be any subsequence of this sequence satisfying $\|u_{k_j} - u_{k_{j+1}}\| \le C\,2^{-j}$ with $C > 0$ that is independent of j. Then, with u_{k_j} also denoting a representative function from the equivalence class $u_{k_j} \in N^{1,p}(X)$, and setting E to be the collection of all points $x \in X$ for which $(u_{k_j}(x))_j$ is not a Cauchy sequence of real numbers, we have that E is a p-exceptional set.*

2.2. Construction of homogeneous Newton-Sobolev spaces $HN^{1,p}(X)$. We set $D^{1,p}(X)$ to be the collection of all functions $f : X \to [-\infty, \infty]$ that have an upper gradient g such that $\int_X g^p\, d\mu$ is finite, that is, g is p-summable. Note that as constant functions have the identically-zero function as an upper gradient, it follows that constant functions belong to $D^{1,p}(X)$, regardless of whether $\mu(X)$ is finite or infinite. For $f \in D^{1,p}(X)$ we set

$$\|f\|_{D^{1,p}(X)} = \inf_g \left(\int_X g^p\, d\mu \right)^{1/p},$$

where the infimum is over all upper gradients g of f. Note that $\|\cdot\|_{D^{1,p}(X)}$ is only a seminorm on $D^{1,p}(X)$, as non-zero constant functions have their seminorm be zero. As in the previous subsection, we say that two functions $f_1, f_2 \in D^{1,p}(X)$ are equivalent, $f_1 \sim f_2$, if $\|f_1 - f_2\|_{D^{1,p}(X)} = 0$.

LEMMA 2.3. *Let $f_1, f_2 \in D^{1,p}(X)$ such that $f_1 \sim f_2$. Then there is a family Γ of non-constant compact rectifiable curves in X with $\mathrm{Mod}_p(\Gamma) = 0$, such that whenever γ is a non-constant compact rectifiable curves in X with $\gamma \notin \Gamma$, we have that $r_1 \circ \gamma = f_2 \circ \gamma + C$. Moreover, $\|f_1\|_{D^{1,p}(X)} = \|f_2\|_{D^{1,p}(X)}$.*

PROOF. Since $f_1 \sim f_2$, for each $n \in \mathbb{N}$ we can find an upper gradient g_n of $f_1 - f_2$ such that $\int_X g_n^p\, d\mu < 2^{-np}$. As before, let $\rho = \sum_n g_n$, and note that $\rho \in L^p(X)$. Let Γ_b be the collection of all non-constant compact rectifiable curves γ in X for which $\int_\gamma \rho\, ds = \infty$. It follows that $\mathrm{Mod}_p(\Gamma_b) = 0$.

For ease of notation, we set $u := f_1 - f_2$. Fix any $\gamma \notin \Gamma_b$, say, $\gamma : [a, b] \to X$. Note that as $\int_\gamma \rho\, ds < \infty$, it follows that for each $s, t \in [a, b]$ with $s < t$ we also have that $\gamma|_{[s,t]} \notin \Gamma_b$. Hence for each $n \in \mathbb{N}$ we have that

$$|u(\gamma(s)) - u(\gamma(t))| \le \int_{\gamma|_{[s,t]}} g_n \, ds \le \int_\gamma g_n \, ds. \tag{2.4}$$

A theorem of Fuglede [**6**, **9**] yields that there is a family Γ_F of non-constant compact rectifiable curves with $\mathrm{Mod}_p(\Gamma_F) = 0$ such that whenever $\gamma \notin \Gamma_F$ is a non-constant compact rectifiable curve in X, we have that

$$\lim_{n\to\infty} \int_\gamma g_n \, ds = 0.$$

Note that with $\Gamma = \Gamma_b \cup \Gamma_F$, we have $\mathrm{Mod}_p(\Gamma) = 0$. So if $\gamma \notin \Gamma$, then by (2.4) we have that for each $s, t \in [a, b]$ with $s < t$,

$$|u(\gamma(s)) - u(\gamma(t))| \le 0.$$

Thus $u \circ \gamma$ is constant for each $\gamma \notin \Gamma$.

The last claim of the lemma follows from the triangle inequality for the seminorms:

$$\|f_1\|_{D^{1,p}(X)} \le \|f_2\|_{D^{1,p}(X)} + \|f_2 - f_1\|_{D^{1,p}(X)} = \|f_2\|_{D^{1,p}(X)},$$

and

$$\|f_2\|_{D^{1,p}(X)} \le \|f_1\|_{D^{1,p}(X)} + \|f_2 - f_1\|_{D^{1,p}(X)} = \|f_1\|_{D^{1,p}(X)}.$$

□

The *homogeneous Newton-Sobolev space* $HN^{1,p}(X)$ is the collection of all equivalence classes of $D^{1,p}(X)$ under the equivalence relationship $\sim$, and we denote $HN^{1,p}(X) := D^{1,p}(X)/\sim$. Thanks to the above lemma, this homogeneous space inherits the seminorm of $D^{1,p}(X)$ as a norm. Note that in general, it is not true that $f_1 \sim f_2$ implies that $f_1 - f_2$ is a constant; for example, if X has no non-constant compact rectifiable curves, then all functions on X are equivalent as zero function would be an upper gradient for it. This is in stark contrast to $\widetilde{N^{1,p}}(X)$, where if f_1 is equivalent to f_2 with equivalence as defined in the previous subsection, then $f_1 = f_2$ almost everywhere. There are some circumstances under which $D^{1,p}(X)$ has such a regulated behavior, and one such circumstance is that in which X supports a p-Poincaré inequality.

REMARK 2.5. We point out here that in some literature the term *homogeneous Sobolev space* in relation to Euclidean spaces stands in for the $HN^{1,p}$–norm-closure of the collection of all compactly supported smooth functions on Euclidean spaces, see for instance [**16**]. We do not consider this analog here, but it would be interesting to know what the relationship between these two spaces are in general. Our definition of homogeneous spaces seem to be more intimately connected to trace theorems for unbounded domains [**5**].

2.3. Poincaré inequalities. As in the previous subsections, we fix an index p with $1 \le p < \infty$. For measurable sets $A \subset X$ with $0 < \mu(A) < \infty$ and $f : X \to \mathbb{R}$ a measurable function, we set

$$\fint_A f \, d\mu := \frac{1}{\mu(A)} \int_A f \, d\mu =: f_A.$$

We say that the metric measure space (X, d, μ) supports a p-Poincaré inequality if there are constants $C > 0$ and $\lambda \geq 1$ such that, whenever f is a measurable function on X and g is an upper gradient of f, and $B(x,r)$ is a ball in X, we have

$$\fint_{B(x,r)} |f - f_B|\, d\mu \leq C\, r \left(\fint_{B(x,\lambda r)} g^p\, d\mu \right)^{1/p}.$$

Since we assume that balls in X have positive and finite μ-measure, it follows that X is connected whenever (X, d, μ) supports a p-Poincaré inequality. Under some additional assumptions (such as doubling property of μ and local compactness of X) it also follows that X is quaisconvex, that is, there is a constant $C_q \geq 1$ such that whenever $x, y \in X$ are two distinct points, there is a rectifiable curve $\gamma_{x,y}$ with end points x and y and with length $\ell(\gamma_{x,y}) \leq C_q\, d(x,y)$, see for instance [**9**, Theorem 8.3.2] or [**2**, **12**]. In this subsection we will not assume that μ is doubling, and so the quasiconvexity property is not necessarily true.

LEMMA 2.6. *Suppose that (X, d, μ) supports a p-Poincaré inequality and $f_1, f_2 \in D^{1,p}(X)$ such that $f_1 \sim f_2$. Then there exists $C = C_{f_1 - f_2} \in \mathbb{R}$ such that $f_2 = f_1 + C$ μ-a.e. in X.*

PROOF. Suppose that $f_1 \sim f_2$. Set $u := f_1 - f_2$. Then as $\|u\|_{D^{1,p}(X)} = 0$, for each $n \in \mathbb{N}$ we can find an upper gradient g_n of u such that $\int_X g_n^p\, d\mu \leq 2^{-np}$.

We fix a point $x_0 \in X$ and $R > 0$ and denote $B = B(x_0, R)$. Then by the Poincaré inequality, we have that

$$\fint_B |u - u_B|\, d\mu \leq C\, R \left(\fint_{B(x_0,\lambda R)} g_n^p\, d\mu \right)^{1/p} \leq \frac{C\, R}{\mu(B(x_0, \lambda R))^{1/p}}\, 2^{-n}.$$

Letting $n \to \infty$ we see that we must have $u = u_B$ μ-a.e. in B.

Now, exhausting X by replacing B by $B(x_0, kR)$ for each $k \in \mathbb{N}$ and applying the above argument, we now conclude that u is to be constant μ-a.e. in X; choosing C to be that constant completes the proof of the lemma. □

Theorem 2.7 was proved in [**6**].

THEOREM 2.7 ([**6**, Theorem 5.1]). *If the measure μ is doubling and supports a p-Poincaré inequality, then there are constants $C_0 > 0$ and $\lambda_0 \geq 1$ such that whenever $f : X \to [-\infty, \infty]$ is a measurable function with upper gradient $g \in L^p(X)$ and $B(x,r)$ is a ball in X, then we have*

$$\left(\fint_{B(x,r)} |f - f_B|^p\, d\mu \right)^{1/p} \leq C_0\, r \left(\fint_{B(x,\lambda_0 r)} g^p\, d\mu \right)^{1/p}.$$

3. Banach space property of $HN^{1,p}(X)$

The goal of this section is to provide a proof of Theorem 1.1. We start with the following useful tool.

LEMMA 3.1. *Suppose the hypotheses of Theorem 1.1 hold. Let $u \in HN^{1,p}(X)$ with upper gradient g, and fix a point $x_0 \in X$. If $0 < R_0 < R < \infty$, then setting*

$B_0 = B(x_0, R_0)$, *we have*

$$\left(\fint_{B(x_0,R)} |u - u_{B_0}|^p \, d\mu\right)^{1/p} \leq C_0 \left(1 + \left(\frac{\mu(B(x_0,R))}{\mu(B_0)}\right)^{1/p}\right) R \left(\fint_{B(x_0,\lambda R)} g^p \, d\mu\right)^{1/p}.$$

PROOF. We have that

$$\left(\fint_{B(x_0,R)} |u - u_{B_0}|^p \, d\mu\right)^{1/p} \leq \left(\fint_{B(x_0,R)} |u - u_{B(x_0,R)}|^p \, d\mu\right)^{1/p} + |u_{B(x_0,R)} - u_{B_0}|.$$

Note that

$$\begin{aligned} |u_{B(x_0,R)} - u_{B_0}| &\leq \fint_{B_0} |u - u_{B(x_0,R)}| \, d\mu \\ &\leq \left(\fint_{B_0} |u - u_{B(x_0,R)}|^p \, d\mu\right)^{1/p} \\ &\leq \left(\frac{\mu(B(x_0,R))}{\mu(B_0)} \fint_{B(x_0,R)} |u - u_{B(x_0,R)}|^p \, d\mu\right)^{1/p}. \end{aligned}$$

Now an application of Theorem 2.7 yields the claim of the lemma. □

Now we are ready to prove Theorem 1.1.

PROOF OF THEOREM 1.1. Let $(u_k)_k$ be a Cauchy sequence in $HN^{1,p}(X)$, and we fix a ball $B_0 = B(x_0, R_0)$. With u_k also denoting any choice of representative function from the equivalence class $u_k \in HN^{1,p}(X)$ for each $k \in \mathbb{N}$, we set the function v_k by

$$v_k := u_k - (u_k)_{B_0},$$

where $(u_k)_{B_0} = \fint_{B_0} u_k \, d\mu$ is the mean of u_k on the ball B_0. Note that the mean of v_k on the same ball B_0 is zero, i.e., $(v_k)_{B_0} = 0$. Note that for each $k \in \mathbb{N}$, v_k belongs to the same equivalence class as u_k, that is, as far as $HN^{1,p}(X)$ is concerned, $v_k = u_k$.

As $(v_k)_k$ is Cauchy in $HN^{1,p}(X)$, we know that for each $m, n \in \mathbb{N}$ with $m > n$, there is an upper gradient $g_{v_m - v_n}$ of $v_m - v_n$ such that $\lim_{m,n\to\infty} \int_X g^p_{v_m - v_n} \, d\mu = 0$. Therefore, by passing to a subsequence if necessary, we may assume that for each $k \in \mathbb{N}$ we have $\int_X g^p_{v_{k+1} - v_k} \, d\mu < 2^{-kp}$. Note that if we can show the convergence of this subsequence in $HN^{1,p}(X)$, then the entire Cauchy sequence also converges.

We fix $R > R_0$, and consider the sequence of *functions* $v_k \in L^p(B(x_0, R))$. By an application of Lemma 3.1, we know that

$$\begin{aligned} &\int_{B(x_0,R)} |v_{k+1} - v_k|^p \, d\mu \\ &\qquad \leq C_0^p \frac{R^p}{\mu(B(x_0,R))} \left(1 + \left(\frac{\mu(B(x_0,R))}{\mu(B_0)}\right)^{1/p}\right)^p \int_X g^p_{v_{k+1} - v_k} \, d\mu \\ &\qquad \leq C_0^p \frac{R^p}{\mu(B(x_0,R))} \left(1 + \left(\frac{\mu(B(x_0,R))}{\mu(B_0)}\right)^{1/p}\right)^p 2^{-kp}. \end{aligned}$$

Setting

$$A(R) := C_0 \frac{R}{\mu(B(x_0,R))^{1/p}} \left(1 + \left(\frac{\mu(B(x_0,R))}{\mu(B_0)}\right)^{1/p}\right),$$

we now have that

$$\|v_{k+1} - v_k\|_{L^p(B(x_0,R))} \le A(R)\, 2^{-k}.$$

It follows that the sequence $(v_k)_k$ is Cauchy in $L^p(B(x_0,R))$, and hence converges to a function v_R there. Moreover, by the above inequality, we also know that this convergence is also *pointwise* μ*-a.e. in* the ball $B(x_0,R)$. Since this happens for each $R > R_0$, it follows that there is a function $v \in L^p_{loc}(X)$ with $v_k \to v$ both in $L^p_{loc}(X)$ and μ.a.e. in X.

It only remains to show that (a) there is a function $\widehat{v} \in D^{1,p}(X)$ with $\widehat{v} = v$ μ-a.e. in X, and that (b) $v_k \to \widehat{v}$ in $HN^{1,p}(X)$. To do so, note that $(v_k|_{B(x_0,R)})_k$ is a Cauchy sequence in $N^{1,p}(B(x_0,R))$, thanks to Lemma 3.1 and the condition on the choices $g_{u_m-u_n}$ which are upper gradients of $v_m - v_n$ on $B(x_0,R)$ as well. Indeed, this sequence also satisfies the second hypothesis of Lemma 2.2 with the entire sequence being considered as its own subsequence. Thus, by Lemma 2.2, we know that there is a function $w_R \in N^{1,p}(B(x_0,R))$ such that $v_k \to w_R$ in $N^{1,p}(B(x_0,R))$. As the $L^p(B(x_0,R))$-norm is part of the $N^{1,p}(B(x_0,R))$-norm, it follows that $w_R = v$ μ-a.e. in $B(x_0,R)$. Moreover, this lemma also tells us that $v_k \to w_R$ pointwise in $B(x_0,R) \setminus E[R]$ for some p-exceptional set $E[R]$. Here we take $E[R]$ to be the set of points x in $B(x_0,R)$ for which $(v_k(x))_k$ is not a Cauchy sequence; thus, we also have that $E[R_2] \cap B(x_0,R_1) = E[R_1]$ whenever $R_2 > R_1 > R_0$. Hence, when $R_0 < R_1 < R_2$, we also have that $w_{R_2} = w_{R_1}$ on $B(x_0,R_1)$, and so we can find a globally defined function $\widehat{v}$ so that the restriction of $\widehat{v}$ to $B(x_0,R)$ is w_R.

For each $k \in \mathbb{N}$, note that $\sum_{j=k}^\infty g_{v_{j+1}-v_j}$ is an upper gradient of $v_l - v_k$ in X for each $l \in \mathbb{N}$ with $l > k$. It follows that $\sum_{j=k}^\infty g_{v_{j+1}-v_j}$ is an upper gradient of $\widehat{v} - v_k$ in $B(x_0,R)$ for each $R > R_0$. It now follows from the fact that $v_k \in D^{1,p}(X)$ that $\widehat{v} \in D^{1,p}(X)$, that is, $\widehat{v}$ belongs to an equivalence class in $HN^{1,p}(X)$.

As

$$\|\sum_{j=k}^\infty g_{v_{j+1}-v_j}\|_{L^p(X)} \le \sum_{j=k}^\infty \|g_{v_{j+1}-v_j}\|_{L^p(X)} < 2^{1-k},$$

it follows that

$$\|\widehat{v} - v_k\|_{D^{1,p}(X)} < 2^{1-k},$$

that is, $v_k \to \widehat{v}$ in $D^{1,p}(X)$ and hence in $HN^{1,p}(X)$. This completes the proof of the first part of the theorem.

To show prove the last claim of the theorem when $1 < p < \infty$, we need to show that bounded sequences in there have a convex combination subsequence that converges to some function in $HN^{1,p}(X)$. To this end, let $(v_k)_k$ be a representative sequence from a bounded sequence in $HN^{1,p}(X)$. We can, by replacing v_k with $v_k - (v_k)_{B_0}$ if necessary, assume that each v_k has mean $(v_k)_{B_0} = 0$. Then, from Lemma 3.1 and by the p-Poincaré inequality, we know that for each $R_0 < R < \infty$ the sequence $(v_k)_k$ is bounded in $N^{1,p}(B(x_0,R))$. By [**9**, Theorem 7.3.12], which uses only the reflexivity property of the spaces L^p, we know that $N^{1,p}(B(x_0,R))$ satisfies the conclusion of Mazur's lemma. Therefore it follows that there is a convex-combination subsequence $w_{R,j} = \sum_{k=j}^{N_j} \lambda_{j,k}\, v_k$ with $0 \le \lambda_{j,k} \le 1$ for each

$k = j, \cdots, N_j$ and $\sum_{k=j}^{N_j} \lambda_{j,k} = 1$, and a function $w_R \in N^{1,p}(B(x_0, R))$, such that $w_{R,j} \to w_R$ in $N^{1,p}(B(x_0, R))$.

Now choosing a strictly monotone increasing sequence of radii R_k with $\lim_k R_k = \infty$, and for each $k \geq 2$ making sure to consider the sequence $(w_{R_{k-1},j})_j$ on $B(x_0, R_j)$, and utilizing a Cantor-type diagonalization argument, we obtain the function w. Note that for each k, j, the function $w_{R_{k-1},j}$ has average 0 on the ball B_0. As in the previous paragraph, we see that w_R can be chosen to be independent of R, and so we obtain a consistent globally defined function w. This completes the proof. □

References

[1] Sara Barile, Silvia Cingolani, and Simone Secchi, *Single-peaks for a magnetic Schrödinger equation with critical growth*, Adv. Differential Equations **11** (2006), no. 10, 1135–1166. MR2279741

[2] J. Cheeger, *Differentiability of Lipschitz functions on metric measure spaces*, Geom. Funct. Anal. **9** (1999), no. 3, 428–517, DOI 10.1007/s000390050094. MR1708448

[3] Sylvester Eriksson-Bique, Pekka Koskela, and Khanh Nguyen, *On limits at infinity of weighted Sobolev functions*, J. Funct. Anal. **283** (2022), no. 10, Paper No. 109672, 40, DOI 10.1016/j.jfa.2022.109672. MR4470734

[4] William G. Faris, *Weak Lebesgue spaces and quantum mechanical binding*, Duke Math. J. **43** (1976), no. 2, 365–373. MR425598

[5] Ryan Gibara, Riikka Korte, and Nageswari Shanmugalingam, *Solving a Dirichlet problem on unbounded domains via a conformal transformation*, Math. Ann. **389** (2024), no. 3, 2857–2901, DOI 10.1007/s00208-023-02705-8. MR4753076

[6] Piotr Hajłasz and Pekka Koskela, *Sobolev met Poincaré*, Mem. Amer. Math. Soc. **145** (2000), no. 688, x+101, DOI 10.1090/memo/0688. MR1683160

[7] Juha Heinonen, *Lectures on analysis on metric spaces*, Universitext, Springer-Verlag, New York, 2001, DOI 10.1007/978-1-4613-0131-8. MR1800917

[8] Juha Heinonen and Pekka Koskela, *Quasiconformal maps in metric spaces with controlled geometry*, Acta Math. **181** (1998), no. 1, 1–61, DOI 10.1007/BF02392747. MR1654771

[9] Juha Heinonen, Pekka Koskela, Nageswari Shanmugalingam, and Jeremy T. Tyson, *Sobolev spaces on metric measure spaces: An approach based on upper gradients*, New Mathematical Monographs, vol. 27, Cambridge University Press, Cambridge, 2015, DOI 10.1017/CBO9781316135914. MR3363168

[10] Ilkka Holopainen, *Volume growth, Green's functions, and parabolicity of ends*, Duke Math. J. **97** (1999), no. 2, 319–346, DOI 10.1215/S0012-7094-99-09714-4. MR1682233

[11] Ilkka Holopainen and Pekka Koskela, *Volume growth and parabolicity*, Proc. Amer. Math. Soc. **129** (2001), no. 11, 3425–3435, DOI 10.1090/S0002-9939-01-05954-8. MR1845022

[12] Riikka Korte, *Geometric implications of the Poincaré inequality*, Results Math. **50** (2007), no. 1-2, 93–107, DOI 10.1007/s00025-006-0237-x. MR2313133

[13] Josh Kline, Pekka Koskela, and Khanh Nguyen, *Large-scale behaviour of Sobolev functions in Ahlfors regular metric measure spaces*, J. Geom. Anal. **35** (2025), no. 7, Paper No. 214, 49, DOI 10.1007/s12220-025-02052-x. MR4917148

[14] Pekka Koskela and Khanh Nguyen, *Existence and uniqueness of limits at infinity for homogeneous Sobolev functions*, J. Funct. Anal. **285** (2023), no. 11, Paper No. 110154, 22, DOI 10.1016/j.jfa.2023.110154. MR4638845

[15] Pekka Koskela, Khanh Ngoc Nguyen, and Zhuang Wang, *Trace and density results on regular trees*, Potential Anal. **57** (2022), no. 1, 101–128, DOI 10.1007/s11118-021-09907-2. MR4421924

[16] Hideo Kozono and Masao Yamazaki, *Exterior problem for the stationary Navier-Stokes equations in the Lorentz space*, Math. Ann. **310** (1998), no. 2, 279–305, DOI 10.1007/s002080050149. MR1602012

[17] Vladimir Maz'ya, *Sobolev spaces with applications to elliptic partial differential equations*, Second, revised and augmented edition, Grundlehren der mathematischen Wissenschaften

[Fundamental Principles of Mathematical Sciences], vol. 342, Springer, Heidelberg, 2011, DOI 10.1007/978-3-642-15564-2. MR2777530

[18] Debanjan Nandi, *A density result for homogeneous Sobolev spaces*, Nonlinear Anal. **213** (2021), Paper No. 112501, 32, DOI 10.1016/j.na.2021.112501. MR4295184

[19] Nageswari Shanmugalingam, *Newtonian spaces: an extension of Sobolev spaces to metric measure spaces*, Rev. Mat. Iberoamericana **16** (2000), no. 2, 243–279, DOI 10.4171/RMI/275. MR1809341

[20] Jussi Väisälä, *Lectures on n-dimensional quasiconformal mappings*, Lecture Notes in Mathematics, Vol. 229, Springer-Verlag, Berlin-New York, 1971. MR454009

Department of Mathematical Sciences, P.O. Box 210025, University of Cincinnati, Cincinnati, Ohio 45221-0025

Email address: shanmun@uc.edu

Selected Published Titles in This Series

828 **Hafedh Herichi, Maria Rosaria Lancia, Therese Basa Landry, Anna Rozanova-Pierrat, and Steffen Winter, Editors,** Fractal Geometry in Pure and Applied Mathematics, 2025

823 **Mikhail Bershtein, Anton Dzhamay, and Andrei Okounkov, Editors,** The Versatility of Integrability, 2025

822 **Ahmad Barhoumi, Roozbeh Gharakhloo, and Andrei Martinez-Finkelshtein, Editors,** Recent Developments in Orthogonal Polynomials, 2025

821 **Bogdan D. Suceavă, Adara M. Blaga, Marian Ioan Munteanu, Cezar Oniciuc, Shoo Seto, Gabriel-Eduard Vîlcu, and Lihan Wang, Editors,** Recent Advances in Differential Geometry and Related Areas, 2025

820 **Sudeb Mitra, Rakesh Kumar Parmar, Saminathan Ponnusamy, and Toshiyuki Sugawa, Editors,** Geometric Function Theory and Related Topics, 2025

819 **Howard S. Cohl, Roberto S. Costas-Santos, and Robert S. Maier, Editors,** Applications and q-Extensions of Hypergeometric Functions, 2025

818 **Howard S. Cohl, Roberto S. Costas-Santos, and Robert S. Maier, Editors,** Classical Hypergeometric Functions and Generalizations, 2025

817 **Mikhail Khovanov, Joshua Sussan, and Anton Zeitlin, Editors,** From Representation Theory to Mathematical Physics and Back, 2025

816 **Luca F. Di Cerbo and Laurenţiu G. Maxim, Editors,** Geometry and Topology of Aspherical Manifolds, 2025

815 **Daniel Orr and Joshua Jeishing Wen, Editors,** Macdonald Theory and Beyond, 2025

814 **Susan Montgomery and Siu-Hung Ng, Editors,** Quantum Groups, Hopf Algebras, and Applications, 2025

813 **Thomas Creutzig, Julia Plavnik, and David Ridout, Editors,** Quantum Symmetries, 2025

812 **Ross Geoghegan, Craig R. Guilbault, and Kim Ruane, Editors,** Topology at Infinity of Discrete Groups, 2025

811 **Krishnendu Gongopadhyay and Shashikant A. Katre, Editors,** Geometry, Groups and Mathematical Philosophy, 2025

810 **Robert J. Berman and Yanir A. Rubinstein, Editors,** Convex and Complex: Perspectives on Positivity in Geometry, 2025

809 **Fabrice Baudoin and Luca Rizzi, Editors,** New Trends in Sub-Riemannian Geometry, 2025

808 **Loring W. Tu, Editor,** Group Actions and Equivariant Cohomology, 2024

807 **Galina Filipuk, Editor,** Recent Progress in Special Functions, 2024

806 **Maximilian Engel, Hildeberto Jardón-Kojakhmetov, and Cinzia Soresina, Editors,** Topics in Multiple Time Scale Dynamics, 2024

805 **Anthony Iarrobino, Pedro Macias Marques, Maria Evelina Rossi, and Jean Vallès, Editors,** Deformation of Artinian Algebras and Jordan Type, 2024

804 **Mahir Bilen Can and Jörg Feldvoss, Editors,** A Glimpse into Geometric Representation Theory, 2024

803 **Peter Gothen, Margarida Melo, and Montserrat Teixidor i Bigas, Editors,** Moduli Spaces and Vector Bundles—New Trends, 2024

802 **Ralph M. Kaufmann, Martin Markl, and Alexander A. Voronov, Editors,** Higher Structures in Topology, Geometry, and Physics, 2024

801 **K. A. Brown, T. J. Hodges, M. Vancliff, and J. J. Zhang, Editors,** Recent Advances in Noncommutative Algebra and Geometry, 2024

For a complete list of titles in this series, visit the
AMS Bookstore at **www.ams.org/bookstore/conmseries/**.

FSC
www.fsc.org
MIX
Papier | Fördert gute Waldnutzung
FSC® C083411

Kolibri 360 GmbH
Ferdinand-Jühlke-Straße 7
99095 Erfurt, Deutschland
produktsicherheit@kolibri360.de